Geoheritage, Geoparks and Geotourism

Conservation and Management Series

Series Editors

Spectacular geo-morphological landscapes and regions with special geological features or mining sites are becoming increasingly recognized as critical areas to protect and conserve for the unique geoscientific aspects they represent and as places to enjoy and learn about the science and history of our planet. More and more national and international stakeholders are engaged in projects related to "Geoheritage", "Geo-conservation", "Geoparks" and "Geo-tourism"; and are positively influencing the general perception of modern Earth Sciences. Most notably, "Geoparks" have proven to be excellent tools to educate the public about Earth Sciences; and they are also important areas for recreation and significant sustainable economic development through geotourism. In order to develop further the understanding of Earth Sciences in general and to elucidate the importance of Earth Sciences for Society, the "Geoheritage, Geoparks and Geotourism Conservation and Management Series" has been launched together with its sister "GeoGuides" series. Projects developed in partnership with UNESCO, World Heritage and Global Geoparks Networks, IUGS and IGU, as well as with the 'Earth Science Matters' Foundation will be considered for publication. This series aims to provide a place for in-depth presentations of developmental and management issues related to Geoheritage and Geotourism in existing and potential Geoparks. Individually authored monographs as well as edited volumes and conference proceedings are welcome; and this book series is considered to be complementary to the Springer-Journal "Geoheritage".

More information about this series at http://www.springer.com/series/11639

Patricia Erfurt

The Geoheritage of Hot Springs

Springer

Patricia Erfurt
Geotourism
Hervey Bay, QLD, Australia

ISSN 2363-765X ISSN 2363-7668 (electronic)
Geoheritage, Geoparks and Geotourism
ISBN 978-3-030-60462-2 ISBN 978-3-030-60463-9 (eBook)
https://doi.org/10.1007/978-3-030-60463-9

Cover Illustrations: Travertine terraces of Pamukkale (Turkey); Caldeira das Furnas, São Miguel (Azores); Roman Bath Complex, Bath (UK); Peninsula Hot Springs, Victoria (Australia—Photo credit: Charles Davidson); Pōhutu Geyser, Rotorua (New Zealand); Umi Jigoku, Beppu (Japan)

This Springer imprint is published by the registered company Springer Nature Switzerland AG
The registered company address is: Gewerbestrasse 11, 6330 Cham, Switzerland

Travertine Terraces at the Hierapolis-Pamukkale World Heritage Site, Turkey

Foreword

The first time I met Patricia she was undertaking research for her book *Health & Wellness Tourism—Spas and Hot Springs* (2009), one of many publications she has produced in her field of passion—Volcanic Tourism and Hot Springs and Spa based Health and Wellness Tourism. It was a sunny autumn day and we sat on the balcony of the Spa Dreaming Centre at our newly opened Peninsula Hot Springs. Immediately it was apparent we shared a deep passion for natural warm water, a gift from mother earth, that had swept us up and become a major force and indeed purpose in our respective lives. Patricia looking through the eyes of an academic observer who had grown up in Germany where she was immersed in the culture of thermal bathing. Myself, a business graduate of Melbourne University Commerce who first experienced hot springs in a snow covered Japanese countryside town of Kusatsu in the spring of 1992. The impact of that seminal experience was profound. An epiphany moment when I knew I would spend my life helping bring the incredible sensation of thermal mineral spring bathing to Australia and indeed the world.

Thermal waters are the great connector. Something that everyone of all ages from every culture can relate to and connect with. In the past 22 years I have had the opportunity to visit thousands of hot springs in 51 countries in a search for the ways cultures of the world embrace the gift of thermal water to be able to share in the experiences we provide our guests at Peninsula Hot Springs. Bathing is so fundamental to life it is often not even considered part of culture and yet it is at the very heart of the way we live and interact with each other when we are at home in our communities. In periods when I am unable to visit hot springs, sometimes weeks at a time, the initial sensation or returning to the thermal waters is always the same, a sense of coming home.

The epiphany moment in Kusatsu was when I realised the power thermal water had to make us feel as one, at home with ourselves, with our community and most importantly with nature around us. That moment was the inspiration to bring back home the gift of thermal bathing and its ability to provide a space where we can experience and embrace cultural differences. Soon after learning about the thermal waters deep underground on the Mornington Peninsula the search for global cultural knowledge began with a research and educational trip to Russia, the Czech Republic and Yemen.

All communities in the world have bathing cultures. Be it for cleansing, religious ceremonies or importantly for social connection, bathing is a practice for all people for their entire lives.

Thermal waters have long been associated with health and wellbeing. In Europe the industry was at the forefront of the medical industry in the time before allopathic (drug and surgery based) medicine became the dominant healing modality. Thermal towns were the places where people went for the Kur (cure) and to rest, recuperate, socialise and live a wellness life. Healthy habits and practices were taught to visitors who participated in multi week programs. Examples of thermal spa towns include: Baden Baden (Germany), Budapest (Hungary), Vichy Spa (France), Saturnia (Italy), Tiberias (Israel), Damt (Yemen), Manikaren (India), Addis Ababa (Ethiopia), Rotorua (New Zealand), Irrwanyere (Dalhousie Springs—Australia), Yampah (Glenwood—USA), Kusatsu (Japan) and many more across the globe.

These were the vanguard and historical precursors to the modern day health retreats like Kamalaya, Gwingana, Canyon Ranch, Six Senses and Golden Door.

Traditional thermal spa towns and modern wellness retreats provide an escape from the mundane world, a place where time is forgotten, and relaxation is the code of harmony. In these environments we slow down enough to give ourselves opportunity to catch up, to rest and be. This allows our bodies to find homeostasis and gives our immune systems time to re-charge enabling our bodies to be our own best doctors. Strong immune systems are our bodies first line of defence against the pathogens that are naturally present in our daily lives.

Patricia's book helps us understand the historical and physical picture of hot springs in an incredibly well researched and thoughtful way. It opens our eyes to the depth and diversity of the industry and reminds us of the importance of respecting the thermal resources and the gifts they have for us all.

Warm regards
Charles Davidson
Co-founder, Chairman
Peninsula Hot Springs
Fingal, Australia

Preface

During long winters, wearing layers of warm clothes and spending much time in heated environments to avoid the freezing cold outside, it is nice to know there are some hot springs not far away.

Having lived in Europe, in an area with several spa towns within a 30-minute drive, the rising mist hovering above the outdoor hot spring pools in winter remains a fond memory. Especially once, when a group of ducks was battling the deep layers of snow around the thermal pools, determined to join the bathers. Growing up with the knowledge that natural hot springs are used for health, wellness and rehabilitation as well as recreation, I have looked for them in every country I have visited, keeping records of each experience, hoping that one day I would have the time to write a book about the rich geodiversity of this remarkable natural resource.

In addition to many personal observations and countless 'immersions' into the research subject, another essential part in the preparation for this book was an extensive review of the literature related to hydrothermal activity worldwide. Taken together this resulted in further valuable insights into research undertaken by geologists, hydrologists, hydrogeologists, geochemists and volcanologists. Their endeavours are greatly appreciated as supporting literature sources, which have added relevant scientific data and region-specific information to several challenging subjects.

A 170-year-old book I came across during the literature review proved to be one of the more interesting sources of information, not just from a scientific viewpoint but also on a personal level. The book's author Charles Daubeny (1795–1867), an Oxford scholar and physician, undertook special studies and fieldwork as a geologist, chemist and botanist while travelling widely. Visiting many hot springs and volcanic locations during his journeys, I realised I myself have visited and explored most of these places nearly 200 years later. Daubeny's precise descriptions of individual hot springs, their geophysical background and their geochemistry would have been one of the most competent scholarly reports at the time. This is one example where I have valued the freedom of not being limited to research from the last ten or twenty years but was able to delve further back to extract information about the earliest use of hot springs. Given the topic of this book, I enjoyed the liberty of searching for written accounts that have recorded how ancient civilisations have benefitted from hot springs for a variety of purposes.

While not every country with natural hot springs was featured in this book, it will hopefully encourage further research and more publications to add to the literature about hot springs. There are many countries rich in hydrothermal resources and an abundance of cultural traditions linked to the use of hot springs, and I hope this inspires other authors to contribute their knowledge and their accounts about the diversity of this unique natural resource. Visiting industrial sites, where geothermal resources are used for various purposes, was especially eye-opening as it demonstrates the remarkable multiplicity of natural hot spring water. We should see more of this in the future.

Writing this book about the geoheritage of hot springs was a challenging task and took longer than planned. At this point, I would like to acknowledge everybody who contributed advice, illustrations and peer reviews. I am especially grateful to my dear friend Charles

Davidson for his time, support and advice as well as the generous contribution of additional text sources and illustrations. We share a similar enthusiasm for hot springs, although with different results—I am writing about hot springs while Charles has created a hot spring paradise—Peninsula Hot Springs in Victoria—and is planning the development of a hot spring bathing trail that connects many hot spring destinations in Australia and across the world.

I would also like to thank Ken McQueen, Professor of geochemistry at the University of Canberra, for giving his time to provide comments and improvements. His peer review has helped to enhance some important points in my writing, and I feel grateful for his kind support.

Particular thanks are given to a number of people and institutions who granted permission to use their images, a valuable contribution that enriched the text where needed:

- Ivan Lapper, a famous artist specialised in historic construction paintings, who has created an impression of the Romans building the hot spring complex in Bath (UK) two thousand years ago, which fits perfectly within the chapter about the history of hot springs.
- Professor Victor Gostin from the University of Adelaide, who took the contemporary photograph of the radioactive Paralana Hot Springs in South Australia.
- Information and images related to Maruia Hot Springs in New Zealand were kindly provided by the owners James White and Charles Davidson.
- Simon Lewis from the Friends of Mound Springs (FOMS) in South Australia generously provided literature resources and photographs to choose from.
- GNS Science (New Zealand), USGS and NOAA (USA), and the MARUM University Bremen (Germany), who make their information and images available to the public.
- Thanks also go to geologist Mark Fisher from Wyoming, for the use of his hot spring pictures from Thermopolis.

For the ongoing editorial support, encouragement and of course for their patience, I would like to thank Annett Büttner and Dörthe Mennecke-Bühler from the staff at Springer in Germany, as well as the helpful team of copy editors in India. And finally, to everybody who told me about hot springs and their personal experiences, whether it was related to a relaxing timeout, resulted in health improvements or simply observing hydrothermal activity like erupting geysers and bubbling mud pools—I am grateful for sharing this with me.

This book is for all hot spring enthusiasts, health and wellness professionals, spa managers, tourism stakeholders, and advanced undergraduate and graduate students in earth science, environmental science and even history. There should be something for everyone—enjoy reading. I would also like to acknowledge GeoSea in North Iceland, who have made their images available for use in this book.

Hervey Bay, Australia Patricia Erfurt

Contents

About the Author

Patricia Erfurt studied geography, geology and soil science at the University of New England (UNE) in Armidale, New South Wales. She received her Ph.D. from James Cook University (JCU) in Cairns, Queensland, for the subject of natural hot springs and their role in health, wellness and recreation. Since retirement from university teaching, she has worked as a research scientist and consultant at GEOTOURISM Australia with a special focus on risk prevention and risk management in geothermal and volcanic environments. Her main research interests include exploring new hot spring locations with an emphasis on sustainable management and conservation of endangered sites. She spends much of her time publishing research findings and has authored and co-authored books, chapters and articles about hot spring tourism, volcano tourism and geotourism. She actively promotes the concept of sustainable geotourism, assists with planning and establishing geotrails and works as advisor for the conservation of natural and cultural resources and their geoheritage.

April 2021 Patricia Erfurt

Acronyms

AFRETH	Association Française pour la Recherche Thermale (French Association for Thermal Research)
AGA	Australian Geothermal Association
Bq	Becquerel
BSL	Below Sea Level
CAM	Complementary and Alternative Medicine
CCS	Carbon Capture and Sequestration
CHPM	Combined Heat, Power and Metal [Extraction]
CSG	Coal Seam Gas
CSIRO	Commonwealth Scientific and Industrial Research Organisation (Australia)
DEC	Department of Environmental Conservation (Alaska, USA)
DMP	Department of Mines and Petroleum (since 2017 Department of Mines, Industry Regulation and Safety, Western Australia)
EIA	Environmental Impact Assessment
ENSO	El Niño–Southern Oscillation
EPA	Environmental Protection Agency
FLA	Free-Living Amoebae
FOMS	Friends of Mound Springs (Australia)
FoRST	Fondazione della Ricerca Scientifica Termale (Italian Association for Thermal Research)
GAB	Great Artesian Basin (Australia)
GAS	Guarani Aquifer System (South America)
GEG	Geothermal Energy and Geofluids (Switzerland)
GHG	Greenhouse Gas
GPS	Global Positioning System
GRC	Global Resources Council
GWI	Global Wellness Institute
HGPS	Hellisheiði Geothermal Power Station (Iceland)
HSNP	Hot Springs National Park (Arkansas, USA)
IDDP	Iceland Deep Drilling Project
IMA	International Mineral Association
IRENA	International Renewable Energy Agency
IVC	Indus Valley Civilisation (Pakistan)
LD	Legionnaires' Disease
MAR	Mid-Atlantic Ridge
ML	Mega Litre
MoE	Ministry of the Environment (Japan)
MOR	Mid-ocean Ridge
MWe	Megawatt electric
NASA	National Aeronautics and Space Administration (USA)
NOAA	National Oceanic and Atmospheric Agency (USA)

NPS	National Park Service (USA)
PAM	Primary Amoebic Meningoencephalitis
pH	Potential of Hydrogen
PHS	Peninsula Hot Springs/Paralana Hot Springs (Australia)
POI	Point of Interest
ppm	Parts per Million
PRM	Physical and Rehabilitation Medicine
QGC	Queensland Gas Company (Australia)
RoF	Ring of Fire
ROV	Remote-Operated Vehicle
SAP	Strategic Action Program (South America)
SMS	Short Message Service
SPA	Sanus per Aquam (Health through Water)
TDS	Total Dissolved Solids
TSS	Total Suspended Solids
TVZ	Taupo Volcanic Zone (New Zealand)
UNESCO	United Nations Educational, Scientific and Cultural Organization
UOG	Unconventional Oil and Gas (Extraction)
USGS	United States Geological Service
USP	Unique Selling Point
VMS	Volcanogenic Massive Sulphide (Ores)
WGC	World Geothermal Congress
WHA	World Heritage Area

List of Figures

List of Tables

List of Examples

Old Faithful eruption depicted in an old oil painting. *Source* A. Bierstadt c.1881–1886) (Public Domain)

1 Hot Springs: A General Perspective

Contents

An idyllic hot spring pond surrounded by tropical vegetation. Bitter Springs, Northern Territory, Australia

P. Erfurt, *The Geoheritage of Hot Springs*, Geoheritage, Geoparks and Geotourism,
https://doi.org/10.1007/978-3-030-60463-9_1

Everywhere in many lands gush forth beneficent waters, here cold, there hot, there both ... in some places tepid and lukewarm, promising relief to the sick ...
Pliny, the Elder, c.77 AD

1.1 Introduction

Hot springs are a remarkable natural resource. For thousands of years they have played a significant role in a wide range of areas related to human society, including culture, history, religion, and most specifically, human health, wellbeing and recreation. In addition, hot springs are also an essential resource for the renewable energy sector providing environmentally responsible solutions based on geothermal energy in many countries (Fig. 1.1).

A significant point is the abundance of hot springs all over the world with only a few countries without direct access to this natural resource. But first things first—what exactly are hot springs? A representative definition from Michael Allaby's (2013:284) 'Dictionary of Geology & Earth Sciences' describes hot springs as *a continuous flow of hot water through a small opening on to the Earth's surface. The water is usually groundwater heated at depth by hot rocks and recycled to the surface by convection.* Based on their temperature, the correct term is *hydrothermal* spring, derived from the Greek language with *hydros* meaning water and *thermos* referring to heat. However, in reference to their subterranean origin, hot springs are more commonly referred to as *geothermal* springs or just *thermal* springs, whereby *geo* refers to earth (Erfurt-Cooper 2018). The difference between geothermal and hydrothermal can be further clarified with hydrothermal as a subset of geothermal, whereas the term geothermal refers to all processes that transfer heat from the interior of the earth to the surface using water, either as a liquid or as steam (Heasler et al. 2009; Keary 1996). Depending on the information source the terms hydrothermal, thermal and geothermal springs may be used to refer to natural hot springs throughout this book, unless a further definition such as geyser or fumarole is required.

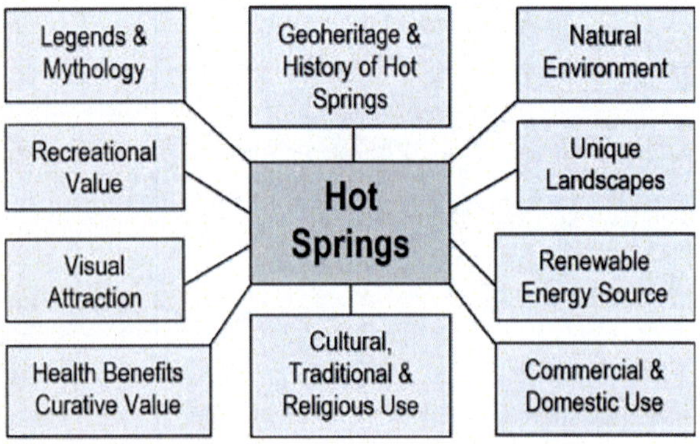

Fig. 1.1 Natural hot springs are of considerable economic and environmental importance and play a central role worldwide. *Source* Author

The use of natural hot springs dates back several thousand years with people using them for various purposes. Apart from their past and present use for washing, cooking and bathing, natural hot springs also play a significant role in various tourism sectors, including health and wellness tourism, geotourism, adventure- and ecotourism. To clarify hot spring tourism, a definition was created based on research related to hot springs, which was collected over more than two decades:

> *Hot spring tourism (or geothermal tourism) involves visiting a destination, location, attraction or facility that takes advantage of geothermal resources in the form of natural hot and mineral springs* (Erfurt 2012).

For health and wellness purposes as well as for recreation, the reputed curative powers of hot springs, mainly based on the mineral contents of individual springs, are generally held in high regard and determine the attractiveness of hot spring destinations. Together with natural surroundings and a pleasant water temperature, hot springs are a sought-after renewable resource for sustainable tourism. Many areas that contain hot springs are either protected in national parks, located in designated World Heritage sites, are part of the geological heritage in national and global geoparks or are protected by local government legislation. For instance, national parks with geothermal resources (e.g. Yellowstone or Banff) have provided attractive tourist destinations with unusual natural environments since their inception (in 1872 and 1885 respectively) for visitor numbers that increase every year.

An abundance of hydrothermal areas is located along the 'Ring of Fire' (Fig. 1.2), which surrounds the Pacific region and is known for its seismic and volcanic activity. Here, not just hot springs for recreational use are found, but also phenomena such as geysers, fumaroles and boiling mud ponds (Appendix 1.1). Outside the Ring of Fire, Yellowstone (Wyoming, USA) is not just the oldest national park in the USA but is also well-known for its estimated 10,000 hydrothermal features, a major attraction for millions of annual visitors [4,020,288 million people in 2019 (Yellowstone NP, 2020)]. A short case study about Yellowstone in Chap. 7 (Hot springs and their Natural Geoheritage) explores this remarkable environment in more detail.

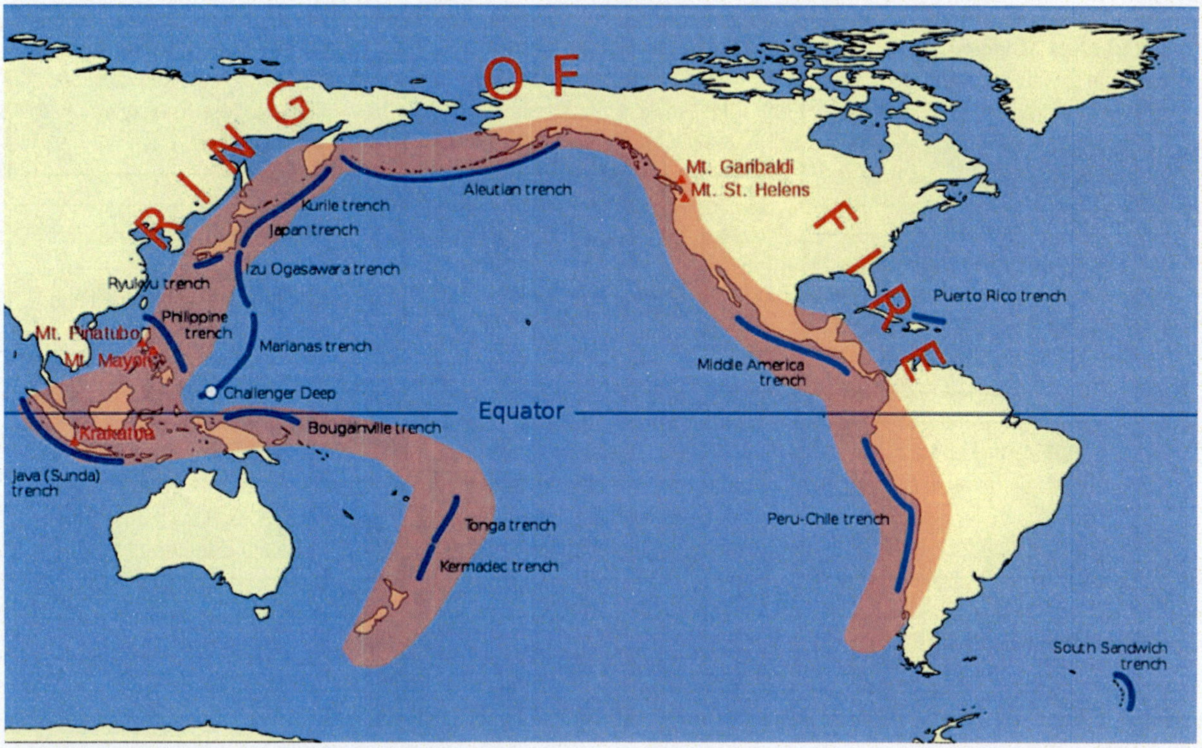

Fig. 1.2 The Pacific Ring of Fire is an area of tectonic activity where tens of thousands of volcanic hot springs are located. *Source* Gringer 2009 (Public Domain)

Hot springs are also located in areas where one would not necessarily expect to find them. Australia for example has no current volcanic activity and many hot springs are fed by the Great Artesian Basin (GAB), where water is heated at depth and enriched with minerals accumulated on the way to the surface. France and Germany also do not have any currently active volcanoes, but many thermal springs originating from deep reservoirs supply a large hot spring-based tourism industry. The origins of hot spring water are discussed in more detail in Chap. 2 (Geology of Hot Springs).

1.2 Why Are Hot Springs So Popular?

Natural hot springs have always captivated people's attention and their popularity is not new. In fact, hot springs have been important sources for various domestic purposes, health treatments, recreational bathing as well as settings for socio-cultural and traditional functions for thousands of years. The variations in settings, usage, culture and extent of development of hot springs is represented in Fig. 1.3, which illustrates the continuum from undeveloped wilderness hot springs to highly developed multipurpose facilities as well as purely visual destinations (Erfurt 2012).

Ancient settlements have been repeatedly discovered adjacent to natural hot springs and for peoples of many countries the occurrence of this resource was a motivation to start communities and build villages, some of which developed into large population centres over time. Archaeological excavations at the spa town *Piešťany* in Slovakia for example reveal a significant amount of Stone Age settlements. The density of these settlements strongly indicates a close correlation to the location of a cluster of ten hot springs that emerge at constant temperatures between 67 and 69 °C (Fendek et al. 1999).

The establishment of Roman military settlements was also often determined by the location of existing hot springs as the Romans were aware of the healing properties of the mineral-laden waters as well as their recreational benefits. The city of *Aachen* in Germany for instance was selected by the Romans for settlement based on several hot springs. The city of *Bath*, where the only genuine hot springs in England were developed for therapeutic purposes over 2000 years ago by the Romans is another example (Erfurt 2012), and is

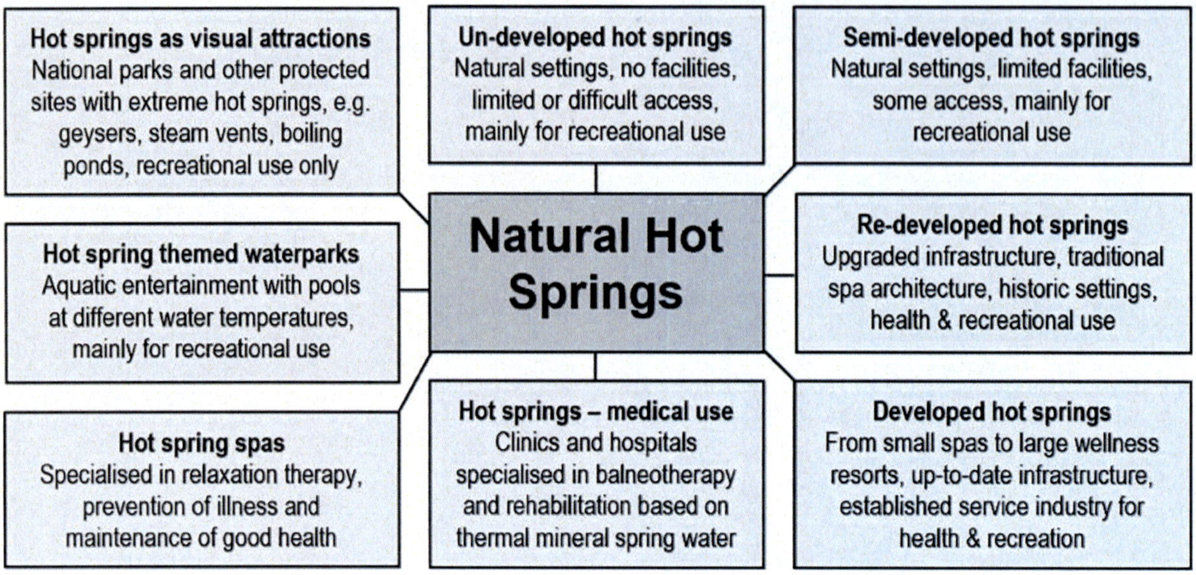

Fig. 1.3 The different stages and possible transitions of hot spring locations from undeveloped natural settings to highly developed hot spring resort destinations. *Source* Author

explored in detail in Chap. 5, which analyses the history of hot springs in more detail.

Other reasons for the establishment of settlements close to natural hot springs have their origin in the legends and mythology of many countries. Countless legends describe how healing springs were initially discovered by hunters following wounded animals that were miraculously healed by immersion in a hot spring. This led people to experiment and when they realised that their health was improving, and injuries were healing faster, they decided to settle in the surrounding area to make use of this valuable resource.

However, not all countries planned their settlements based on the proximity of thermal water. According to the Icelandic sagas and land-taking records, the first settlers did not build their homesteads near hot springs, despite the fact, that hundreds of hot springs all over the country would have provided some form of comfort and hygiene, especially in a cold climate (Fridleifsson 1999). Yet today Iceland is famous for its unusual hot springs, which attract more visitors annually than the country's overall population.

In more recent times, tourists all over the world are enjoying the increasing ease of access to active hydrothermal environments. Various reasons for visiting a hot spring include health, wellness and recreational purposes as well as well as observing the remarkable surface features such as erupting geysers, steaming fumaroles and boiling lakes or mud ponds. Adventure minded travellers are looking increasingly for authentic experiences that include opportunities to explore interesting and unusual destinations. One of the most fascinating aspects of hydrothermal activity undoubtedly is the visual appeal (Fig. 1.4; Appendix 2.1), and many of these attractions were subsequently developed into exceptional tourist destinations such as National Parks, Geoparks or World Heritage sites. However, there are also thousands of hot springs in wilderness settings, which may require long cross-country hikes to reach.

Hot springs can be found just about anywhere; on islands or along the beach, in hot climates or in freezing conditions like **Antarctica**, where *Deception Island* has become a distinctive tourist attraction (Fig. 6.7). Weather permitting, cruise ships sail into the centre of Deception Island, which is a volcanic crater that is accessible through a narrow passage by sea. Cruise ship passengers can take the 'polar plunge' and experience the freezing Antarctic waters followed by a soak in volcanic hot springs that are heated by a shallow magma chamber, and which are emerging in certain areas along the beach. Assisted by the ship's crew, small ponds are dug into the black volcanic sand where water seeps out of the ground. While Deception Island is not a developed hot spring destination as such, this unparalleled adventure is part of the trip agenda of several cruise ships travelling the Antarctic. A note of caution is needed here though—Deception Island is an active volcano, which last erupted in

Fig. 1.4 The Pohutu Geyser is located in the Whakarewarewa Thermal Village and erupts at regular intervals. Rotorua, New Zealand

Fig. 1.5 The Blue Lagoon in the south of Iceland with the Svartsengi Geothermal Power plant in the background

1970 and has shown periods of unrest as recently as 2015 (Geyer et al. 2019).

Looking at more developed hot spring areas, Iceland's capital Reykjavik was designated a 'spa city' due the proximity to high-temperature geothermal fields, which provide an abundant supply of hot spring water. Several active volcanoes are responsible for countless hot springs all over the country and most towns in Iceland have geothermally heated public swimming pools. One of the greatest tourist attractions in Iceland is the *Blue Lagoon* south of Reykjavik and close to the international airport in Keflavik, which makes the Blue Lagoon a popular stop-over either after arrival or before departure. The distinct atmosphere of the Blue Lagoon is not only due to the surrounding ancient lava flows, which are devoid of vegetation, but also to the co-existence with the Svartsengi geothermal power station. The white steam clouds rising into the air from the power plant create a surreal backdrop (Fig. 1.5).

Environments that contain different types of hot springs are without doubt an invaluable contribution to the tourism industry, as they represent important areas for leisure and recreational pursuits. Based on their high visual impact, especially sectors such as geotourism, ecotourism and adventure tourism value hot springs as key features to attract outdoor enthusiasts.

During the last decade of the twentieth century new consumer trends focusing on a more holistic health awareness emerged. This affected health resorts and spas as the main stakeholders, as the ensuing wellness movement

Fig. 1.6 The white travertine cliffs of Hierapolis (Pamukkale, Turkey) are a major tourist attraction. Hot spring water cascades over limestone sills and has created terraced shallow pools filled with thermal water

contributed to the growth of hot spring spa tourism worldwide (Erfurt-Cooper 2009). This consumer driven development added new confidence for hot spring destinations to take advantage of their natural resources as an integrated part of this growing trend for health and wellness. With an overall preference for natural environments in combination with lifestyle and recreation as well as health and medical aspects, these trends continue to stimulate the growth of hot spring tourism (Erfurt 2012).

In Japan for example, health and recreation is to a large extent based on thousands of volcanic hot springs throughout the country, resulting in an industry that attracts an annual average of over 150 million overnight visitors (Erfurt-Cooper 2014). In Germany, many regions with health resorts and spas based on hot and mineral springs, which offer natural remedies, special facilities and a typical local ambience combined with thermal spa treatments (Erfurt 2012). Other areas with hydrothermal resources such as Pamukkale in Turkey not only offer hot spring facilities and resorts for health and recreational use but benefit from the unique attraction of the famous white travertine terraces formed by mineral deposits from the constant flow of hot spring water (Fig. 1.6).

Hot spring connoisseurs with an interest in authentic experiences generally know what they are looking for; however, consumers who are planning to visit spa resorts that offer thermal bathing, need to be aware of some subtle differences in the terminology.

Note Hot spring spas and thermal or mineral spas are not the same.

Because of the popularity of natural hot springs for health and recreational purposes, spa resorts and wellness facilities without access to hydrothermal resources are finding a middle ground by offering thermal or mineral spas or pools. If the natural hot spring source is not specified, the water is artificially heated and promoted as thermal bathing, because every facility with access to natural hot springs will clearly state this in their promotional literature, on their websites and at the entrance to their facilities. While mineral or thermal spas do not have access to genuine hot springs, they might be able to use cold mineral spring water sourced locally. After heating the water the term 'thermal mineral water' is not incorrect per se, although this may be misleading spa visitors who are led to believe that they are enjoying a genuine natural hot spring, especially when outdoor pools are designed to look natural. While the water may originate from a mineral-rich spring, in reality the water is heated to a pleasant temperature and possibly also treated with chemicals.

In most countries it is not against the law to advertise heated pools as hot spring pools; however, consumers should be aware that at some 'thermal' facilities this could very well refer to cold spring water or even glacier water that has been heated, with customised blends of minerals added. In fact, *hot pool* or *thermal pool* establishments that heat up cold spring water and chemically treat it, still advertise this as a natural experience, while at the same time pointing out that 'real' hot springs may carry the risk of contracting *Naegleriasis* (amoebic meningoencephalitis), a rare, but frequently fatal infection of the brain. Ironically, this infection can also be contracted in poorly maintained public and private swimming pools. Because health and safety are such important issues, more details about risk avoidance and risk management at natural hot springs are discussed in Chap. 9.

To indicate a genuine natural hot spring, the term geothermal water is frequently used to inform visitors, usually along with a current geochemical water analysis of the mineral content, temperature range, rate of discharge and other relevant facts about the origin of the hot spring. Therefore, any hot spring facility with access to a naturally heated water source will clearly state this in their promotional literature and on their online marketing platforms.

1.3 Aim and Structure of the Book

The content of this book is based on the research of natural hot springs in many countries and intends to cover as many aspects as possible related to hot springs, a renewable resource that deserves to be protected for future generations. Considering the worldwide distribution and the popularity of hot springs, a surprisingly small number of academic publications acknowledge the significance of this natural resource. While there are books about individual hot spring areas in different countries, these are predominantly guide books for hot spring connoisseurs. Their main focus is to encourage people to explore different regions and their unique hot springs. Also, there appears to be a scarcity of reference books that include a global overview of natural hot and mineral springs and their role in different industry sectors supported by up to date research.

While many research articles have been published about geothermal resources associated with specific topics such as geothermal energy, hydrothermal alteration or the existence of subaqueous hydrothermal vents or thermophiles and extremophiles, there is a general lack of consideration of natural hot springs and hot spring destinations in relation to their purpose in human societies. Existing information is scattered and/or outdated, which indicates potential gaps in the current literature and therefore a need for more comprehensive facts and updated figures based on active scholarly research in this field. The aim of this book is to bring together important aspects related to hot spring environments, their distinctive history and their contemporary use as well as their potential as a sustainable resource and their conservation. Not every facet can be included because the subject is too extensive; however, the ensuing chapters aim to contribute a collection of thought-provoking topics directly associated with natural hot springs and their geothermal origin.

One of the key objectives of this book is to take a broader view by using a cross-disciplinary approach to allow for a diversity of perspectives. As the following chapters' outlines show, the individual themes focus not only on clarifying the different origins of hot spring use in many countries, but also support the significance of hot springs as a renewable natural resource. Apart from a general introduction, this introductory chapter shares details why hot springs are so popular and provides the foundation for the content of the book. The next paragraphs briefly outline the topics of the following chapters to present an overview of the subjects addressed and how the book is structured. Also, the chapters do not require sequential reading—they can be explored as individual segments, complete with references and in some cases appendices.

Following the introduction, Chap. 2 (Geology of Hot Springs) examines the key geophysical processes related to the origin of natural hot springs and their occurrence. Parts of the chapter content are focusing in more detail on the different types of hot springs, their definitions and specific classifications.

In Chap. 3 the fundamental geochemistry of hot springs is explained with special references to their mineral content, while different sections of the chapter are looking at hydrothermal processes as well as unusual environments with extraordinary thermotolerant lifeforms.

Chapter 4 (Conservation of Hot Springs) raises awareness of environmental issues affecting hydrothermal resources and addresses the need for the conservation of hot springs, using global examples.

Chapter 5 (History of Hot Springs) delves into the past use of hot springs and provides a timeline leading back to ancient peoples. Records of early civilisations and their use of hot springs as well as their remaining geoheritage are discussed and supported with examples from various countries.

A different approach to the geoheritage of hot springs is taken in Chap. 6 (Hot Springs and their Cultural Heritage). Here the cultural and religious customs closely related to hot springs are analysed and compared. The mythology and regional legends that are interwoven with the traditional use of these natural resources are discussed with examples from a range of social and cultural settings.

In Chap. 7 (Hot Springs and their Natural Heritage) a more specific focus is used to explore the natural geoheritage of hot springs and their geographic settings. Topics discussed include the appeal of hot springs and active hydrothermal areas for various tourism sectors and their potential for future sustainable development. Protected sites with hot springs and other hydrothermal features are compared to aquatic theme parks based on natural hot springs with special consideration of their environmental impact in the near future.

Hot Springs and their curative value are investigated in detail in Chap. 8 (Hot Springs, Health and Wellbeing), by exploring the connection between hot springs and their application in the health sector. To provide an overview of the role of natural hot springs for medicinal purposes, examples from countries with hot spring spas and clinics are introduced.

Chapter 9 (Visitor Expectations and Risk Management at Hot Spring Destinations) analyses the expectations of hot spring visitors based on their reasons and motivations. This chapter also explains the main risks and hazards in relation to active hydrothermal features and points out the importance of relevant safety guidelines, as well as the need for suitable risk management strategies as potential hazards in active hydrothermal environments can have a direct impact on visitor safety.

Chapter 10 (Glossary of Hot Spring Related Terms) is a collection of definitions and technical terms that are central to the subject of natural hot springs, their geoheritage and other directly related subject areas. Because some of the terms may not be familiar to all readers, definitions have been selected to present a quick reference as well as to clarify similar sounding terminology.

Chapter 11, the concluding chapter (Hot Springs—A Final Overview) draws together several essential viewpoints discussed in this book and summarises some of the research related to different aspects of hot springs. The potential of natural hot springs as a renewable resource is briefly explored with examples from several countries to provide an overview of the versatility of this natural resource. In addition, the need for conservation of the cultural and natural geoheritage of hot springs is considered with an emphasis on sustainability.

The aim of this book is to provide useful and constructive insights into the geodiversity of hot springs combined with an understanding of the global distribution of this amazing natural resource. While this volume is contributing a comprehensive acknowledgement of the overall geoheritage of hot springs, it can and should be used as a stepping-stone for further research. As a final point, this text is intended for a wide audience interested in hot springs, including students, researchers, environmental managers, tourism planners, spa managers and other professional stakeholders, and of course anyone who enjoys a nice soak in a hot spring and would like to know more about them.

1.4 Suggested Readings

For readers interested in a general overview, the following titles provide some examples of the available hot spring literature. Most books are illustrated and focus on a specific areas or countries (Fig. 1.7a, b). Academic journals are another source of information about hot springs; however, research articles are not always easy to access, or they require to be purchased.

Death in Yellowstone: Accidents and Foolhardiness in the First National Park, 2nd Edn. by L.H. Whittlesey (2014). Amazon Digital Services, Inc. Roberts Rinehard.

Healing Springs: The Ultimate Guide to Taking the Waters – From Hidden Springs to the World's Greatest Spas by N. Altman (2000). Rochester, Vermont: Healing Arts Press.

Healing Waters—Missouri's Historical Mineral Springs and Spas by L. Bullard. University of Missouri Press, Columbia, MO.

Health and Wellness Tourism: Spas and Hot Springs by P. Erfurt-Cooper, and M. Cooper (2009). Channel View Publications, Bristol, UK.

Holistic Holidays in South Africa: Health spas, hot springs, magical places and sacred spaces by S. Spicer, and J. Nepgen (2005). Cape Town, SA: Human & Rousseau.

Japan's Hidden Hot Springs by R. Neff (1995). Tokyo: Tuttle Publishing.

Springs and Bottled Waters of the World – Ancient History, Source, Occurrence, Quality and Use by P. E. LaMoreaux, & J.T. Tanner (Eds). (2001). Berlin-Heidelberg: Springer.

Stories from a Heated Earth—Our Geothermal Heritage by R. Cataldi, S.F. Hodgson, and J.W. Lund (1999). Sacramento, California: Geothermal Resources Council, International Geothermal Association.

Taking the Waters in Texas: Springs, Spas and Fountains of Youth by J.M. Valenza (2000). Austin, TX: University of Texas Press.

The Japanese Spa: A Guide to Japan's Finest Ryokan and Onsen by A. Seki, and E.H. Brooke (2005). Tokyo: Tuttle Publishing.

Touring New Mexico Hot Springs. A Falcon Guide by M. Bishoff (2001). Guilford, CT: Globe Pequot Press.

Volcano and Geothermal Tourism: Sustainable Geo-Resources for Leisure and Recreation by P. Erfurt-Cooper, and M. Cooper (2010). Earthscan, London.

a

Fig. 1.7 **a** A selection of books about natural hot springs from various countries. **b** The majority of the literature about natural hot springs focuses either on a specific region or country

b

Fig. 1.7 (continued)

1.5 Summary

The objective of this chapter was to introduce some of the topics covered in this book to raise awareness about the abundant distribution of hot springs and the diversity of this natural resource (Fig. 1.8). By using a global approach, the content of the book aims to contribute different perspectives of hot springs and their individual place in many cultures and societies with overviews of the traditional and contemporary use of natural hot springs supported by examples from different regions.

On reviewing the literature and by visiting hot spring destinations in many countries, it became obvious that the sheer number of hot springs worldwide is overwhelming and without doubt beyond the scope of this book. However, while their diversification and abundance may present limitations, taking a multi-disciplinary approach intends to cover as many aspects related to natural hot springs and their geoheritage as possible. The vast amount of hot springs worldwide, together with their distinctive characteristics and their local significance, means these hydrothermal manifestations require more than one volume to do their diversity justice. Regardless, the objective of this book is to close some existing gaps in the current literature with the main focus on providing interesting insights and discussing key aspects of the geoheritage of hot springs.

Fig. 1.8 Natural hot spring in the Caldeira Velha, São Miguel, Azores

Private onsen at a hot spring resort in Hokkaido, Japan

1.6 Appendix

Appendix 1.1 Various Mud Pots

1, 2—Bubbling mud pools in Rotorua, New Zealand (left) and Japan (Beppu); 3, 4—Boiling mud pots in Iceland (Námafjall) and in Japan (Beppu); 5, 6—Furna do Enxofre, Graciosa, Azores and Mt Garandake, Kyushu, Japan; 7, 8—Dried mud pools in the Caldeira Velha, São Miguel, Azores and Krýsuvík, Iceland

References

Allaby M (2013) A Dictionary of Geology and Earth Sciences, 4th edn. Oxford University Press, Oxford

Bierstadt A (c.1881–1886) Old faithful. Oil painting. Retrieved from https://commons.wikimedia.org/wiki/File:Bierstadt_Albert_Old_Faithful.jpg

Erfurt-Cooper P, Cooper M (2009) Health and wellness tourism: spas and hot springs. Channel View Publications, Bristol

Erfurt P (2012) An assessment of the role of natural hot and mineral springs in health, wellness and recreational tourism. Unpublished Doctoral Dissertation, School of Business, James Cook University, Cairns

Erfurt-Cooper P (2014) Health and wellness tourism—an integrated approach. In: Pforr C, Voigt C (eds) Wellness tourism—a destination perspective. Routledge, London, pp 235–254

Erfurt-Cooper P (2018) Active hydrothermal features as tourist attractions. In: Fearnley CJ, Bird DK, Haynes K, McGuire WJ, Jolly G (eds), Observing the volcano world: volcano crisis communication. Advances in volcanology series. Springer, Berlin, pp 85–105. https://doi.org/10.1007/11157_2016_33

Fendek M, Rebro A, Fendeková M (1999) A spell cast: historical aspects of thermal spring use in the Western Carpathian Region. In: Cataldi R, Hodgson SF, Lund JW (eds) Stories from a heated Earth—our geothermal heritage. Geothermal Resources Council, Sacramento, pp 251–264

Fridleifsson IB (1999) Historical aspects of geothermal utilisation in Iceland. In: Cataldi R, Hodgson SF, Lund JW (eds) Stories from a heated Earth. Geothermal Resources Council, International Geothermal Association, Sacramento, pp 306–319

Geyer A, Álvarez-Valero AM, Gisbert G, Aulinas M, Hernández-Barreña D, Lobo A, Marti J (2019) Deciphering the evolution of Deception Island's magmatic system. Sci Rep 9(373). https://doi.org/10.1038/s41598-018-36188-4

Gringer (2009) Pacific ring of fire. Public domain. Retrieved from https://commons.wikimedia.org/wiki/File:Pacific_Ring_of_Fire.svg

Heasler HP, Jaworowski C, Foley D (2009) Geothermal systems and monitoring hydrothermal features. In: Young R, Norby L (eds) Geological monitoring: Boulder. Geological Society of America, Colorado, pp 105–140

Keary P (1996) The new Penguin dictionary of geology. Penguin Books, London

Pliny, the Elder (c.77 AD) Natural history, vol XXXI, pp i–ii

Yellowstone NP (2020) Recreation visitators (1904—Last Calendar Year). Retrieved from https://tinyurl.com/kgk5xh8

Fumarole in the Whakarewarewa Geothermal Reserve, New Zealand

The Geology of Hot Springs

2

Contents

P. Erfurt, *The Geoheritage of Hot Springs*, Geoheritage, Geoparks and Geotourism,
https://doi.org/10.1007/978-3-030-60463-9_2

A small boiling pond is shrouded in mist on a cold and rainy day. Located at the edge of the pond and covered with large rocks is a high temperature degassing steam vent (Kyushu, Japan)

2.1 Introduction

Hot springs are a remarkable natural resource and have fascinated civilisations all over the world for thousands of years. While most people are aware of the existence of hot springs, not everybody is familiar with the key geological processes that lead to the presence of these springs at so many locations. As mentioned in the first chapter, hot springs are naturally occurring hydrothermal resources that arrive at the surface of the Earth in various forms and at various temperatures. Not all springs are hot though; some are warm or merely tepid, especially when compared to the ambient air temperature at the source.

Note Natural hot springs, independent of their temperatures or whether they are of volcanic or non-volcanic origin, are frequently referred to as geothermal or thermal springs. The correct definition for a hot spring nevertheless is hydrothermal spring (see Chaps. 1 and 10).

Warm water, regardless of its origin, is lighter than cold water and readily rises to the surface if unconfined, with very few restrictions where hot springs can emerge. In **Greenland** near *Uunartoq* hot springs occur along the shoreline, in stark contrast to icebergs floating in the coastal waters. Hot springs also emerge from the bottom of lakes, feed into rivers and streams or cause mud pots to boil. Some of the more common hydrothermal surface and subsurface manifestations are briefly described in Table 2.1 followed by more details throughout this chapter.

Wherever feasible, hot spring water is used for domestic and industrial purposes. It is therefore essential to establish management strategies to oversee and regulate the sustainable extraction of geothermal groundwater. To accomplish this challenging task, the characteristics of suitable aquifers are usually identified through studies of the geological processes, assessments of geochemical analyses of the water, and most importantly by closely monitoring discharge against recharge (Guillou-Frottier et al. 2013).

2.2 Hot Springs and Their Origin

Hot springs are a natural source of water that is heated underground near active magma reservoirs or cooling igneous bodies where it becomes less dense and rises under pressure to the surface (Corbett 2001; Erfurt 2012). This process depends on a heat source at depth, as well as areas of recharge and discharge. If the heat source is derived from an active magma chamber, the subterranean networks are classed as volcanic or magmatic hydrothermal systems (Delmelle et al. 2015; Fischer and Chiodini 2015). This includes about 70% of all hydrothermal vents on the seafloor (De Ronde et al. 2015). In comparison, geothermal systems consist of subterranean rock formations that can host hot

Table 2.1 Hydrothermal manifestations that are common on the Earth's surface as well as underwater

Hydrothermal features	Description	Examples
Explosion craters Hydrothermal eruption craters	Caused by violent eruptions of steam, hot water, mud and rocks—the craters can be small in areas of hydrothermal activity or large volcanic crater lakes	**Craters of the Moon,** Taupo, New Zealand **Krýsuvík**, Seltún, Iceland **Porkchop Geyser,** Yellowstone **Raupo Pond,** Waimangu Rift Valley, NZ **Sileri Crater,** Dieng Plateau, Indonesia **Dallol Crater**, Danakil Depression, Ethiopia
Fumaroles Solfataras Figs. 2.7, 2.9 and 2.10	Hydrothermal steam or gas vents that commonly occur in or near active volcanic environments—can also be present around dormant and extinct volcanoes—high temperatures of emissions	**Campi Flegrei** (Phlegraean Fields), Italy **Námafjall,** Krafla, Iceland **Stefanos crater**, Nisyros volcano, Greece **Caldeiras das Furnas,**São Miguel, Azores **Xiaoyoukeng,** Yangmingshan NP, Taiwan **White Island,** New Zealand
Geysers Figure 2.7	Hot spring that discharges columns of hot water and steam at and above boiling point—popular tourist attraction due to its visual appeal	**Norris Geyser Basin,** Yellowstone, USA **Valley of the Geysers**, Kamchatka, Russia **El Tatio Geyser Field,** Chile **Haukadalur** Geyser Field, Iceland **Whakarewarewa** Geyser Flat, Rotorua, NZ **Umi Jigoku,** Beppu, Japan
Hot springs (health and recreational use) Figure 2.3	Natural springs that are fed by a constant flow of groundwater at temperatures that allow for the direct use such as rehabilitative and recreational bathing	**Peninsula Hot Springs,** Victoria, Australia **Polynesian Spa,** Rotorua, New Zealand **Blue Lagoon**, Iceland **Hanmer Springs**, New Zealand **Cebu,** Philippines
Hot springs (viewing only) Figs. 2.2 and 2.7	Hot springs that emerge under pressure and at extreme temperatures—commonly used as visual attractions in volcanic areas	**Yellowstone** National Park, USA **Dallol**, Danakil Depression, Ethiopia **Caldeiras das Furnas,**São Miguel, Azores **Wai-O-Tapu** Thermal Wonderland, NZ **Kinryu Jigoku,** Beppu, Japan
Hot springs (geothermal cooking, baking and washing)	Hot springs at boiling point, steam vents and hot sand pits are used in many countries for traditional domestic purposes and as tourist attractions	**Kannawa Onsen District,** Beppu, Japan **Caldeiras das Furnas,**São Miguel, Azores **Ngāraratuatara**—Māori cooking, New Zealand **Semliki** hot springs, Uganda, Africa **Rachel Spring,** Rotorua, New Zealand
Hot rivers and streams Figure 2.14	Watercourses fed hot springs at varying temperatures—hot springs can be the source of the stream or discharge into the stream elsewhere	**Rio Tabacon,** Costa Rica **Kerosene Creek,** New Zealand **Shanay-Timpishka** - Boiling River of the Amazon, Peru **Mainit Sulphuric Hot Springs,**Philippines **Hot Creek Gorge,** USA
Mud ponds, mud pots (also known as mud volcanoes) Figs. 2.5 and 2.16	Boiling or bubbling mud ponds contain a mixture of hot spring water, dissolved rock and hydrogen sulphide. The mud can reach temperatures up to boiling point depending on the source	**Lake Oyunuma**, Hokkaido, Japan **Hells Gate** mud volcano, Rotorua, New Zealand **Námafjall,** Krafla, Iceland **Furna do Enxoffre,** Graciosa, Azores **Caldeiras das Furnas,**São Miguel, Azores **Mt Garandake,** Kyushu, Japan
Travertine formations Sinter terraces Figure 2.8	Limestone/sinter terraces formed by the deposition of calcium carbonate or silica evaporating from thermal spring water	**Mammoth Hot Springs**, Yellowstone NP, USA **Pamukkale** WHA, Turkey **Badab-e Surt** Sinter Terraces, northern Iran **Hammam Mesqoutine,** Algeria

(continued)

Table 2.1 (continued)

Hydrothermal features	Description	Examples
Warm, hot or boiling lakes Figs. 2.12 and 7.13	Volcanic crater lakes, lakes or ponds which are fed by submerged hot springs that are emitting large volumes of extremely hot water	**Lake Hévíz,** Hungary **Lake Kinrin**, Yufuin, Japan **Lake Bogoria,** Kenya **Frying Pan Lake,** Waimangu Rift Valley, NZ **Boiling Lake**, Dominica (Caribbean) **Inferno Crater Lake,** Waimangu Rift Valley, NZ
Waterfalls and hot springs Figs. 2.14 and 2.15	Warm or hot streams cascading over natural rock formations or travertine terraces	**Saturnia** Natural Spa, Tuscany, Italy **Mainit Sulphuric Hot Springs,**Philippines **Finca el Paraiso**, Guatemala **Namtok Ron Khlong Thom**, Krabi, Thailand **Tikitere** (Hells Gate), New Zealand

Source Compiled by Author

fluids and steam but can also consist of only 'dry hot rock'. This shows that while some geothermal systems are also hydrothermal, others are not (Arnórsson et al. 2015).

Hot springs, geysers and fumaroles are typical surface characteristics of geothermal heat (Branney and Acocella 2015), but their occurrence is not limited to specific regions, or to a particular rock type. While natural springs of different temperatures can emerge in the same area, they may be discharging from separate hydrologic systems.

Springs also differ greatly in their rate of discharge, which depends on factors such as water pressure in the aquifer, size of the reservoir basin and the rate of replenishment through atmospheric water and infiltration from lakes, rivers and oceans (Fig. 2.1). Three original water sources are responsible for the flow of natural hot springs, although in most hydrothermal systems *meteoric water* is the dominant fluid, which circulates underground after it enters the groundwater system primarily from rainfall and meltwater of snow and ice.

Another source is *connate water*, which refers to fluids that are entrapped in the interstices of sedimentary or extrusive igneous rocks at the time of their formation. Connate fluids often have a high content of chemical constituents or can consist of highly saline groundwater if formed in marine environments. The third source is *magmatic* or *juvenile water*, the result of volcanic processes and located in large volumes within deep magma bodies. It is *a residual fluid formed during the late stages of crystallisation of the body*, has never been in the atmosphere and is only transferred by *episodic magmatic fluid expulsion* to contribute directly to hydrothermal systems, although this contribution may be too small to be discovered (Allaby 2013:288; Stimac et al. 2015:807). Fluid expulsion takes place during volcanic eruptions, when juvenile fluids are released into the atmosphere for the first time and enter into the hydrologic cycle. Magmatic or juvenile water can form in very large quantities and can remain underground indefinitely.

Apart from these sources of recharge another method of aquifer replenishment is well injection, whereby water is re-injected into an artificial well that reaches down into fractured rocks; a common practice in countries where over-exploitation has led to a reduced supply of 'natural' hot water (Erfurt 2012; Kralj et al. 2009; Seibt et al. 2005).

Other factors differentiating hot springs are their heat source and the depth from which they rise. Volcanic hot springs can acquire their temperature closer to the surface if an active heat source in the form of a magma chamber or a body of cooling igneous rock is present. Artesian springs derive their heat from depth, pressure, friction and time spent in circulation underground before ascending. The content of dissolved components in hot springs depends on the fluid temperature in the hydrothermal system as well as on the composition of the surrounding host rock. Based on these factors and the duration underground, the composition of hot spring water and its content of total dissolved solids (TDS) can vary significantly.

2.2.1 Hot Springs of Volcanic Origin

Volcanic hot springs are primary hot springs that are common in low- and high-temperature geothermal fields associated with active volcanic environments (Fig. 2.2). In high-temperature geothermal fields (>180 °C), partly molten rock heats up groundwater that is passing through permeable rock formations following fractures and faults. Based on their close proximity to active volcanism, these hot springs contain solids, gases and trace elements generated by

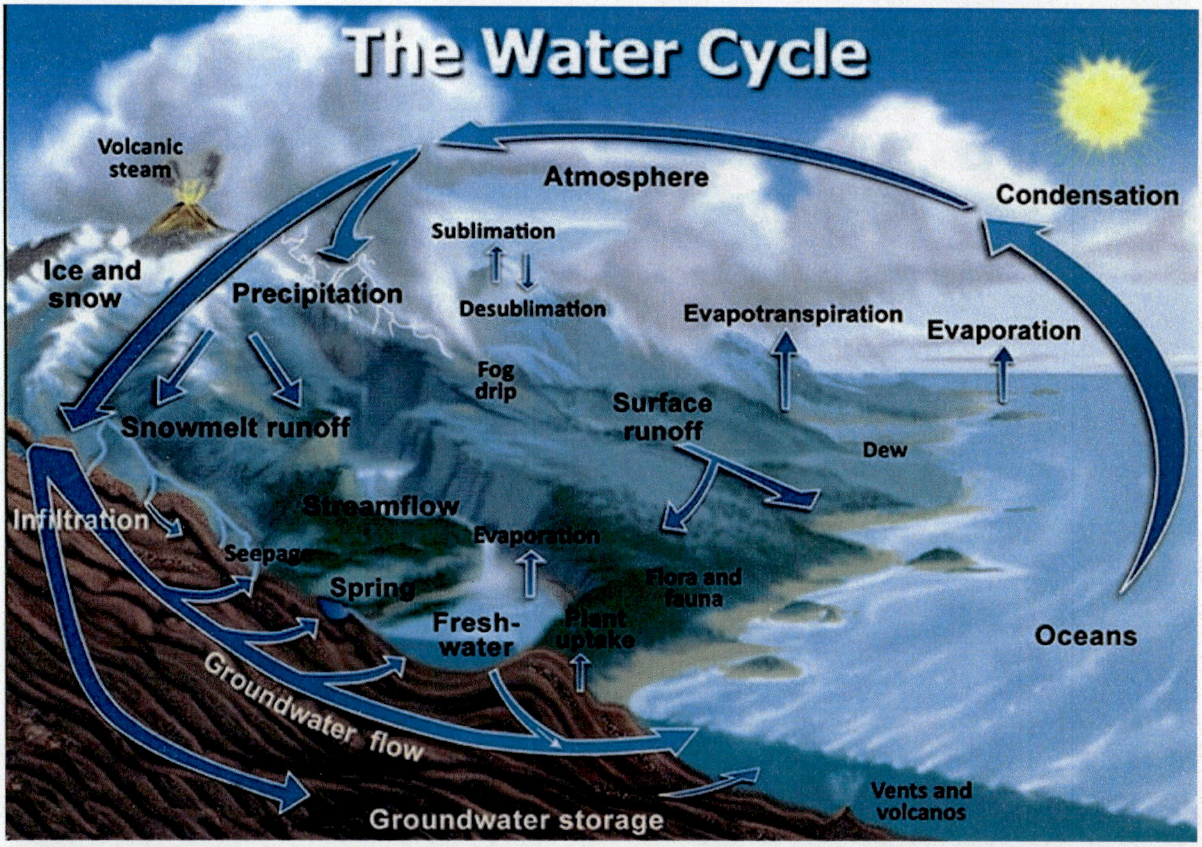

Fig. 2.1 The hydrologic cycle is the perpetual natural circulation of water involving evaporation, condensation and precipitation. *Source* Evans and Perlman 2013

magmatic processes that influence the characteristics of hot springs and determine their geochemistry (see also Chap. 3—The Geochemistry of Hot springs). Many hydrothermal resources never reach the surface and are only discovered during exploration drilling for other natural resources such as gas or oil (Fridleifsson et al. 2008; Stimac et al. 2015).

Example 2.1 Yellowstone, Wyoming, USA

In the case of Yellowstone, hydrothermal fluids originate from meteoric water and circulate within thick layers of rhyolite and ignimbrite, where temperatures can reach up to and above 350 °C at depths of over 500 m.

Source Gibson and Hinman (2013); Lowenstern and Hurwitz (2008)

2.2.2 Hot Springs of Non-volcanic Origin

In non-volcanic regions hot springs emerge from artesian basins where the groundwater derives its heat through convection. These artesian reservoirs produce hot springs similar to those emerging in active volcanic areas (Fig. 2.3). Depending on depth and pressure, the temperature of subsurface water can vary between 25 and >100 °C. This is due to the fact that water circulating to a depth of 2–3 km can be heated substantially above the ambient surface temperature at a given location.

Geothermal Gradient

In 1963 the US Geological Survey Bulletin 1172 determined geothermal gradients as *due to dissipation of subsurface heat, which is not everywhere the same, and gradients vary from place to place because of the differences both in rock and in regional and local heat sources* (Lovering and Goode 1963:4). Since then the knowledge about subterranean heat transfer has increased; although, what happens out of our reach is still to some extent left to speculation and assumption based on currently available data.

Contained within the Earth's crust, natural thermal energy generates significant amounts of heat, which is transferred to the surface generally by conduction, causing a change in temperature of approximately 25–30 °C per kilometre depth

Fig. 2.2 A boiling spring rises to the surface under pressure at the Caldeiras das Furnas, São Miguel Island, Azores

(Fridleifsson et al. 2008). This temperature change is known as the *temperature gradient* or *geothermal gradient* and increases with depth, although extrapolating downward based on an averaged regional temperature gradient can lead to considerable miscalculations of the temperature variations in the heat transfer. Deviations can result from fluctuations in thermal conductivity and thermal convection or from excessive heat flowing from 'hot rocks' (Guillou-Frottier et al. 2013). In volcanic areas this variation in temperature mainly depends on the size and/or proximity of a magma chamber or on the residual heat from a slowly cooling granite intrusion, which can heat subsurface water at rather shallow depths.

Throughout the world, artesian basins contain heated water stored in sedimentary rock formations. The largest artesian basins are the Great Artesian Basin (GAB) in Australia (Habermehl and Pestov 2002) (largest worldwide) and the Guarani Aquifer in South America (Fig. 2.4). These reservoirs are the source of countless cold, warm and hot springs that are free flowing from the ground based on varying degrees of pressure in the underlying aquifer. But not all artesian springs are free flowing. Bores or wells are often used to access the aquifer and to cause the water to rise until it reaches its hydrostatic equilibrium with a constant flow velocity.

Other non-volcanic thermal water sources are circulating in deep carbonate rock aquifers in karst drainage systems, where they move by gravity and follow topographic gradients (Goldscheider et al. 2010). Until the water discharges as warm or hot springs, various sources of heat and transport systems influence the formation and functioning of the hydrologic karst system. In those aquifers thermal water follows fault lines in highly permeable limestone formations by using subterranean conduits and passages as a support network, which essentially controls the water flow within a subterranean system (Luhmann et al. 2011). Natural hot springs all over the world are associated with karst aquifers and are used as an important resource for health and recreational purposes. For example, the hot springs of *Budapest* (Hungary), one of

Fig. 2.3 The Mataranka Thermal Pools are located in a tropical bush setting, where the thermal water rises from a subterranean reservoir at 34 °C. Northern Territory, Australia

Europe's largest natural hot spring systems, are supplying numerous baths and health facilities with thermal water originating in carbonate rock formations. In countries such as Algeria, Canada, China, Germany, France, Italy, Jordan, Switzerland, Tunisia, Turkey, as well as in many other regions worldwide hot springs emerge from subterranean thermal karst systems (Goldscheider et al. 2010).

Hydrothermal systems in non-volcanic areas can also obtain their heat from radioactive decay taking place in deep seated impermeable crystalline granite formations. For example, the *Paralana Hot Springs* in South Australia are described by several scientists as deriving their heat from the decay of radium and uranium (Anitori et al. 2004; Greene et al. 2016; Wright et al. 2012). However, the Paralana Hot Springs are not recommended for recreational use due to concerns about the elevated levels of their natural radioactivity (see also Chap. 3). Other hot springs containing low levels of radioactive elements are known around the world where they are used for health and recreational purposes.

The following examples describe three different hot spring locations that derive their hot water from artesian reservoirs.

Example 2.2 Germany

Throughout Germany hydrothermal resources of non-volcanic origin are accessible in porous, fractured or cavernous rocks, with Northern Germany offering the most favourable geological conditions made up of sandstones, clay and carbonates. Aquifers contain productive horizons bearing 40–120 °C hot water at depths ranging from 1000 to 3000 m and are located in the North German Sedimentary Basin, the Molasse Basin in southern Germany, and along the Upper Rhine Graben. These natural reservoirs hold sufficient hydrothermal resources with adequate flow rates, discharge temperatures of up to 72 °C (Aachen), and high concentrations of mineral elements to supply countless

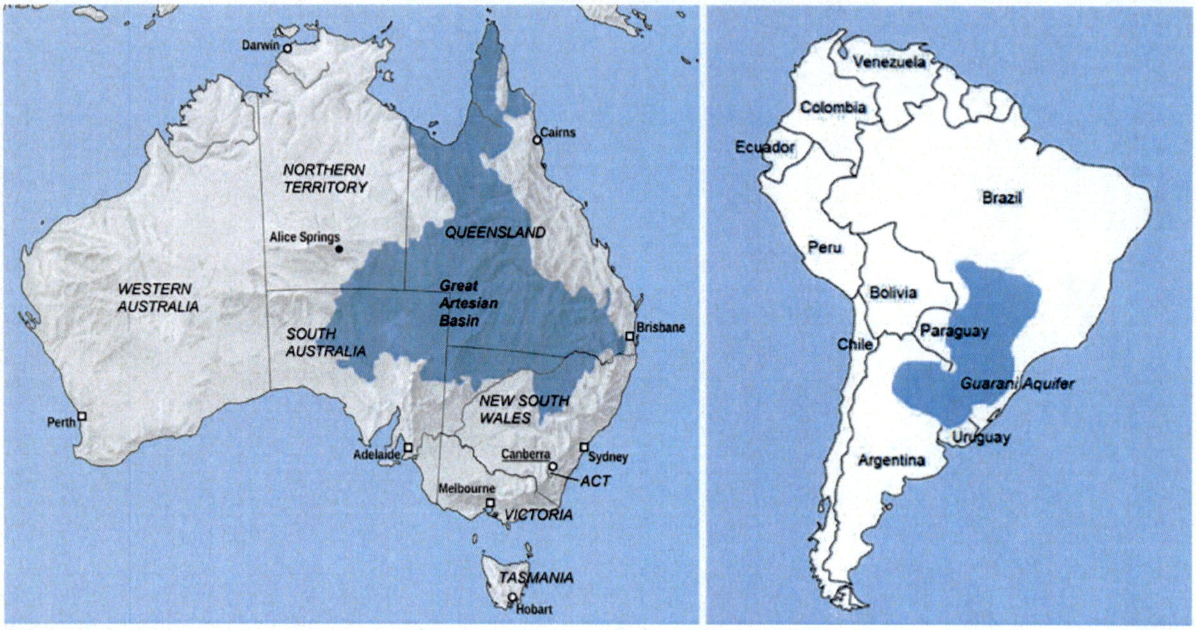

Fig. 2.4 The approximate size and location of the Great Artesian Basin in Australia and the South American Guarani Aquifer. *Source* Public Domain. Maps modified by the Author

health resorts and spa towns in almost every region in Germany with natural hot spring water.

In the city of *Baden Baden*, twelve sodium chloride springs (68 °C) with a combined flow rate of approximately 285 million litres per year (nearly 800,000 l/d) provide the health spa industry with hot water, while in the classic spa city of *Wiesbaden,* natural hot springs are used to supply large public baths, resorts and many specialised rehabilitation clinics and healthcare centres. The temperature of Wiesbaden's hot springs ranges between 46 and 66 °C with a flow rate of approximately 2 million litres per day, which makes Wiesbaden the second highest hot water producing health resort town after *Aachen* with 3.5 million litres per day (Table 2.3).

Source Erfurt (2012), Rockel et al. (1997), Schellschmidt et al. (2010), Seibt et al. (2005), Wiesbaden (2010)

Example 2.3 Australia

In the southern state of Victoria, the *Peninsula Hot Spring Spa* on the Mornington Peninsula is another example of natural hot spring water sourced from an artesian reservoir and has become a major hot spring destination for domestic and international visitors since the opening of the first stage in 2005. Peninsula Hot Springs is a filtration hot spring of meteoric origin which rises from an aquifer 637 m below the ground. The water arrives at 54 °C and a near neutral pH of 6.8 before it is distributed into many different hot spring pools, which vary in temperature between 34 and 42 °C. Based on its high content of dissolved mineral components (TDS >3500ppm), the water is classified as a genuine hot spring.

Source Erfurt (2012), Peninsula Hot Springs (2019)

Example 2.4 South America

The South American transboundary *Guarani Aquifer System* (GAS) derived its name from the Guaraní Indigenous Nation, although in Uruguay and Argentina it is also known as the *Tacuarembó Aquifer* (Fig. 2.4). As one of the largest artesian basins worldwide it covers an overall area of approximately 1.2 million square kilometres, with the water bearing rock formations of the GAS located in the Paraná sedimentary basin underneath southern Brazil, Uruguay, Paraguay and Argentina. According to hydrogeochemical research in some areas of the GAS the groundwater displays great compositional variations, allowing the water to be used for public supply (80%), industrial and agricultural (15%), as well as for recreational purposes (5%). In the border region between Uruguay and Argentina several thermal aquifers were identified during periods of oil exploration and regarded as a highly productive and sustainable resource if managed responsibly. By the mid-1990s the development of a health and recreational spa industry had started in Uruguay, which was followed by similar developments on the Argentinian side. To supply the many hot spring facilities, artesian wells were drilled into different levels of the low temperature sedimentary basin to extract the water at between 30 and 50 °C from three thermal aquifers at an alarming rate. What was originally thought as a resource with a large potential for future extraction, started to diminish after a few years as the GAS was progressively more exploited (see also Chap. 4).

Source Bonotto and Santos (2007), Foster et al. (2009), Gomez et al. (2010), Hirata et al. (2011), International Waters Governance (2018), Kern et al. (2007), Montaño and Peel (2003), Pesce (2002), Rabelo and Wendland (2009), Wendland et al. (2007)

2.3 Hydrothermal Phenomena

2.3.1 Surface Manifestations

Wherever they occur, hot springs, geysers or fumaroles are attracting people because of their unique visual impact. Most of these natural phenomena are associated with volcanic environments and are common in many countries (e.g. Americas, China, Iceland, Indonesia, Italy, Japan, New Zealand, Portugal, Turkey) where they have been developed into tourist destinations. The more powerful displays usually take place in the vicinity of active volcanoes, where hydrothermal manifestations are observable expressions of underlying tectonic forces. Other remarkable features include boiling lakes, bubbling mud pots (Fig. 2.5), hot rivers and streams, travertine formations and degassing, steaming ground (Di Napoli et al. 2009; Erfurt-Cooper 2010; see Table 2.1). Most hydrothermal areas present a combination of several different features based on similar geological processes.

To identify the location of hot springs and other hydrothermal manifestations the following definitions clarify whether these settings are above or below the Earth's surface:

- **Subaerial**—located and/or occurring, existing, formed or taking place on or close to the Earth's surface;
- **Subaqueous (subaquatic)**—located and/or occurring, existing, formed or taking place under water—refers to the ocean as well as to lakes and rivers;
- **Sublacustrine**—located and/or occurring, existing, formed or taking place below the surface of a lake and/or at the bottom of a lake;
- **Submarine**—located and/or occurring, existing, formed or taking place below the surface of the ocean and/or at the bottom of the seafloor.

All hot springs, geysers and fumaroles emerge from hydrothermal systems, where the temperature range is estimated to be typically between 50 °C and up to over 400 °C in deeper reservoirs (Haase et al. 2009), depending on whether they are magmatic systems or rock/water-based systems. Hydrothermal systems present a continuum of resource temperatures that is relatively open-ended (Heasler et al. 2009), with high-temperature areas of special interest for the generation of geothermal energy. In comparison to the energy sector, the temperature range of hot spring water used for health and recreational facilities generally lies between 37 and 100 °C, often with the need to cool the water down before use. This can be done in several ways; either by adding cold water, which dilutes the amount of mineral components or by aerating the hot water to preserve the authenticity of the spring water (Fig. 2.6).

Geysers

Hot springs come in many forms and sizes (Example 2.5). And while some are flowing quietly from the ground, others, such as geysers, appear with a noisy splash and great visual effect, discharging water and steam at extreme temperatures. Most geysers erupt at intervals, but there are some that continuously discharge steam and water (Fig. 2.7). Their eruptive activity is commonly based on large quantities of water in underlying hydrothermal systems (Ladd and Ryan 2016), where one or more water filled chambers are connected through a main conduit to a hydrothermal network. The water contained underground is heated by conduction and rises under pressure. Depending on the width of the conduit, the pressure of the hot water increases the temperature even more, und the geyser discharges into the air in a column of steam and scalding water.

Example 2.5 Iceland

One of the most famous hot springs of this type is the original *Geysir* in Iceland, which provided the name for all hot springs of this kind. Today *Geysir* only erupts occasionally, induced by treatments with a surfactant (detergent, soap). Not surprisingly, unique natural phenomena such as geysers are of great interest for tourists and are highly sought-after must-see destinations worldwide (See Figs. 9.10 and 9.11; Appendix 7.4).

Source Erfurt (2012), Hróarsson and Jónsson (1992)

Fig. 2.5 Mud pots at Námafjall derive their heat from the Krafla volcano, a high-temperature area in the northeast of Iceland

The complex processes and subsurface conditions that cause geysers to erupt at intervals have been the subject of extensive studies for over two centuries. More recent research into the eruptivity of spouting springs proposes that CO_2 may also play a role as a necessary component in triggering eruptions of certain geysers as reported about the *Spouter Geyser* of Yellowstone National Park. While geysers are generally viewed as systems that are simply based on heat and water, this observation might be too general without recognising other processes potentially involved in geyser eruptions (Ladd and Ryan 2016).

Travertine Formations

Travertine or sinter terraces consist of siliceous precipitates of hydrothermal origin, also known as siliceous sinter or concretionary limestone, depending on their texture, composition and porosity. Travertine terraces, dome-shaped mounds and other depositional features build up as the hot spring water cools and mineral elements such as calcium carbonate are precipitated, and form silicified layers that cover entire hillsides (Fig. 1.6). Some of these formations grow into large, raised mounds (Fig. 2.8) where the hot springs emerge and appear like petrified waterfalls (e.g. Hammam Meskhoutine, Algeria—Fig. 3.10; Hierve el Agua, Mexico) (Bonny and Jones 2003; Boudreau and Lynne 2012; Chafetz and Guidry 2011; Jones and Renaut 2003).

Travertine terraces are also located in countries including Hungary, Iran (Appendix 7.4), Turkey (Appendices 3.2, 5.2) and the USA (Chap. 3, Table 3.2, Appendix 7.4b). Their formations, their hydrothermal origin and geochemical characteristics are discussed in more detail in Chap. 3, which covers the geochemistry of hot springs.

Fumaroles

Fumarolic activity is a common sight in active volcanic environments, where open vents release steam and gas at temperatures up to and above 100 °C from underground hydrothermal systems into the atmosphere (Fig. 2.9). Gases

Fig. 2.6 Hot springs emerging at high temperatures are cooled until the water is suitable for health and recreational purposes. This can be achieved either by using high-tech cooling systems or with the simple process of aerating the hot water. Observed at the Suginoi Hotel Complex, Kyushu (Japan)

such as carbon dioxide (CO_2), hydrogen chloride (HCl), hydrogen sulphide (H_2S) and sulphur dioxide (SO_2) are an indication of active volcanism and present a potential danger to the public, especially if a person suffers from respiratory conditions.

Around vents where cooling vapour contains sulphur elements, layers of yellow crystal deposits are formed, which can be mined as an industrial resource. In the east of **Java** (Indonesia), active fumaroles on the flanks of the *Kawah Ijen* volcano complex are producing large amounts of pure sulphur, which is harvested under hazardous conditions in an extremely hostile environment. Sulphur mining also used to take place on **White Island** (*Whakaari*), New Zealand (Figs. 2.10 and 2.11), until a landslide in 1914 killed ten mine workers. A few years later the operation was shut down permanently (Te Ara 2019).

Note On the 9 December 2019 the volcano Whakaari (White Island) erupted unexpectedly and many tourists close to the crater did not survive. It was one of the most tragic volcanic events in recent times, resulting in 22 fatalities and dozens of people receiving emergency treatment in intensive care units.

Fumaroles are frequently identified as solfataras (Italian for *sulphur place*). *Solfatara* is also the name of a fumarolic crater in the Phlegraen Fields (Campi Flegrei) near **Naples** (Italy); a densely populated area. Close to a million people live here in the neighbourhood of extensive fumarolic activity and ongoing tectonic unrest (Mayer et al. 2016; Petrillo et al. 2013; Todesco et al. 2003; Troiano et al. 2014). As a result, the Phlegraean Fields are located in a region of complex geological processes, which have created remarkable hydrothermal surface manifestations. Based on the level of hydrothermal and possible future volcanic activity of Mount Vesuvius there is a potential danger for the surrounding urban developments in this vast metropolitan region (Todesco et al. 2003; Fig. 2.12).

Despite the hazardous aspects of active volcanic environments, fumarolic activity is also one of the main attractions in the *Yellowstone National Park* (USA), where

Fig. 2.7 This continuously spouting geyser at the Umi Jigoku (Kyushu, Japan) is covered with a concrete roof to contain the scalding water column and for the protection of park visitors

thousands of hydrothermal features draw millions of visitors every year (4,020,288 in 2019; NPS Stats 2020). Steaming fumaroles also exist in Chile, Mexico, Costa Rica, Japan, New Zealand, Iceland, China, the Azores and Kamchatka; all countries or regions with an abundance of hydrothermal manifestations based on volcanic activity.

2.3.2 Submerged Hot Springs

Hot Lakes

Hot or sometimes even boiling lakes are usually related to volcanic activity such as crater lakes but can also include former freshwater lakes or ponds in areas of hydrothermal activity. These lakes are fed by submerged hot springs or fumaroles, which are discharging large volumes of hot water and steam into the body of water. The temperature of hot lakes can vary based on activity levels in the underlying hydrothermal system, and such lakes are always related to active volcanism and can appear after a hydrothermal or volcanic eruption. In New Zealand the hot lake *Frying Pan* (45–55 °C) was created when Mount Tarawera erupted in 1886 and 'rearranged' the surrounding landscape (Fig. 2.13).
According to various sources the *Frying Pan Lake* in the Waimangu Valley is one of the largest hot lakes in the world, followed closely by the aptly named *Boiling Lake* in Dominica, although the size, temperature and water level of both lakes seems to fluctuate over time (GNS Science 2019; Scott 1992; Vandemeulebrouck et al. 2008; Virtual Dominica 2018). The Boiling Lake is located in the *Morne Trois Pitons National Park*, which was recognised as a UNESCO World Heritage site in 1997 for its *rare combination of natural features of World Heritage value,* including not just the Boiling Lake, but also 50 fumaroles, numerous hot springs, and five volcanoes (UNESCO 2019).

Another example of sublacustrine hydrothermal activity is *Lake Rotomahana*, a crater lake in the *Okataina Volcanic Zone* in northern New Zealand (Stucker et al. 2016). In fact, there are several lakes in New Zealand where hydrothermal venting occurs with the *Horomatangi System* in *Lake Taupo* considered as a sublacustrine equivalent to areas of hydrothermal venting on the seafloor (De Ronde et al. 2002;

Fig. 2.8 Iron-rich travertine mound created by the continuous flow of hot spring water. Karahayıt, Denizli Region, Turkey

Jones et al. 2007). Lake-floor hydrothermal activity is also reported from *Lake Baikal* in Siberia (Granina et al. 2007; Crane et al. 1991), from *Lake Tanganyika* in East Africa (Elsgaard and Prieur 2011), as well as from *Lake Yellowstone* in Wyoming, USA (Fournier 1989; Yang et al. 2011). Not all lakes are warm or hot throughout though but comprise areas where hot springs emerge at the lake-floor and generate warm plumes. In cold climate areas this can be enough to prevent a lake from completely freezing over (Fig. 2.14).

Hot Streams and Waterfalls

Depending on their location, hot springs can either add to a watercourse by discharging large volumes of heated water from subterranean hydrothermal systems, or they can start a hot stream if the topographic gradient allows the runoff to follow a natural channel. Depending on the topography, hot spring-fed waterfalls are not uncommon in regions rich in volcanic hot springs.

In the Waikite Valley in New Zealand the *Otamakokore Stream* flows from the boiling *Te Manaroa* spring (98 °C), to create a natural watercourse fed by nearly 2.6 million litres of steaming hot water per day. *Hot Creek Gorge* in the Long Valley Caldera (California, USA) is another example of hydrothermal processes related to active volcanism. Here, hot springs emerge alongside the stream known as Hot Creek and contribute over 20.7 million litres of hot water per day to rock pools and into the waterway (USGS 2019).

While hot or boiling rivers and streams are primarily related to active volcanic environments, one boiling river, hidden in the rainforest of Peru in South America, is located hundreds of kilometres away from volcanic activity. This river, locally known as *Shanay-Timpishka,* reaches temperatures between 49 and 91 °C and, according to Andrés Ruzo, the scientist who discovered this river, there are sections of the river that are actually boiling (Ruzo 2016). Generally though, the water temperature of hot rivers decreases relatively quickly downstream as cold water from other sources mixes with the hot water and the geothermal influence declines (Duggan et al. 2007). Hot waterfalls are only found in connection with hot spring-fed water courses, when these encounter terraced rock formations or concrete barriers and continue their flow as steaming cascades (Figs. 2.15 and 2.16).

Mud Pools

Mud pots or mud pools are a different kind of hot spring where the boiling and bubbling of the mud is caused by hydrothermal vents degassing below the surface. Heat combined with just enough water to keep the mud liquid cause the water to react with hydrogen sulphide and form

Fig. 2.9 A slowly degassing fumarole in São Miguel's Caldeiras das Furnas, Azores

sulphuric acid, which dissolves the surrounding rock material into very fine particles of silica and clay. These particles mix with hot water and steam and create the appearance of boiling and bubbling mud (Fig. 2.5). Apart from all shades of white, grey and black, mud pots can be coloured bluish-grey, yellow, pink, orange or even purple, depending on the presence of different oxides (Fig. 2.17; Appendix 1.1).

The viscosity of bubbling mud pots can differ greatly. Some are very 'slow', with air bubbles struggling through the thick clayey mud until they reach the surface, where they burst into ever changing forms and shapes (Appendix 1.1). Other, more liquid mud pots boil vigorously, and splash mud droplets everywhere. If dry-degassing fumaroles get flooded, they can turn into boiling ponds or mud pots, depending on their geophysical preconditions.

2.4 Submarine Hot Springs

2.4.1 Hydrothermal Vents

Apart from hydrothermal springs submerged in hot streams, lakes and mud pots, a rather unique type of spring is found on the seafloor in geologically active regions such as at mid-ocean ridges (MOR). Comparable to hot springs and geysers at the Earth's surface these vents are located in the abyssal zones (4000–6000 m BSL[1]) and hadal zones (6000 m BSL), where plate boundaries are undergoing constant change as a result of seafloor spreading. Intensive volcanic activity on the seafloor, especially along the Pacific Ring of Fire (Fig. 1.2), has created high-temperature fields, where hydrothermal vents such as *black and white smokers* with their chimney-like structures were discovered.

[1]Below Sea Level.

Fig. 2.10 Fumarolic activity with sulphur deposits inside the volcanic crater of White Island (Whakaari, New Zealand)

In fact, it is not that long ago on the geological time scale, that '*submarine hydrothermal venting and their accompanying chemosynthetically based communities*', were first detected in 1977 near the Galápagos Islands and acknowledged as a remarkable form of hydrothermal circulation (German and Von Damm 2006:182). Systematic seafloor exploration and surveys with remote operated vehicles (ROV) has resulted in documenting the geomorphology of the seafloor (Stix 2015; Thal et al. 2016), with a special focus on areas hosting hydrothermal vents.

2.4.2 Black Smokers

Since their original discovery, submarine hot springs and their hydrothermal vents are commonly known as *black smokers*, which are conduits or chimneys that emit hydrothermal fluids at extreme temperatures enriched with mineral components and metallic trace elements (Figs. 2.18, 2.19 and 11.16; Example 2.6). During the process of hydrothermal circulation at the seafloor, cold seawater seeps through fractures into the oceanic crust. Here the water is heated and undergoes chemical alterations while reacting with the surrounding rock, and becomes an acidic superheated solution enriched with mineral elements. During descending, hydrothermal fluids can reach supercritical conditions at extreme temperatures up to and in excess of 400 °C, at which point the fluids start to rise again and discharge from hydrothermal vents on the seafloor (German and Von Damm 2006; Passarella et al. 2017; Schmincke 2006).

Based on the rock composition underneath the seafloor, the geochemistry as well as the temperature of the chemical laden fluids can vary significantly (Fouquet 2011; German and Von Damm 2006). This includes pH levels, which have been measured at various locations and recorded as changing between as low as pH 0.9 to pH 6.0 by way of increased seawater mixing (De Ronde and Stucker 2015; Ding et al. 2005).

Note Despite the fact that hydrothermal fluids reach temperatures of several hundred degrees Celsius, they do not reach boiling point due to the pressure at the seafloor. Submarine hydrothermal systems can also

Fig. 2.11 Remnants of the destroyed sulphur mining facility on White Island approximately a decade ago. Prolonged exposure to acidic gas emissions from the nearby crater and occasional eruptions have further damaged the structure over time

remain active at lower temperatures, even if volcanic activity decreases over time.

Source Cashman and Scheu (2015), McPhie and Cas (2015), Staudigel and Koppers (2015), WHOI (2018).

Besides their occurrence at mid-ocean ridges, hydrothermal vents also exist in back-arc basins and island arcs associated with subduction zones (Fouquet 2011; Passarella et al. 2017; Stix 2015; Wu et al. 2016). Submarine hydrothermal activity has also been discovered in the shallow waters (200 m) of the *Aira Caldera* in the south of Kyushu, Japan. At the *Wakamikenu* hot venting site in the *Wakamiko Crater* several (at least three) hydrothermal vents were discovered with two of the vents described as having formed talc chimneys and are emitting hydrothermal fluids at temperatures of up to 200 °C (Ishibashi et al. 2008; Yamanaka et al. 2013).

Example 2.6 New Zealand

In a media release from 2010, GNS Science (New Zealand) described a black smoker chimney that was retrieved from the bottom of the Pacific Ocean. After drying out, it was cut into half and put on display in the Te Papa Museum in Wellington, and the GNS Science headquarters. The mineral layers that line the inside of the chimney enable a rare view of the path the super-heated fluids would have taken on their way from below the seafloor. As per GNS Science, the minerals analysed inside this particular black smoker chimney include sulphides of iron, lead, zinc, and copper. The inside layers are also used to determine when and how fast the chimney formed. In fact, some black smokers can reach up to and over five metres, growing several centimetres per day.

Source GNS Science (2019)

Fig. 2.12 Mount Vesuvius is surrounded with residential areas, which are encroaching on the mountain side with obvious disregard for the potential impact any future eruptions may have on these densely populated suburbs

2.4.3 White Smokers

Depending on their chemical components, hydrothermal vents are divided into *black and white smokers*. White smokers are found in similar environments at ocean ridges, where they discharge plumes of high-pH (alkaline) hydrothermal fluids at temperatures between 260 and 300 °C. The solution from white smokers is rich in light coloured precipitates of particles such as baryte and silica and their chimneys are created similarly to black smokers through accumulation of dissolved mineral particles contained in the hydrothermal plumes. Worth mentioning is that although their temperatures and pH levels are different, black and white smokers can indeed occur in close proximity of each other.

2.5 Hot Springs—Definitions and Classifications

Natural hot springs occur in nearly every country and are not limited by region or rock type, which means that their classification has always been challenging. Most countries have their own classification system to categorise natural hot springs together with laws and regulations for their use and their protection. To this day however, there appears to be no comprehensive classification system that covers and categorises the vast range of *all* natural springs. Over time, numerous hydrologists have attempted to establish classification systems for springs, some of which are still applicable and incorporated in current classifications of natural springs. In 1919, Kirk Bryan from the US Geological Survey commented on the lack of a complete classification of all springs because *'Only springs with unusual characteristics have been thought worth of study'* (Bryan 1919). While Bryan's work may be over a century old, his classification of springs still holds much value today.

Based on their hydrologic and geophysical diversity, the development of a classification system is a complex task that goes beyond the geophysical characteristics. Generally, hot springs are categorised according to their content of dissolved mineral elements, temperature, magnitude of discharge from the source (flow rate) and/or by their pH level (acidic, neutral, alkaline). These are the common criteria shared by all types of springs, including cold springs. To further categorise natural springs becomes increasingly challenging when different geologic conditions and hydrothermal processes are involved (Pitts and Alfaro 2001),

Fig. 2.13 Frying Pan Lake in the Waimangu Volcanic Rift Valley, New Zealand

although additional spring categories can include the location and geologic setting, e.g. rock type, structure and embedded mineral deposits.

For a detailed hot spring classification the collection of physical data, together with chemical analyses is required, while at the same time taking into consideration that the flow rates, elements in solution, the temperature and as well as the pH level can vary. Such changes can be due to seasonality or the dynamic forces of the geothermal and/or volcanic environment where the springs are located.

In Central Europe, research into nearly one hundred springs was carried out over several years, which confirmed that discharge rates for springs can fluctuate throughout the year, mainly based on differences in the seasonal recharge of aquifers (Moniewski 2015). These findings are supporting to some degree a historical classification of springs proposed in 1923 by O.E. Meinzer, who used the average annual volume of discharge of a spring, which he described as the *spring magnitude*. Depending on the 'size' of a spring, the classification specifies the *First Magnitude* as having a very high flow rate, down to the *Eighth Magnitude* with the lowest flow rate (Pitts and Alfaro 2001). The magnitude of discharge is still one of the main categories used to assess the sustainability of a natural spring, although the flow rate is dependent on the aquifer recharge, which varies based on the amount of precipitation and infiltration that replenish the reservoir. If the discharge exceeds the recharge, the flow volume of the spring will ultimately decrease or stop altogether (Chap. 4).

A review of the various historical classifications of natural hot springs for this book has led to the conclusion that there is still no principal internationally recognised classification system available for hot springs. Apart from classifying springs based on their discharge (Meinzer 1923), other suggestions made in the past include the classification of all springs by the pathways they follow to the surface (Keilhack 1912), by their geochemical factors (Bliss 1983), by the specific type of their flow (Shuster and White 1971), and by criteria (Clarke 1924) that consist of the geological origin of hot springs, their physical as well as their chemical

Fig. 2.14 The frozen lake *Panteko* with submerged hot springs melting the thick ice where plumes of warm water rise (Akan National Park, Hokkaido)

Fig. 2.15 The steaming waterfalls of the Mainit Sulfuric Hot Springs are a protected landscape area in Mindanao, Philippines

Fig. 2.16 The temperature of this thermal waterfall in Japan can vary from steaming hot to barely lukewarm, depending on rainfall and other seasonal and/or tectonic factors

properties. These earlier attempts of establishing a classification system all focus on different factors. While describing the hydrogeology of a dozen springs, Springer and Stevens (2008:83) point out the need for a *'consistent and comprehensive classification system'*. Perhaps the already existing proposals together with contemporary databases could be linked and consolidated to create a comprehensive and structured classification of hot springs. Table 2.2 presents an overview of different hot spring types and is *not* intended as a classification system. From the descriptions of individual spring types, it becomes obvious that there is room for expanding this list; however, the specified hot springs offer the reader a starting point for further exploration.

Another type of classification can be based on the widespread use of hot springs for their medicinal benefits and curative value (also frequently referred to as healing springs or medicinal springs), which is commonly associated with a specific chemical content. In Chap. 8 (Hot Springs, Health and Wellbeing) this topic is discussed in detail. For more in-depth information about historic spring classifications, the book *Springs and Bottled Waters of the World—Ancient History, Source, Occurrence, Quality and Use* edited by LaMoreaux and Tanner (2001) provides interesting information about some of the early classification attempts.

2.5.1 Hot Spring Temperatures and pH Levels

It appears that there is a general belief that hot springs of volcanic origin are hotter than springs emerging from artesian basins or karst formations. This may be true in high-temperature hydrothermal systems in active volcanic environments, but it merely reflects a basis for categorising the heat source, not the spring itself. Apart from the origin of the water and the heat source, natural hot springs can be categorised according to the following broad temperature ranges, which appear to be generally accepted in many countries:

Fig. 2.17 One of many small mud 'volcanoes' at the 'Shaved Monks Head Hell' (Bouzo Jigoku) in Beppu, Kyushu (Japan)

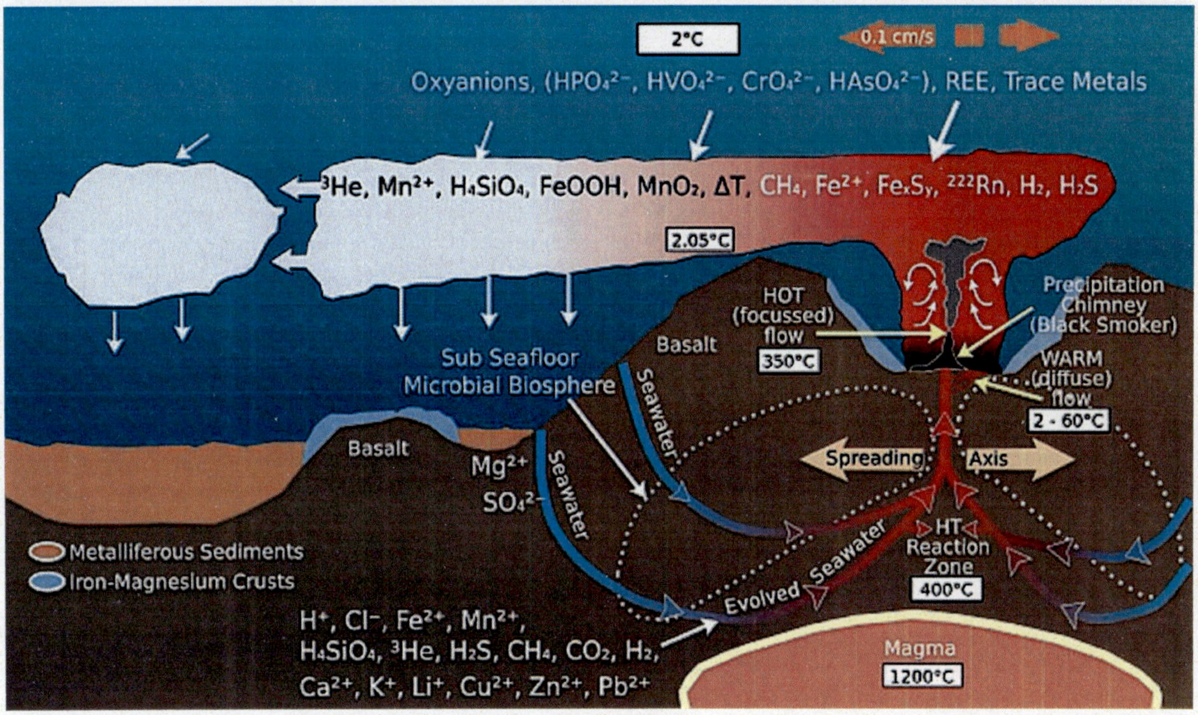

Fig. 2.18 Diagram of a hydrothermal vent and the circulation of hydrothermal fluids at a mid-oceanic ridge (MOR) system. *Source* NOAA (2018). Public Domain

Fig. 2.19 Hydrothermal vents are an unusual phenomenon, which discharge particle-laden superheated water containing large amounts of dissolved mineral elements from the Earth's interior. Mixing with cold seawater, suspended components such as black metal-sulphide, oxide and sulphur start to precipitate and cause the dark colour of the fluid and the black 'smoke' columns rising from the hydrothermal vents. *Source* NOAA (2018)

- cold springs <25 °C;
- tepid springs—between 25 and 34 °C;
- warm springs—between 34 and 42 °C;
- hot springs >42 °C; and
- extreme hot springs >100 °C.

More differentiations are certainly possible, although the above temperatures provide the more common distinctions between cold and hot springs. Depending on their origin, hot springs can range from barely tepid to very hot. For example, to be considered as a hot spring in Germany, the water temperature must be at or above 20 °C, while in Japan the stipulated minimum temperature is 25 °C (Erfurt 2012). In New Zealand the Resource Management Act 1991 (Ministry for the Environment 2019:605) defines geothermal water as *water heated within the earth by natural phenomena to a temperature of 30 °C or more; and includes all steam, water, and water vapour, and every mixture of all or any of them that has been heated by natural phenomena.* Water below 30 °C is classed as 'mere' groundwater.

While scientific and legal classifications vary from one country to another, the Earth science literature has provided several definitions over the years, in which springs are generally classed as hot springs if the water temperature is higher than that of the human body (37 °C or 98°F). By comparing various definitions there appears to be a broad consensus that natural springs at temperatures below 25 °C should not be classed as a genuine hot spring (Allaby 2013; Heasler et al. 2009; Erfurt-Cooper 2018).

Different definitions of hot springs have been reviewed and assessed to contribute suitable criteria to the classification of various types of springs (Pentecost et al. 2003). The findings resulted in two major benchmarks for natural hot springs—the human body temperature and the mean annual local air temperature, which led the researchers to the position, that while no system can provide an objective and unbiased classification, water emerging at temperatures in excess of the core human body temperature of 36.7 °C should be defined as a 'hot spring' (Pentecost et al. 2003). For this reason, health and rehabilitative facilities and hot

Table 2.2 Hot spring categories with a description of their key elements such as dissolved mineral components, location, flowrate and origin

Category	Description of key characteristics
Acidic hot springs	These springs exist in extreme, mainly volcanic environments, and can support acidophilic and thermophilic microbes known as extremophiles (thermophiles and hyperthermophiles). The pH level of acidic hot springs can be as low as 1.2 (*Tamagawa* Hot Spring, Japan) and 1.4 (*Rehai*, China). Submarine hydrothermal vents such as Black Smokers are also acidic with pH levels reported as low as pH 0.9
Alkaline hot springs	This type of spring occurs worldwide in active volcanic areas, and like acidic hot springs, alkaline springs are also able to host a diversity of thermotolerant lifeforms, regardless of the hostile conditions caused by extreme pH levels (up to 12) and high temperatures. Submarine hydrothermal vents such as White Smokers are also alkaline with elevated pH levels due to their specific chemical content and the mixing with seawater
Artesian springs	Springs at a range of temperatures that rise from the ground due to varying degrees of pressure in the underlying artesian basin. Not all artesian springs are free flowing, but require bores or wells drilled into the aquifer (Fig. 2.20), which causes the water to rise until reaching a hydrostatic equilibrium. Artesian hot springs are generally not associated with volcanic activity and occur worldwide. They can be used for used for health and recreational purposes as well as for irrigation and other direct uses including the generation of energy
Brine springs Chloride Springs Saline Springs	Brine springs or saline springs contain high concentrations of sodium chloride ($NaCl$) originating from subterranean rock salt deposits or seawater intrusions and are also known as chloride springs. They can be of either artesian or volcanic origin with a wide temperature range. Cold brine springs are generally heated for health and recreational use and can be used externally (bathing) and internally (drinking, inhaling of vapour; Fig. 8.6; Appendix 8.1)
Calcic springs Calcium springs Calcium chloride springs	Natural hot springs containing calcium carbonate (CaO_3), calcium hydroxide ($Ca(OH)$) or silicon dioxide (SiO_2)—depending on the water temperature and the elements in solution this can cause the formation of calcite deposits such as travertine or tufa. Examples: Pamukkale (Turkey), Mammoth Hot Springs (USA) or Huanglong (China)
Carbonated springs Carbonic or carboniferous springs Bicarbonate Springs	Warm and hot springs containing carbon dioxide (CO_2), nitrogen (N) and oxygen (O) at varying levels. Carbonic springs occur worldwide, many of them in volcanic regions (Fig. 3.4), while others rise from artesian reservoirs and are used internally (drinking cure) as well as externally (bathing cure)
Chalybeate springs Ferruginous springs	Natural springs containing salts of iron in high concentrations. Chalybeate spring water has a reputation for therapeutic value and is considered an important natural resource for health and recreational spa treatments. However, these springs are usually cold water springs and require heating before use other than for drinking. Example: Chalice Well in Glastonbury, UK (Fig. 10.2)
Fissure springs Fracture springs	Springs discharging from faults or joints in the ground, where the water has followed a natural course of openings or fissures in underlying non-porous rock formations. These springs are warm to hot and usually contain high concentrations of dissolved mineral components
Filtration springs Seepage springs	Natural springs that emerge in depressions or at the bottom of hillsides where groundwater is flowing or seeping through sand and gravel. Temperatures vary from cold to hot
Geyser Spouting springs	A natural hot spring that discharges columns of water and steam at extreme temperatures from an underlying hydrothermal system (Fig. 1.4). Most geysers erupt at intervals, while some continuously discharge steam and water. Geysers create spectacular sights and are popular tourist attractions. However, cold spouting springs also exist
Gravity springs	Springs formed by gravity are mostly found on hillsides, where water moves horizontally after encountering confining layers until it finds an outlet from which it can emerge. Unless the water is circulating near a magma body, these springs are usually cold springs
Hydrothermal springs Geothermal springs Hydrothermal vents	Includes all types of hot springs and extreme hot springs discharging water that is heated naturally while circulating through subterranean faults and fractures. Often referred to as geothermal or thermal springs Hydrothermal vents, also known as Black or White Smokers (Figs. 2.19, 11.16), are located on the sea floor at mid-ocean ridges where super-heated fluids are released at temperatures of up to 400 °C. These springs transport chemicals from the Earth's interior to the seafloor, where they accumulate into potentially valuable mineral deposits
Intermittent springs	A spring that discharges only periodically, sometimes depending on seasons, weather conditions and sufficient recharge from rainfall and/or snowmelt. By definition, intermittent springs can include geysers as these erupt at certain intervals

(continued)

Table 2.2 (continued)

Category	Description of key characteristics
Medicinal springs Healing springs	Natural hot springs considered as beneficial in the treatment of various medical conditions. These springs are rich in essential mineral and trace elements and have been used for therapeutic purposes for thousands of years
Mineral springs Hot or cold springs	Naturally occurring springs containing dissolved mineral components and metallic trace elements that are considered to have therapeutic value. To be classed as true mineral water a spring usually must contain more than 1 g of dissolved mineral elements per litre. Mineral springs can be cold, warm, hot, extremely hot as well as artificially heated and are used for health and recreational purposes
Mound springs	Mound springs are formed by the precipitation of mineral components contained in the water. Other methods of growth such as build-up of organic matter and sediments can also lead to the formation of mound springs (Fig. 10.14; Appendix 4.2)
Natural hot springs	Generic term for hydrothermal springs naturally discharging from the ground and above the ambient temperature—often co-existing with other hydrothermal features such as geysers and/or fumaroles (Figs. 1.8, 2.3, 2.21, 7.15 and 10.14). A valuable resource for the hot spring tourism industry
Perennial springs	Natural hot water springs with a constant flow rate all year round
Primary hot springs Volcanic Hot Springs	Closely related to volcanic activity, primary hot springs are containing mineral and trace elements generated predominantly by volcanic processes. The majority of volcanic spring water is of meteoric origin heated in the vicinity of a magma reservoir
Radioactive springs Radium springs Radon springs	Springs containing traces of radioactive substances such as radon gas at low levels are used for health and recreational purposes with no known negative effects. In the case of the Australian Paralana Hot Springs however, the level of the natural radioactivity does not encourage their use for any purposes at all (Fig. 3.6)
Thermal springs	Includes all warm and hot springs at temperatures generally above 25 °C but can also refer to natural springs that are artificially heated and used for health and recreational purposes
Thermo-mineral spring	Collective term for hydrothermal springs and artesian springs at temperatures between 25 and 100 °C—used for a variety of purposes including health and recreation as well as industrial applications
Subaerial hot springs	All natural springs that are located on or close to the Earth's surface—not underground
Sublacustrine hot springs	Springs that are located below the surface of a lake or at the bottom of a lake—commonly in volcanic areas where some form of activity takes place. Examples: Lake Taupo and Lake Rotomahana, New Zealand; Lake Akan (Fig. 2.14), Japan; Lake Baikal, Russia; Lake Tanganyika in East Africa; Lake Yellowstone, USA
Submarine hot springs Hydrothermal Vents	Springs or hydrothermal vents that are located at the bottom of the ocean in active volcanic areas (Figs. 2.19 and 11.16)
Sulphur springs Sulphate springs Sulphurous hot springs	Natural hot springs containing sulphur or sulphur compounds—usually of volcanic origin. Sulphur springs tend to be acidic with pH levels ranging between <1.6 up to 4.6—compare Acidic Hot Springs
Volcanic hot springs	Natural hot springs heated by volcanic activity or a cooling magma body underground and which contain mineral components, gases and metallic trace elements of volcanic origin (e.g. Figs. 3.4, 3.18, 4.2, 6.1, 7.2 and 9.9)

Source Compiled from various sources

spring spas prefer temperatures of natural hot springs to be in the range of 25–48 °C for bathing, medical treatments and for recreational purposes (Erfurt 2012; Pentecost 2005; Pentecost et al. 2003).

A further defining factor in the classification of natural hot springs is their pH level, which indicates the degree of acidity or alkalinity of a liquid. In the case of hot springs there is a remarkable range in pH levels, a factor of importance for health and recreational use, where acceptable pH levels are required for medical treatments and bathing. Examples from Japan show that low pH levels (Tamagawa Onsen, pH 1.2; Tsukahara Onsen, pH 1.4; and Kusatsu Onsen, pH 1.6), as well as higher levels (Blue Lagoon—Iceland, pH 7.5; Fairmont Hot Springs—Canada, pH 8.4) are all tolerable for the human body, depending on the temperature and time of immersion in highly acidic or highly alkaline waters. Comparing hot spring water to other liquids; battery acid and stomach acid, for example, have a pH level of 2.0, while pure water, blood and saliva are neutral at pH 7.0 and seawater is slightly alkaline at pH 8.0. Antacids have

Fig. 2.20 Beel's bore in Cunnamulla, Queensland, was one of the original artesian bores drilled into the Great Artesian Basin underlying much of Australia's outback. At the time, the water from this bore rose from a depth of 520 m at a temperature of 42.7 °C with an enormous discharge rate of 15.2 million litres per day. *Source* Kerry, c. 1884–1917

a pH of 10.0, the pH of bleach is 13.0 and caustic soda (sodium hydroxide—NaOH) tops the scale at pH 14.0. At the other end of the scale we can find hydrochloric acid (HCl) with a pH of 0.0.

2.5.2 Discharge Volume of Hot Spring Water

The viability of most hot spring destinations and their facilities relies heavily on the volume of hot spring water discharged from the source. Are the flow rates low, inconsistent or unsustainable, it rules out large-scale hot spring resort developments that use millions of litres of hot water every day. For instance, the *Spa Resort Hawaiians* in the Fukushima Prefecture, Japan, can use up to 5 million litres of geothermal water per day for their many themed pools (Spa Resort Hawaiians 2018; see Chap. 7). In fact, the daily discharge of all hot spring water in Japan combined adds up to more than 4 billion litres. Due to the demand for genuine hot spring water in Japan, around 70% of the overall discharge in the order of 2.77 million litres per minute does not flow from natural springs but must be pumped out of the ground (Erfurt 2012; Nippon Onsen Research Institute 2020). To illustrate the amount of hot spring water that is discharged at the source on a daily basis in different countries, Table 2.3 uses examples from 36 hot spring locations with a flow volume of more than one million litres per day.

Varying definitions of natural hot springs and their inconsistent usage of existing classifications, including by scientists and researchers from different disciplines and different countries, have led to some discrepancy in the quest for a universally agreed classification system. Temperature categories such as cold, warm and hot water are relative terms used according to an individual's perception of what is hot or cold. In the case of higher temperatures not suitable for thermal bathing, the term extreme hot spring may be used

Table 2.3 Hot springs with discharge volumes above one million litres per day

Country	Hot spring	Temperature (°C)	Flowrate (l/d)
Algeria	Hammam Meskhoutine	70–98	6.9 million l/d
Argentina	Federaçion Thermal Spa	41	10.8 million l/d
Australia	Dalhousie Springs Complex	38–43	**>1.46 billion l/d**
Australia	Mataranka Springs	34	22.4 million l/d
Australia	Moree Artesian Bore Baths	41.6	13.5 million l/d
Australia	Peninsula Hot Springs	45	4.3 million l/d
Austria	Baden	36–47	1 million l/d
Brazil	Caldas Novas	34–57	28.8 million l/d
Canada	Miette Hot Springs	37–40	2.2 million l/d
Canada	Fairmont Hot Springs	32–39	7.5 million l/d
China	Huaqing Hot Springs	43–49	2.7 million l/d
Costa Rica	Tabacon Hot Springs	50	6.9 million l/d
England (UK)	Bath Hot Springs	43–49	>1.2 million l/d
France	Aix-les-Bains	43–45	4 million l/d
Germany	Aachen	45–75	3.5 million l/d
Germany	Wiesbaden	46–66	2 million l/d
Hungary	Budapest	21–78	70–80 million l/d
Iceland	Blue Lagoon	40	21.6 million l/d
Iceland	Deildartungukver	100	12.9–17 million l/d
Israel	Hamat Gader	20–42	12–16.8 million l/d
Italy	Abano Terme	65–87	86.4 million l/d
Italy	Terme di Saturnia	37.5	1.15 million l/d
Japan	Beppu City—combined sources	Up to 150	>50 million l/d
Japan	Kusatsu	55–79	>46 million l/d
Japan	Tamagawa Hot Spring	98	12.9 million l/d
Japan	Iwaki—Spa Resort Hawaiians	58	>5 million l/d
Jordan	Wadi Afra Springs	47–49	>38 million l/d
Mexico	Agua Hedionda	26.6–27.7	>73.4 million l/d
New Zealand	Polynesian Spa	33–43	5 million l/d
New Zealand	Te Manaroa Spring	98	~2.6 million l/d
Switzerland	Baden	35–47	>4.5 million l/d
Turkey	Balcova	62–80	>2.4 million l/d
Turkey	Pamukkale	35–38	>31.5 million l/d
Turkey	Yalova Hot Springs	57–60	1.3 million l/d
United States	Hot Creek—Long Valley Caldera	93	20.7 million l/d
United States	Thermopolis, Wyoming	22–56	8.2–10.4 million l/d

For some locations, the flowrates as well as the temperatures vary depending on the information source or if there are clusters of hot springs. The highest discharge rate per day is attributed to the Dalhousie hot springs complex in Australia
Source Compiled by Author

to differentiate from thermal waters used for health and recreational purposes.

Natural hot spring water, while circulating underground, undergoes changes in its composition caused by interaction with the surrounding rock through heat, pressure and time, thus acquiring loads of dissolved mineral elements of up to 50% TDS. Because chemical elements are an important component in the classification of hot springs, this topic is

Fig. 2.21 Natural hot spring pool at Landmannalaugar in the southern highlands of Iceland

explored in more detail in Chap. 3 (Hot Springs and their Geochemistry). In Chap. 8 (Hot Springs, Health and Well-being) a number of mineral and trace elements of importance for the hot spring health tourism sector are discussed and the reasons why they are such a highly sought-after resource.

2.6 Summary

The basic geological processes that determine the existence of different categories of natural hot springs of volcanic and non-volcanic origin were outlined in this chapter. Important surface manifestations were described, and their physical characteristics explained. The geophysical diversity of hot springs includes not only springs found on the Earth's surface, but also takes into account hydrothermal vents submerged on the bottom of lakes, rivers and oceans. The remarkable multitude of submarine hydrothermal vents presents a relatively new (first discovered in 1977) and exciting area of scientific study. Another section of this chapter highlighted the varying definitions of hot springs; a subject that has generated a number of different interpretations over time; mainly based on sets of physical key indicators including temperature, dissolved mineral content, host rock formation and rate of discharge at the source. Although there is still more to be discovered about the geophysical aspects of natural hot springs, one of the challenges is to establish and agree on a comprehensive classification system for all springs that can be applied internationally. At this point, a standard classification of hot springs remains undecided and existing categories vary from one country to another.

Such is the geodiversity of natural hot springs that it is hard to incorporate the many factors that influence their existence and emphasise the versatility of this natural resource. This chapter has presented a general overview of the different types of hot springs and their key characteristics with examples from selected locations. To further advance the geosciences, natural hot springs and their many related hydrothermal features present one of the most exciting and challenging fields for further study.

Travertine terraces in Karahayıt, Turkey

2.7 Appendix

Appendix 2.1 Hydrothermal Activity—Steam Vents (Fumaroles)

1—Caldeira das Furnas, a hydrothermal area in São Miguel, Azores (left); 2—Steam vents on the volcanic island Whakaari (White Island), New Zealand; 3, 4, 5—Degassing high temperature fumaroles in Kyushu, (Japan); 6—Degassing hillside in Terceira, Azores; 7—Fumaroles in the crater wall of White Island, New Zealand (left); 8—Steaming vents that are used by locals for cooking. Furnas, São Miguel, Azores

References

Allaby M (2013) A Dictionary of Geology and Earth Sciences, 4th edn. Oxford University Press, Oxford

Anitori RP, Trott C, Saul DJ, Bergquist PL, Walter MR (2004) The microbial community of a radon hot spring. In: Norris R, Stootman F (eds) Bioastronomy 2002: life among the stars, proceedings of IAU symposium #213. Astronomical Society of the Pacific, San Francisco, p 374

Arnórsson S, Thórhallsson S, Stefánsson A (2015) Utilization of geothermal resources. In: Sigurdsson H et al (eds) The encyclopedia of volcanoes, 2nd edn. Academic Press. London, pp 851–864

Bliss JD (1983) Basic data for thermal springs and wells as recorded in geothermal-part A, Idaho, U.S.G.S. Open file report 83-0431-A, p 64

Bonny S, Jones B (2003) Relict tufa at Miette Hot Springs, Jasper National Park, Alberta. Canada. Can J Earth Sci 40(11):1459–1481. https://doi.org/10.1139/e03-050

Bonotto DM, Santos TO (2007) Health risk related to radon in a thermal spa utilizing waters from Guarani aquifer. WIT Trans Biomed Health 11:137–146. https://doi.org/10.2495/EHR070151

Boudreau AE, Lynne BY (2012) The growth of siliceous sinter deposits around high-temperature eruptive hot springs. J Volcanol Geoth Res 247–248:1–8. https://doi.org/10.1016/j.jvolgeores.2012.07.008

Branney M, Acocella V (2015) Calderas. In: Sigurdsson H et al (eds) The encyclopedia of volcanoes, 2 edn. Academic Press, London, pp 299–315

Bryan K (1919) Classification of Springs. USGS Staff—published research. J Geol 27(7):522–561. Retrieved from https://digitalcommons.unl.edu/usgsstaffpub/493

Cashman KV, Scheu B (2015) Magmatic Fragmentation. In: Sigurdsson H et al (eds) The encyclopedia of volcanoes, 2nd edn. Academic Press, London, pp 459–471

Chafetz HS, Guidry SA (2011) Deposition and diagenesis of mammoth hot springs Travertine, Yellowstone National Park, Wyoming, USA. Can J Earth Sci 40(11):1515–1529. https://doi.org/10.1139/e03-051

Clarke FW (1924) The data of geochemistry. Chapter VI: mineral wells and springs, pp 181–217

Corbett T (2001) The making of American resorts—Saratoga Springs, Ballston Spa, Lake George. Rutgers University Press, New York

Crane K, Hecker B, Golubev V (1991) Hydrothermal vents in Lake Baikal. Scientific correspondence Nature 350:281. https://doi.org/10.1038/350281a0

Delmelle P, Henley RW, Bernard A (2015) Volcano-Related Lakes. In: Sigurdsson H et al (eds) The encyclopedia of volcanoes, 2nd edn. Academic Press, London, pp 851–864

De Ronde CEJ, Stoffers P, Garbe-Schönberg D, Christenson BW, Jones B, Manconi R, Brown PRL, Hissman K, Botz R, Davy BW, Battershill CN (2002) Discovery of active hydrothermal venting in Lake Taupo, New Zealand. J Volcanol Geoth Res 115:257–275. 10.1016/S0377-0273(01)00332-8

Di Napoli R, Aiuppa A, Bellomo S, Brusca L, D'Alessandro W, Gagliano Candela E, Longo M, Pecoraino G, Valenza M (2009) A model for Ischia hydrothermal system: evidences from the chemistry of thermal groundwaters. J Volcanol Geoth Res 186:133–159. 10.1016/j.jvolgeores.2009.06.005

Ding K, Seyfried WE Jr, Zhang Z, Tivey MK, Von Damm KL, Bradley AM (2005) The in situ pH of hydrothermal fluids at mid-ocean ridges. Earth Planet Sci Lett 237(1–2):167–174. https://doi.org/10.1016/j.epsl.2005.04.041

Duggan IC, Boothroyd IKG, Speirs DA (2007) Factors affecting the distribution of stream macroinvertebrates in geothermal areas: Taupo Volcanic Zone, New Zealand. Hydrobiologia 592:235–247. https://doi.org/10.1007/s10750-007-0748-9

Elsgaard L, Prieur D (2011) Hydrothermal vents in Lake Tanganyika harbor spore-forming thermophiles with extremely rapid growth. J Great Lakes Res 37(1):203–206. https://doi.org/10.1016/j.jglr.2010.11.019

Erfurt-Cooper P (2010) The importance of natural geothermal resources in tourism. In: Proceedings world geothermal congress, Bali, Indonesia, 25–29 April 2010

Erfurt P (2012) An assessment of the role of natural hot and mineral springs in health, wellness and recreational tourism. Unpublished doctoral dissertation, School of Business, James Cook University, Cairns, Queensland, Australia

Erfurt-Cooper P (2018) Active hydrothermal features as tourist attractions. In: Fearnley CJ, Bird DK, Haynes K, McGuire WJ, Jolly G (eds) Observing the volcano world: volcano crisis communication. Advances in volcanology series. Springer, Berlin, Heidelberg, pp 85–105. https://doi.org/10.1007/11157_2016_33

Evans J, Perlman H (2013) The water cycle (Natural water cycle). Credits: Howard Perlman, USGS and John Evans, USGS. Public domain. Retrieved from https://tinyurl.com/vwyvg3s

Fischer TP, Chiodini G (2015) Volcanic, magmatic and hydrothermal gases. In: Sigurdsson H et al (eds) The encyclopedia of volcanoes, 2nd edn. Academic Press, London, pp 779–797

Foster S, Hirata R, Vidal A, Schmidt G, Garduño H (2009) The Guarani aquifer initiative—towards realistic groundwater management in a transboundary context. Worldbank—Water Partnership Program

Fouquet Y (2011) Black smoker. In: Gargaud M et al (eds) Encyclopedia of astrobiology. Springer , Berlin, Heidelberg. https://doi.org/10.1007/978-3-642-11274-4

Fournier RO (1989) Geochemistry and dynamics of the Yellowstone National Park hydrothermal system. Annu Rev Earth Planet Sci 17:13–53. https://doi.org/10.1146/annurev.ea.17.050189.000305

Fridleifsson IB, Bertani R, Huenges E, Lund JW, Rybach L (2008) The possible role and contribution of geothermal energy to the mitigation of climate Change. In: Hohmeyer O, Trittin T (eds) IPCC scoping meeting on renewable energy sources—proceedings. IPCC special report

German CR, Von Damm KL (2006) Hydrothermal processes. In: Elderfield H (ed) The oceans and marine geochemistry. Treatise on geochemistry series. Elsevier, Amsterdam

Gibson ML, Hinman NW (2013) Mixing of hydrothermal water and groundwater near hot springs, Yellowstone National Park (USA): hydrology and geochemistry. Hydrogeol J 21:919–933. https://doi.org/10.1007/s10040-013-0965-4

GNS Science (2019) Microorganisms inhabiting submarine volcanoes. Retrieved from https://tinyurl.com/yahrur8k

Goldscheider N, Mádl-Szönyi J, Eröss A, Schill E (2010) Review: thermal water resources in carbonate rock aquifers. Hydrogeol J 18 (6):1303–1318. https://doi.org/10.1007/s10040-010-0611-3

Gomez AA, Rodriguez LB, Vives LS (2010) The Guarani Aquifer System: estimation of recharge along the Uruguay-Brazil border. Hydrogeol J 18(7):1667–1684. https://doi.org/10.1007/s10040-010-0630-0

Granina LZ, Klerkx J, Callender E, Leermakers M, Golobokova LP (2007) Bottom sediments and pore waters near a hydrothermal vent in Lake Baikal (Frolikha Bay). Russ Geol Geophys 48(3):237–246. https://doi.org/10.1016/j.rgg.2006.04.001

Greene AC, Wright MH, Aldosary HA (2016) Bacterial diversity and metal reducing bacteria in Australian thermal environments. In: Méndez-Vilas A (ed) Microbes in the spotlight: recent progress in the understanding of beneficial and harmful microorganisms. BrownWalker Press, Boca Raton

Guillou-Frottier L, Carrè C, Bourgine B, Bouchot V, Genter A (2013) Structure of hydrothermal convection in the Upper Rhine Graben as inferred from corrected temperature data and basin-scale numerical models. J Volcanol Geoth Res 256:29–49. https://doi.org/10.1016/j.jvolgeores.2013.02.008

Haase KM, Koschinsky A, Petersen S, Devey CW, German C, Lackschewitz KS, Melchert B, Seifert R, Borowski C, Giere O, Paulick H (2009) Diking, young volcanism and diffuse hydrothermal activity on the southern Mid-Atlantic Ridge: the Lilliput field at 9°33'S. Mar Geol 266(1–4):52–64. 10.1016/j.margeo.2009.07.012

Habermehl R, Pestov I (2002) Geothermal resources of The Great Artesian Basin, Australia, Bureau of Rural Sciences. GHC Bulletin, June 2002. Geo-Heat Center, Oregon

Heasler HP, Jaworowski C, Foley D (2009) Geothermal systems and monitoring hydrothermal features. In: Young R, Norby L (eds) Geological monitoring. Geological Society of America, Boulder, pp 105–140

Hirata R, Gesicki A, Sracek O, Bertolo R, Giannini PC, Aravena R (2011) Relation between sedimentary framework and hydrogeology in the Guarani Aquifer System in São Paulo state, Brazil. J S Am Earth Sci 31:444–456. https://doi.org/10.1016/j.jsames.2011.03.006

Hróarsson B, Jónsson SS (1992) Geysers and hot springs in Iceland. Mál og Menning, Reykjavik, Iceland

International Waters Governance (2018) Guaraní Aquifer System. Retrieved from https://www.internationalwatersgovernance.com/guaraniacute-aquifer-system.html

Ishibashi J, Nakaseama M, Seguchi M, Yamashita T, Doi S, Sakamoto T, Shimada K, Shimada N, Noguchi T, Yamanaka T (2008) Marine shallow-water hydrothermal activity and mineralization at the Wakamiko crater in Kagoshima bay, south Kyushu, Japan. J Volcanol Geoth Res 173:84–98. 10.1016/j.jvolgeores.2007.12.041

Jones B, De Ronde CEJ, Renaut RW, Owen RB (2007) Siliceous sublacustrine spring deposits around hydrothermal vents in Lake Taupo, New Zealand. J Geol Soc 164:227–242. https://doi.org/10.1144/0016-76492005-102

Jones B, Renaut RW (2003) Hot spring and geyser sinters: the integrated product of precipitation, replacement, and deposition. Canadian Journal of Earth Science 40(11):1549–1569. https://doi.org/10.1139/e03-078

Keilhack K (1912) Lehrbuch der Grundwasser und Quellenkunde, 3rd edn. Geb. Borntraeger, Berlin

Kern ML, Vieiro AP, Machado G (2007) The fluoride in the groundwater of Guarani Aquifer System: the origin associated with black shales of Paraná Basin. Environ Geol 55(6):1219–1233. https://doi.org/10.1007/s00254-007-1067-1

Kerry CH (c. 1884–1917) Beels bore. Powerhouse Museum Collection, Sydney. No known restrictions on publication. Retrieved from https://ma.as/27744

Kralj P, Eichinger L, Kralj P (2009) The Benedikt hydrothermal system (north-eastern Slovenia). Environ Geol 58(8):1653–1661. https://doi.org/10.1007/s00254-008-1631-3

Ladd BS, Ryan MC (2016) Can CO_2 trigger a thermal geyser eruption? Geology 44(4):307–310. https://doi.org/10.1130/G37588.1

Lovering TS, Goode HD (1963) Measuring geothermal gradients in drill holes less than 60 feet deep, East Tintic district, Utah. Government Printing, Office, Washington

Lowenstern JB, Hurwitz S (2008) Monitoring a supervolcano in repose: heat and volatile flux at the Yellowstone Caldera. Elements 4:35–40. https://doi.org/10.2113/gselements.4.1.35

Luhmann AJ, Covington MD, Peters AJ, Alexander SC, Anger CT, Green JA, Runkel AC, Alexander EC Jr (2011) Classification of Thermal Patterns at Karst Springs and Cave Streams. Ground Water 49(3):324–335. https://doi.org/10.1111/j.1745-6584.2010.00737.x

Mayer K, Scheu B, Montanaro C, Yilmaz TI, Isaia R, Aßbichler D, Dingwell DB (2016) Hydrothermal alteration of surficial rocks at Solfatara (Campi Flegrei): Petrophysical properties and implications for phreatic eruption processes. J Volcanol Geoth Res 320:128–143. https://doi.org/10.1016/j.jvolgeores.2016.04.020

McPhie J, Cas R (2015) Volcanic successions associated with ores deposits: facies characteristics and ore-host relationships. In: Sigurdsson H et al (eds) The encyclopedia of volcanoes, 2nd edn. Academic Press, London, pp 865–879

Meinzer OE (1923) Outline of ground-water hydrology, with definitions. Government Printing Office. US Geological Survey Water Supply Paper 494, Washington, D.C, 69 pp

Ministry for the Environment (2019) Resource management act 1991. Reprint as at 11 July 2018. New Zealand. Retrieved from https://www.legislation.govt.nz/act/public/1991/0069/latest/DLM230265.html

Moniewski P (2015) Seasonal variability of discharge from selected springs in Central Europe. Episodes 38(3):189–196. oai:CiteSeerX.psu:10.1.1.831.616

Montaño J, Peel E (2003) Geothermalism in the Guarani Aquifer System, Uruguay. Geothermal Resources Council Transactions 27, 12–15 Oct 2003

Nippon Onsen Research Institute (2020) Data on hot springs in Japan. Retrieved from https://www.onsen-japan.info/data/you.html

NOAA (2018) Deep sea vent chemistry diagram. U.S. National Oceanic and Atmospheric Administration. Public Domain. Retrieved from https://commons.wikimedia.org/wiki/File:Deep_Sea_Vent_Chemistry_Diagram.svg

NPS Stats (2020) Recreation visitators (1904—Last Calendar Year). Retrieved from https://tinyurl.com/kgk5xh8

Passarella M, Mountain BW, Seward TM (2017) Basalt-seawater interaction at near-supercritical conditions (400°C, 500 bar): Hydrothermal alteration in the sub-seafloor. In: Proceedings 39th New Zealand Geothermal Workshop, 22–24 Nov 2017 Rotorua, New Zealand

Peninsula Hot Springs (2019) Our water. Retrieved from https://www.peninsulahotsprings.com/our-water/minerals/

Pentecost A (2005) Hot springs, thermal springs and warm springs. What's the difference? Geol Today 21(6):222–224. https://doi.org/10.1111/j.1365-2451.2005.00536.x

Pentecost A, Jones B, Renaut RW (2003) What is a hot spring? Can J Earth Sci 40(11):1443–1446. https://doi.org/10.1139/e03-083

Pesce A (2002) Thermal Spas: an economic development alternative along both sides of the Uruguay River. In: GHC bulletin. September 2002. Geo–Heat Center, Oregon

Petrillo Z, Chiodini G, Mangiacapra A, Caliro S, Capuano P, Russo G, Avino R (2013) Defining a 3D physical model for the hydrothermal circulation at Campi Flegrei caldera (Italy). J Volcanol Geoth Res 264:172–182. 10.1016/j.jvolgeores.2013.08.008

Pitts MW, Alfaro C (2001) Geologic/hydrogeologic setting and classification of springs. In: LaMoreaux PE, Tanner JT (eds) Springs and bottled waters of the world—ancient history, source, occurrence, quality and use. Springer, Berlin

Rabelo JL, Wendland E (2009) Assessment of groundwater recharge and water fluxes of the Guarani Aquifer System, Brazil. Hydrogeol J 17:1733–1748. https://doi.org/10.1007/s10040-009-0462-y

Rockel W, Hoth P, Seibt P (1997) Charakteristik und Aufschluß hydrogeothermaler Speicher. Geowissenschaften 15(8):244–252

Ruzo A (2016) The boiling river: adventure and discovery in the Amazon. Simon & Schuster, New York

Schellschmidt R, Sanner B, Pester S, Schulz R (2010) Geothermal energy use in Germany. In: Proceedings world geothermal congress 2010. Bali, Indonesia. 25–29 April 2010

Schmincke HU (2006) Volcanism. Springer, Berlin

Scott BJ (1992) Waimangu: a volcanic encounter. Waimangu Volcanic Valley, Rotorua

Seibt P, Kabus F, Hoth P (2005) The Neustadt-Glewe geothermal power plant—practical experience in the reinjection of cooled thermal waters into sandstone aquifers. In: Proceedings world geothermal congress 2005. Antalya, Turkey, 24–29 April 2005

Shuster ET, White WB (1971) Seasonal fluctuations in the chemistry of limestone springs: a possible means for characterizing carbonate aquifers. J Hydrol 14(2):93–128. https://doi.org/10.1016/0022-1694(71)90001-1

Spa Resort Hawaiians (2018) General profile of Spa Resort Hawaiians. Retrieved from https://www.hawaiians.co.jp/english/about/srh.html

Springer AE, Stevens LE (2008) Spheres of discharge of springs. Hydrogeol J 17:83–93. https://doi.org/10.1007/s10040-008-0341-y

Staudigel H, Koppers AAP (2015) Seamounts and island building. In: Sigurdsson H et al (eds) The encyclopedia of volcanoes, 2nd edn. Academic Press, London, pp 405–421

Stimac J, Goff F, Goff CJ (2015) Intrusion-related geothermal systems. In: Sigurdsson H et al (eds) The encyclopedia of volcanoes, 2nd edn. Academic Press, London, pp 799–822

Stix, J (2015) Volcanic interactions. In H. Sigurdsson et al (eds) The encyclopedia of volcanoes, 2nd edn. Academic Press, London, pp 777–778

Te Ara (2019) Sulfur mining, White Island. Retrieved from https://teara.govt.nz/en/photograph/5258/sulfur-mining-white-island

Thal J, Tivey M, Yoerger DR, Bach W (2016) Subaqueous cryptodome eruption, hydrothermal activity and related seafloor morphologies on the andesitic North Su volcano. J Volcanol Geoth Res 323:80–96. https://doi.org/10.1016/j.jvolgeores.2016.04.041

Todesco M, Chiodini G, Macedonio G (2003) Monitoring and modelling hydrothermal fluid emission at La Solfatara (Phlegraean Fields, Italy). An inter-disciplinary approach to the study of diffuse degassing. J Volcanol Geothermal Res 125(1–2):57–79. https://doi.org/10.1016/S0377-0273(03)00089-1

Troiano A, Di Guiseppe MG, Patella D, Troise C, De Natale G (2014) Electromagnetic outline of the Solfatara-Pisciarelli hydrothermal system, Campi Flegrei (Southern Italy). J Volcanol Geoth Res 277:9–21. https://doi.org/10.1016/j.jvolgeores.2014.03.005

UNESCO (2019) Morne Trois Pitons National Park. Retrieved from https://whc.unesco.org/en/list/814/

USGS (2019) Long valley Caldera field guide—Hot Creek geologic site. Retrieved from https://www.usgs.gov/media/images/temperature-hot-creek

Vandemeulebrouck J, Hurst AW, Scott BJ (2008) The effects of hydrothermal eruptions and a tectonic earthquake on a cycling crater lake (Inferno Crater Lake, Waimangu, New Zealand). J Volcanol Geoth Res 178:271–275. https://doi.org/10.1016/j.jvolgeores.2008.06.021

Virtual Dominica (2018) The boiling lake. Retrieved from www.avirtualdominica.com/the-boiling-lake/

Wendland E, Barreto C, Gomes LH (2007) Water balance in the Guarani Aquifer outcrop zone based on hydrogeologic monitoring. J Hydrol 342:261–269. https://doi.org/10.1016/j.jhydrol.2007.05.033

WHOI (Woods Hole Oceanographic Institution) (2018) Hydrothermal vents. Retrieved from https://www.whoi.edu/main/topic/hydrothermal-vents

Wiesbaden (2010) Wiesbaden—Thermalquellen, Kurbetrieb und Gesundheitswesen. Retrieved from https://tinyurl.com/uulayv3

Wright MH, Patel BKC, Greene AC (2012) Thermophilic bacteria from paralana hot springs in the northern flinders ranges of South Australia. Conference paper 10.13140/RG.2.1.2765.5525

Wu S-F, You C-F, Lin Y-P, Valsami-Jones E, Baltatzis E (2016) New boron isotopic evidence for sedimentary and magmatic fluid influence in the shallow hydrothermal vent system of Milos Island (Aegean Sea, Greece). J Volcanol Geoth Res 310:58–71. https://doi.org/10.1016/j.jvolgeores.2015.11.013

Yamanaka T, Maeto K, Akashi H, Ishibashi J, Myoshi Y, Okamura K, Noguchi T, Kuhwara Y, Toki T, Chiba H (2013) Shallow submarine XE "Submarine" hydrothermal activity with significant contribution of magmatic water producing talc chimneys in the Wakamiko Crater of Kagoshima Bay, southern Kyushu, Japan. J Volcanol Geoth Res 258:74–84. 10.1016/j.jvolgeores.2013.04.007

Yang T, Lyons S, Aguilar C, Cuhel R, Teske A (2011) Microbial communities and chemosynthesis in Yellowstone Lake sublacustrine hydrothermal vent waters. Frontiers in Microbiology 2(June). https://doi.org/10.3389/fmicb.2011.00130

Thermophilic bacteria at a degassing fumarole. Tatsumaki Jigoku in Beppu, Japan

The Geochemistry of Hot Springs

3

Contents

P. Erfurt, *The Geoheritage of Hot Springs*, Geoheritage, Geoparks and Geotourism,
https://doi.org/10.1007/978-3-030-60463-9_3

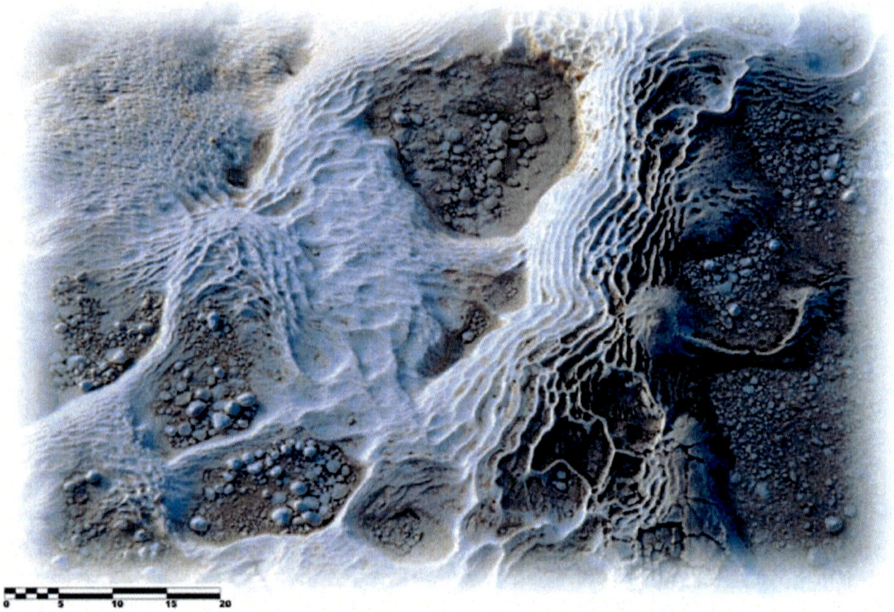

Micro-terracettes with small potholes on a travertine slope (Pamukkale, Turkey). Scale in centimetres

3.1 Introduction

Geochemistry is the study of the chemical composition of the Earth including of its minerals, rocks, soils, water and the atmosphere and the processes that control their distribution and formation. This chapter explores some of the basic geochemistry that relates to different types of hot springs and their mineral content, beginning with a brief overview regarding the overall volume of water on our planet.

Over 70% of the Earth's surface is covered by water, 96.5% of which is saline water contained in oceans and/or lakes. While this may seem plentiful, only between three and 4% is fresh water, which exists as vapour in the air, in lakes and rivers, locked up in glaciers and ice caps, as moisture in the soil and stored in subterranean aquifers (Fig. 2.1, Chap. 2). Although not all of the groundwater it accessible, numerous and extensive artesian basins contribute vast amounts of water at varying temperatures toward the hydrologic cycle.

Water is unique as it is the most flexible of natural compounds and can be used in each of its individual phases, as a liquid, as vapour or gas, and ice as a crystalline solid. In its solid state water is a mineral, whereas the liquid is not (Klein 2002). Pure water (H_2O) is colourless, tasteless and odourless, but with unique physical properties that are primarily temperature dependant. For example, water reaches its boiling point of 100 °C at sea level and freezes at 0 °C. While in a frozen state, water takes on distinct structures, which allow ice to float on water due to its lesser density (Daintith 2000; Park and Allaby 2013).

The main aspect of hot springs is their natural origin. Whether they are volcanic or rising from artesian aquifers, hot springs can reach similar temperatures at discharge (up to boiling point) and contain similar minerals. Surface hot springs can be separated into three broad groups under consideration of the water chemistry and the resulting mineral deposits:

- alkaline, silica-dominated systems
- carbonate-dominated systems and
- acid sulphate systems (Breckenridge and Hinckley 1978; White et al. 1988).

The pH level of hot spring water indicates the degree of either their acidity or alkalinity and hot springs are categorised as acidic, basic or neutral, according to their pH (hydrogen ion content or potential of hydrogen).

- acidic waters <7.0 pH
- neutral waters 7.0 pH and
- basic/alkaline waters >7.0 pH.

Even highly acidic springs are still used for bathing and health treatments at pH levels around 1.4 (Tsukahara Onsen, Japan) and 1.6 (Kusatsu Onsen, Japan). To put this into perspective, vinegar has a pH of 2.0, pure water is neutral at pH 7.0 and seawater is slightly alkaline at pH 8.0. Legal classifications of hot springs based on their chemical composition vary from country to country, although in many jurisdictions a genuine hot spring is expected to contain >1000 mg/l (ppm) of total dissolved solids (see Chap. 2—Geology of Hot Springs). In order to manage the utilisation of hot springs the analysis of the geochemistry of the hydrothermal water is a vital factor (Guo and Wang 2012).

3.2 Hydrothermal Processes

The occurrence of hot springs and other hydrothermal features is dependent on the geophysical and geochemical processes, which create liquid-dominated hydrothermal systems below the Earth's surface. The hot solutions in these systems are residual fluids that contain dissolved elements, compounds, steam and volcanic gases and vary according to their geochemistry, their temperature and pH levels (Stix 2015). Typically, the higher the temperature of the water, the more material is dissolved from the host rock and carried by convection along fractures and faults and through permeable rock layers. Minerals are precipitated in void spaces, where they form areas of deposition including mineral veins (Delmelle et al. 2015; Lagat 2009).

The chemical composition of hydrothermal fluids is defined as '*highly variable with respect to gases and dissolved solids ranging from very dilute to hypersaline brines*' (Arnórsson et al. 2015:1237), with the dominant source of water of meteoric origin derived from rain, snow melt or seawater (Stimac et al. 2015). Then again, meteoric water can also mix with juvenile water from magmatic sources or water released during metamorphic events. Apart from water, the composition of hydrothermal fluids can contain as much as 50% dissolved solids in solution. Depending on the interaction with surrounding rock material these aqueous solutions serve as a transporting agent through highly permeable rocks (McPhie and Cas 2015), with the potential to cause hydrothermal alteration. The pH levels of hydrothermal fluids can range from strongly acidic to strongly alkaline, which makes them effective solvents of ore components, including silver, gold and other valuable metals. Circulation and deposition from such fluids is an important source of metallic ore deposits.

Over time the conditions in subsurface hydrothermal systems can change, which can affect surface features such as hot springs, geysers and fumaroles. This can lead to variations of the water temperature, fluctuating discharge rates or alterations in the geochemistry of hot spring water from acidic to alkaline (Lynne et al. 2017). As waters of hot springs and geysers resemble their hydrothermal host system, they are also subject to modifications caused by changes in temperature, dilution with groundwater and by reacting with the surrounding rock on the way to the surface (Stimac et al. 2015). Owing to their often close relationship to volcanic environments, monitoring of hydrothermal systems is an essential tool for the surveillance of active volcanoes (Rodriguez et al. 2016).

3.3 Hot Springs and Their Contents

A mineral is a naturally occurring solid with a highly ordered atomic arrangement and a definite (but not fixed) chemical composition. It is usually formed by inorganic processes. (Klein 2002:3)

Minerals are defined as elements or chemical compounds that are crystalline and have formed under specific natural conditions. They are characterised by chemico-physical properties with the chemical composition of a mineral conveyed using a specific formula (Ghersetich and Lotti 1996). The most common minerals are feldspars, quartz, pyroxenes, micas, amphiboles, and olivine, all classed as rock forming minerals (Lagat 2009). As of January 2021, the International Mineralogical Association (IMA) has approved 5673 currently valid mineral species (IMA 2021).

During the transport from deep aquifers water is filtered through different rock layers and heated at depth. While the water is circulating underground, the concentration of trace elements and other dissolved components is determined by the composition of the subsurface rock environment as well as the temperature of the solution, the flow rate, the length of the flow path and the time of rock/water contact underground (Example 3.1). Depending on the water temperature the mineral content varies, with hot springs generally containing more minerals than cold springs as their higher temperatures increase the release of minerals and trace elements in greater concentrations resulting in higher amounts of minerals in solution (Erfurt 2012; McGeary et al. 2001).

The concentration of dissolved components in hydrothermal systems can be further influenced by any fluids contained within the host rocks with which the water interacts, including freshwater, pockets of ancient seawater or intruding saltwater in coastal areas (Hunt 1998). Other causes affecting hydrothermal systems include the quality of the water recharging the aquifer as well as any potential anthropogenic contamination (Erfurt-Cooper and Cooper 2009), a subject that is further explored in Chap. 4 (Conservation of Hot Springs).

Example 3.1 Germany

Geological conditions in Germany have created various combinations of dissolved trace elements and other components, which influence the different compositions of hot springs. For instance, the hot spring waters of the northern rim of the Rhine Massif (*Rheinische Schiefergebirge*) and the southern rim of the *Taunus* mountain range derive their salt content from Permian Zechstein, carried by waters that have travelled considerable distances from their area of recharge and salt dissolution to the area of discharge.

Source Erfurt (2012), Rockel et al. (1997), Schellschmidt et al. (2010) and Seibt et al. (2005)

Natural hot springs rich in minerals and trace elements are universally acknowledged as medicinal springs based on their special geochemistry and their physical properties, which make them a much sought-after commodity in the development of hot spring destinations worldwide. Health spas and recreational facilities usually inform their visitors about the specific minerals and trace elements that are contained in the water (Erfurt 2012). This topic is discussed in detail in Chap. 8 (Hot Springs, Health and Wellbeing).

3.3.1 Common Hot Spring Minerals

Depending on their location, temperature, mineral content and flow rates, hot springs are an important natural and economic resource (Erfurt 2012; Hodgson 2004; Zuniga et al. 2003). Some of the common dissolved components that determine the significance of this resource for the hot spring health and recreation industry are listed in Table 3.1, although mineral deposits produced in volcanic and geothermal environments are predominantly utilised for industrial purposes, e.g. the exploitation of ore minerals.

Deposited minerals associated with hydrothermal features such as natural hot springs include the more common mineral calcite ($CaCO_3$) or silica (SiO_2), which forms travertine (sinter) deposits like the terraces at Pamukkale in Turkey (Fig. 3.1). Other minerals deriving from hydrothermal systems are geyserite ($SiO_2 \cdot nH_2O$), a siliceous sinter type of opal (Appendix 3.4), and haematite (Fe_2O_3), one of the principal iron ores. Haematite can form as a deposit around fumaroles and hot springs in high-temperature geothermal areas or by oxidation of magnetite (Fe_3O_4), a primary mineral in igneous rocks that also occurs in veins in geothermal areas (Sæmundsson and Gunnlaugsson 2002). Magnesium (Mg) is another mineral determined by the surrounding host rock and is derived from dolomite, while calcium (Ca) is dissolved out of limestone.

Note

Travertine deposits are influenced by climate, terrain, temperature, discharge rates, bacteria and anthropogenic activities (Pentecost and Viles 1994).

Sulphur, (from the Latin *sulfur*), is a non-metallic element common in many volcanic and geothermal areas worldwide, where it is formed at high-temperature degassing fumaroles (Appendix 3.5). Sulphur deposits range in colour from pale to bright yellow (Figs. 3.2, and 3.3) and are mined for their considerable commercial value. The mineral is used to produce sulphur compounds including sulphuric acid (Daintith 2000), as a component for making fireworks, fertiliser, and is also used within the pharmaceutical industry. Under certain conditions (e.g. humidity) sulphur forms hydrogen sulphide (H_2S), a gas with a rotten egg odour not uncommon around hot springs and volcanoes at levels that can greatly fluctuate.

Hydrogen sulphide (H_2S) contained in hydrothermal solutions underground transforms into sulphuric acid (H_2SO_4) on contact with oxygen-rich water, and thereby contributes to the sulphate content of hot springs (Goldscheider et al. 2010). High concentrations of sulphate (SO_4^{2-}) frequently occur in hot springs discharging from carbonate aquifers where sulphates originate either from the oxidation of sulphide minerals such as pyrite (FeS_2), or from the dissolution of anhydrite ($CaSO_4$) and gypsum (Goldscheider et al. 2010). The mineral pyrite is common at hydrothermal springs, especially in high-temperature areas, where pyrite

Table 3.1 Common components found in natural hot springs (Erfurt 2012)

Common components found in hot springs	Other substances present in hot springs		Main gases in solution
Calcium (Ca) Carbonate (CO_3^{2}) Chloride (Cl–) Fluoride (F–) Iron (Fe) Magnesium (Mg) Sodium (Na) Sulphate (SO_4^{2-}) Sulphide (S^{2-})	Aluminium (Al) Ammonium (NH_4) Arsenic (As) Bicarbonate (HCO_3) Bromine (Br) Caesium (Cs) Cobalt (Co) Copper (Cu) Fluorine (F)	Iodine (I) Lithium (Li_3) Manganese (Mn) Potassium (K) Radium (Ra) Rubidium (Rb) Silica (SiO_2) Zinc (Zn)	Carbon Dioxide (CO_2) Hydrogen Sulphide (H_2S) Nitrogen (N) Oxygen (O) Radon (Rn) Argon (Ar) and Helium (He) can occur in some sulphur springs

Fig. 3.1 A wall of botryoidal travertine deposits is reflected in one of the countless shallow hot spring-fed basins at Pamukkale, Turkey

crystals can be frequently noticed as a grey film floating on thermal mud pots. Pure gypsum ($CaSO_4 \cdot 2H_2O$), from the Greek *gypsos* for chalk, is a comparatively soft evaporite mineral that forms near fumaroles and hot springs in high-temperature areas from calcium and gases rich in sulphur. The colour of gypsum varies from white or grey to colourless depending on its purity (Sæmundsson and Gunnlaugsson 2002).

Common in volcanic areas with extinct hydrothermal systems are metallic carbonates, which are formed by the weathering of sulphide minerals. Other minerals that form in high-temperature (250 °C and above) hydrothermal systems are epidote, prehnite, chlorite, garnet, actinolite, wollastonite and hedenbergite (Sæmundsson and Gunnlaugsson 2002).

Hot spring water of volcanic origin can also contain high concentrations of sodium (Na) and bicarbonates, resulting in a natural effervescence of the spring water (Erfurt-Cooper 2009; Erfurt 2012; Fig. 3.4a, b). Many other minerals and metallic elements are found in or near hot springs at varying concentrations—too many to include here; however, the main objective is to demonstrate the great diversity of minerals and their association with natural hot springs.

3.3.2 Radioactive Hot Springs

It is not uncommon for natural hot springs to contain radioactive elements. This has led to a somewhat debateable reputation of these hot springs and the inevitable question of whether they are safe to use or should rather be avoided. On the other hand, they have been used (mainly for health and recreational purposes) in many different countries without any known negative effects. In most cases the radioactivity is at very low levels without a discernible risk to humans, especially where these are used for health and recreational purposes. To present an overview about hot springs and their radioactive content, some of their minerals and their presence in hot springs are briefly discussed here.

Radioactive elements in hot spring environments can include uranium (U), potassium (^{40}K), Radon (Rn), radium

Fig. 3.2 Sulphur deposits are forming around a degassing fumarole in the crater of the volcano White Island (Whakaari), New Zealand

Fig. 3.3 Sulphur in its solid state is relatively soft and friable. Sample size: circa 5 cm wide

(Ra) and thorium (Th). Long term use of hot springs containing traces of naturally occurring radioactive substances has shown that they are generally considered as safe by hot spring spas at various destinations:

- Montecatini Terme, Italy;
- Radium Hot Springs, Canada;
- Rudas Bath, Budapest, Hungary;
- Hot Springs, Arkansas, United States.

Hot springs containing low levels of radioactivity are regularly tested and analysed for any potential impact on human health. This is of particular concern for staff members at hot spring facilities, who are exposed to radon levels for extended periods of time compared to visitors (Bonotto and Santos 2007).

Radioactive gases that occur in hot spring environments include radon (Rn) or Radon-222 (^{222}Rn), one of the noble

Fig. 3.4 **a** Natural effervescence in one of the hot spring ponds in the Caldeira Velha, São Miguel, Azores. **b** Escaping gas bubbles are clearly visible in this natural spring. Furnas, São Miguel, Azores

gases with a half-life of 3.8 days and a product of decaying radium-226 (^{226}Ra). Radon is a colourless, odourless and tasteless gas and occurs naturally. The concentration of dissolved radon in hot spring water is thought to be related to variables such as the flow rate, water temperature, and multi-year cycles (Yan et al. 2017). Radon concentrations can be an indicator of non-mobile radium contained in the host rocks (Girault et al. 2016), as well as a precursor for seismic activity such as earthquakes and volcanic eruptions (Montazeri et al. 2011). As a trace gas radon is used to detect fluid movements during surveys and/or exploration of geothermal fields (Woith et al. 2011). Some natural hot springs that contain radon, but are perhaps less well known than the locations mentioned above, include:

- Banglazhang Hot Springs, Yunnan-Tibet Geothermal Province, south west China (Guo et al. 2017; Yan et al. 2017);
- East Kerman Ranges (Jowshan Hot Spring), southeast Iran (Montazeri et al. 2011);
- Londrina city, Parana State in Brazil (Bonotto and Santos 2007);
- Tiberias Hot Spring, Israel (Woith et al. 2011);
- Various hot spring sites in northern Venezuela (Horvath et al. 2000);
- Paralana Hot Springs, Australia (Wright et al. 2012; see Example 3.2); as well as approximately
- 30 radon springs located throughout Russia (Voronov 2003).

Radium (Ra) is a radioactive metallic element which occurs in uranium ores, and its most stable isotope Radium-226 (^{226}Ra) decays to radon. Hot springs containing varying concentrations of radium/radon:

- Karlovy Vary, Czech Republic;
- Tamagawa Onsen, Akita, Japan;
- Hot Springs National Park, Arkansas, USA;
- Yellowstone National Park, Wyoming, USA;
- Hammat Gader, Israel;
- Bath hot springs, UK (Gallois 2006).

Some radioactive springs have deposits of rare radioactive minerals, such as hokutolite ($(Ba,Pb)SO_4$), a lead-rich baryte compound of barium sulphate ($BaSO_4$) and lead sulphate ($PbSO_4$) that forms light coloured crystals (Fig. 3.5). Minute quantities of radium (Ra), iodite (Io), polonium (Po) and bismuth (Bi) account for the radioactivity of hokutolite. Hokutolite deposits, also termed Beitou Stone, were initially discovered in 1905 in the Beitou River in the Thermal Valley of Beitou, Taiwan. The small amounts of the radioactive element radium has led to Beitou's Thermal Valley to be also known as Radium Hot Spring. The only other place where hokutolite has been discovered to date are the Tamagawa Hot Springs in Senbuko, Akita Prefecture, in Japan.

For readers interested in more detailed data relevant to hokutolite the following research articles are recommended.

Hokutolite collected from riverbed at Peitou Hot Spring in Taiwan: With emphasis on radiochemical studies in the Journal of Radioanalytical and Nuclear Chemistry 7(3) by J. Tomita, A. Sakaguchi, and M. Yamamoto (2006).

Chemical composition and radioactivity in hokutolite (-plumbian barite) collected at Peito hot spring, Taiwan in the Journal of Environmental Radioactivity 37(1) by N. Momoshima, J. Nita, Y. Maeda, S. Sugihara, I. Shinno, N. Matsuoka, and C.-W. Huang (1997).

Example 3.2 Australia

The radioactive Paralana Hot Springs (PHS) are located on the Paralana fault near Arkaroola in the Flinders Ranges, South Australia; a region with one of the world's largest uranium reserves. PHS consists of two pools, with the hot springs emerging mainly at the primary pool connected to a secondary pool with the water draining into a small creek (Fig. 3.6a, b). The Paralana Hot Springs are non-volcanic, and the low-temperature hydrothermal system is heated by the natural decay of radium and uranium with high Rn concentrations at the springs (10,952 Bq/m^3).

The PHS are fed by meteoric water, which is near neutral (pH range 7–8), contains low levels of dissolved solids (TDS 1144 mg/L) and emerges at temperatures around 56–70 °C. The discharge rate is estimated at 16 L/s, which translates into 1.3 million litres a day. Samples taken at PHS are reported to contain toxic metals such as uranium, vanadium, molybdenum and arsenic with low levels of chromium, cobalt and selenium also present.

There have been debates about the possible source of the meteoric water, which is commonly attributed to the Great Artesian Basin. However, based on the geochemistry of the springs the water could also originate from the nearby Mt. Painter Domain, although the lack of precipitation in this area does not appear to match the current discharge rates of the hot springs.

Source Brugger et al. (2005) and Wright et al. (2012)

In the **Middle East** various hot springs are located in areas with high background radiation and include the Afra

Fig. 3.5 Hokutolite is a rare mineral and occurs only at a few places worldwide, one of them is Beitou (Taiwan). These Hokutolite samples are from Beitou's Thermal Valley and are on display at the Hot Spring Museum in Beitou

Hot Springs in Jordan (Ajlouni et al. 2010; Saqan et al. 2001) as well as a number of other springs in the Dead Sea Rift Valley (Ilani et al. 2006). In the eastern part of Algeria (North Africa) studies have been carried out at thermal springs with natural radionuclide concentrations to determine their potential effect on human health (Kebir and Boucenna. 2017).

3.3.3 Hot Spring Gases

Over 170 years ago Daubeny (1848:557) wrote about hot springs that *'natural thermal springs are in general accompanied by the same gases which volcanoes commonly emit'*. Daubeny's writings are interesting when compared to our current knowledge about hot spring gases, because the scientific information from the mid-1800s still holds a lot of truth.

Daubeny describes for example nitrogen gas as *'a constant concomitant of springs partaking of a high temperature'* (1848:558). As a *'third kind of air'* he explains carbonic acid as enabling thermal waters to hold in solution more minerals *'than they could otherwise do'* (Daubeny 1848:562). Sulphuretted hydrogen, defined as *'generated by the action of organic matter upon alkaline and earthy sulphates'* (Daubeny 1848:561) today is labelled hydrogen sulphide (H_2S). Despite the somewhat archaic language and considering the state of development of science at the time, Daubeny's reports are extremely precise.

Common Hydrothermal Gases

Apart from the already mentioned radioactive gas components hydrothermal fluids contain various other gases, their amounts depending on the physical conditions within individual hydrothermal systems (Fridleifsson et al. 2008). In volcanic areas these are closely associated with magmatic systems, which supply heat, magmatic or juvenile water and various types of hydrothermal gases. The most common of these gases include (in alphabetical order): argon (Ar), ammonia (NH_3), carbon dioxide (CO_2), carbon monoxide

a

b

Fig. 3.6 **a** Paralana Hot Springs in South Australia. *Source* Victor Gostin (2020) (with permission). **b** Paralana Hot Springs in a photo taken by the Rev. Robert Mitchell circa 1898. *Source* State Library of South Australia (2019)

(CO), helium (He), hydrochloric acid (HCl), hydrogen fluoride (HF), hydrogen sulphide (H_2S), methane (CH_4), nitrogen (N_2), oxygen (O_2), sulphur dioxide (SO_2), and water vapour (H_2O). Some fumarolic trace gases include carbonyl sulphide (COS) and carbon disulphide (CS_2).

The chemical composition of hot spring gas emissions depends on a number of variables including their location of discharge, water temperature and pH, as well as the reaction of rock minerals with circulating groundwater (Fischer and Chiodini 2015). While emitting from different sources, degassing from hydrothermal systems can be observed at fumarolic steam vents, extreme hot springs, geysers, boiling mud pots or hot lakes and streams (Figs. 3.7 and 3.8). For example, solfataras (Italian for *sulphur place*) are emitting sulphur-rich gases, while a mofette (French for *noxious fume*) emits gas that is rich in carbon dioxide (Rouwet et al. 2017). The temperature of gases discharging from fumaroles can vary considerably with high temperature fumaroles reaching up to and above 900 °C near active volcanoes (e.g. Sakurajima, Japan) (Tsunogai et al. 2016), where they produce acidic gases such as SO_2, HCl, and HF. Gas emissions at low to medium temperature ranges are commonly observed in less dynamic environments, where hydrothermal processes determine their composition (Fischer and Chiodini 2015; Giggenbach 1987; Shinohara et al. 2011; Symonds et al. 2001). Hydrothermal gases such as CH_4 and H_2S are emitted from low-temperature fumaroles at around 100 °C (Fischer and Chiodini 2015).

Changes in the amount of degassing from a hydrothermal system are indicated by increasing fumarolic activity together with other warning signs like seismic swarms and ground deformation, which can be a sign of transition from quiescence to eruptive activity in volcanic environments (Rouwet et al. 2017). Similar to using radon concentrations as a precursor for seismic activity (Montazeri et al. 2011), variations in the chemical composition of fumarolic gases such as increasing H_2S levels or changes in their discharge characteristics may be indicative of imminent changes in the

Fig. 3.7 High-temperature hot spring with a fumarole degassing under high pressure (Kyushu, Japan)

Fig. 3.8 Degassing takes place at a more leisurely pace at this low-pressure steam vent (Kyushu, Japan)

level of volcanic activity (Italiano and Nuccio 1992; Kiyosu and Okamoto 1998).

In New Zealand the town of Rotorua presents an excellent example of hydrogen sulphide dispersion in the air. Anyone who has visited Rotorua surely remembers the distinctive rotten egg smell of H_2S that lingers in certain areas. Fatal in high concentrations, the intensity of this volatile, corrosive gas can be influenced by humidity, rainfall, wind direction, barometric pressure and air temperature. H_2S emissions from geothermal power stations have the potential to cause serious problems for the surrounding environment (Horwell et al. 2005; Stefánsson et al. 2011). Degassing of high-temperature hydrothermal systems can cause vegetation die-back, which results in erosion and the loss of natural habitats in adjacent areas. As a measure to prevent the release of H_2S into the atmosphere, some countries (e.g. Italy) require its removal prior to any emissions from geothermal power plants (Fridleifsson et al. 2008). Additional Examples 3.3 and 3.4 refer to the analysis of gas emissions at hot springs in North America.

Example 3.3 Wyoming, North America

Research carried out in the Yellowstone National Park indicates that gas emissions at the Hot Spring Basin, compared to other active hydrothermal areas in Yellowstone, are rich in crustal gases such as helium (He) and gaseous hydrogen (H_2), and to a lesser extent enriched with H_2S and CH_4.

Source Werner et al. (2008)

Example 3.4 Rocky Mountains, North America

An analysis of several hot springs in the Rocky Mountains and their volatile gases also indicated that gas components such as Helium-3 (^3He) and Helium-4 (^4He) isotopes, as well as a fraction of dissolved carbon dioxide (CO_2) may be escaping from the underlying mantle. Assisted by fluid circulation in hydrothermal systems these gases make their way to the surface, where they are released into the atmosphere.

Source Karlstrom et al. (2013)

New findings are continuously advancing this highly specialised field of study, and scientific research of hydrothermal and magmatic gases has progressed considerably over recent decades (Fischer and Chiodini 2015). For this reason, a more detailed description of the geochemistry of hydrothermal gases is left to the experts in this field of science, as the scope of this chapter covers only fundamental aspects of geochemistry related to natural hot springs along with suggestions for further reading.

3.4 Mineral Deposition in Hot Spring Areas

3.4.1 Mineralised Landscapes and Vegetation

Hydrothermal mineral deposition is another complex topic that has been explored by scientists for decades, although the classification of mineral deposits remains a point of discussion. While these particular mineral accumulations are often referred to alternately as travertine, tufa, siliceous or calcareous sinter, there is a difference that requires clarification. Siliceous precipitates are composed of silica and associated minerals as well as biological and lithological components (Jones and Renaut 2003). Despite a comparable chemical composition, tufa carbonates or calcareous sinter are precipitates from calcium-bicarbonate waters and are terms usually reserved for mineral deposits from cold springs or water emerging at ambient temperatures, while travertine or siliceous sinter are usually precipitates of hydrothermal origin (Capezzuoli et al. 2014; Cowan et al. 2012; Özkul et al. 2013). This differentiation is based on a much higher content of silica or silicon dioxide (SiO_2) in solutions up to and above boiling point than in colder mineral springs.

Note
The word **travertine** is a variation of the Latin *tiburtinus* and refers to Tibur near Rome in Italy, which is today called Tivoli or Bagni di Tivoli. Large travertine deposits up to 90 m thick have been mined in this area as building material for over 2000 years. In ancient times travertine was known as *lapis tiburtinus*, which means tibur stone and over time developed into travertine.

Source De Filippis et al. (2013) and Faccenna et al. (2008)

The term **tufa** is derived from the Latin *tufus* or *tophus,* which refers to porous limestone and not to be confused with the word *tuff,* which refers to a certain type of igneous rock.

Siliceous deposits of carbonate minerals (travertine) occur in hydrothermal areas where carbon dioxide-rich thermal fluids discharge at the surface, either as hot springs or geysers (Boudreau and Lynne 2012; Dilsiz 2006; Jones and Renaut 2003; Konhauser et al. 2003; Lynne 2007; Sæmundsson and Gunnlaugsson 2002). The constant flow of warm and mineral-rich hot spring water creates travertine features such as silicified hillsides, terraced pools, petrified waterfalls (Fig. 3.9), travertine mounds and microterracettes (Fig. 3.10). Where the water cools and evaporates, mineralised crusts are formed (Fig. 3.11), which indicates that the process of mineral deposition does not take long (McGeary et al. 2001; Renaut 2004). Under favourable geochemical conditions the growth rate of travertine can be tens of centimetres annually, while the rate of growth for tufa is suggested as being considerably slower reaching only a few millimetres per year (Andrews 2006; Kano et al. 2019a).

Note
On reaching the surface, the water temperature changes and causes the silica to precipitate out of solution, thereby forming travertine layers that cover surrounding areas including existing vegetation (Sæmundsson and Gunnlaugsson 2002).

Another interesting point is that mineral accumulations can create diverse depositional features and textures that vary from one hot spring location to another or even between springs connected to the same hydrothermal system (Jones and Renaut 2003). Over time the conditions of hydrothermal discharge can vary (Lynne et al. 2017), a process that is recorded in travertine layers as different periods of precipitation or variations in the deposited mineral content (Fig. 3.12a, b).

Note
Researchers in Japan refer to travertine sites at no less than 30 hot springs throughout the country, which are developing calcareous deposits (Kano et al. 2019b).

Travertine formations with their different silica textures are a distinctive landscape feature. For example, the white sinter terraces of **Pamukkale** (Turkey) consist of vast accumulations of carbonate precipitates resulting from the degassing of CO_2-rich hot spring water (Dilsiz 2006). Several decades ago, anthropogenic pollution caused the discolouration of the siliceous sinter and forced the government to take protective measures in an attempt to restore the terraces to their original white colour (Simsek et al. 2000) (see

Fig. 3.9 This travertine mound is part of the extensive Hammam Mesqoutine area in Guelma, Algeria. *Source* Habib Kaki (2016) (Public Domain)

Fig. 3.10 Micro-terracettes observed at the travertine hillside of Pamukkale, Turkey. Scale in centimetres

Fig. 3.11 Silicified leaves at a high-temperature fumarole (Kyushu, Japan)

also Chap. 2). In some areas unsightly darker precipitates are still visible underneath the newer layers of white travertine (Fig. 3.13).

Another depositional feature at Pamukkale/Hierapolis are areas of mineralised vegetation, caused by hot spring water flowing over cliffs without forming terraces. Instead, the constant water flow has covered exposed root systems and plants with dissolved minerals as they solidified during the cooling process (Appendix 3.2) (Hierapolis).

Apart from Pamukkale, travertine formations are located in other, less well-known areas in **Turkey**. The *Karakoçan-Yoğunağaç* (Elazığ) and *Mazgirt-Dedebağ* (Tunceli) travertines in eastern Anatolia originate from hot springs (24.5–44.4 °C) and have formed white calcite deposits with different morphological features including banded travertines, terraced pools as well as travertine ridges. Analyses of ancient and recent travertine samples indicate that the water temperature may have decreased over time by as much as 17 °C (Kalender et al. 2015).

In the southern hemisphere in **New Zealand,** geysers and hot springs located in the Taupo Volcanic Zone (TVZ) represent a selection of mineral deposits with a wide variety of characteristics (Jones and Renaut 2003) including the deposition of siliceous sinter or geyserite (Fig. 3.14; Appendix 3.4).

Other examples of travertine deposits include the terraces and mounds at *Mammoth Hot Springs*, located in the Yellowstone National Park in **Wyoming (USA)**, which is considered one of the largest sites worldwide with active travertine accumulation, continuously forming new layers (Chafetz and Guidry 2011). A selection of hot spring travertine deposits and their various characteristics, both from a geochemical as well as from a morphological perspective, are briefly described in Appendix 3.1.

Not included here are cold water tufa deposits such as the petrified formations of *Huanglong* (**Sichuan, China**), which are famous for their attractive terraced pools. Chinese researchers have compared the Huanglong formations with Yellowstone and concluded that, unlike the travertine depositions at Mammoth Hot Springs, the Huanglong terraces do not originate from hydrothermal activity (Li et al. 2012; Sun et al. 2014; Wang et al. 2010). The water temperature of the Huanglong spring has been measured and reported by scientists (Liu et al. 2013) as 6.2 °C, which confirms that the terraced pools are not a product of hydrothermal mineral deposits, even though many sources, including the

a

b

Fig. 3.12 **a** Silica deposits are removed on a regular basis from hot spring pools to avoid excessive build-up. Clearly visible are the different layers indicating variations in the mineral content and seasonal changes (Kyushu, Japan; compare Appendix 3.4, Image 6). **b** Enlarged view of the individual layers in one of the silica blocks in the image above

Fig. 3.13 Discolouration caused by pollution several decades ago still remains visible in some sections of the travertine terraces and pools at Pamukkale/Hierapolis (Turkey)

UNESCO world heritage list, refer to Huanglong's terraces as travertine while also mentioning hot springs. The water temperature of the Huanglong spring is around 5 °C higher than the area's annual mean air temperature of 1.1 °C (at over 3500 m ASL), which means this is an extremely cold environment. It is therefore not clear, whether this is a reference to other, genuine hot springs such as the *Jifei* hot spring in Yunnan, south of the Sichuan Province.

There are some disadvantages caused by mineralisation, particularly when hot spring water is used for domestic or industrial purposes. Directly sourced hydrothermal solutions with high concentrations of dissolved components are often responsible for scaling problems in pipes, valves and tap ware. Entire water tanks can be covered with mineral encrustations inside and outside (Fig. 3.15), which requires ongoing maintenance to counteract the consequences of excessive mineralisation. Hot spring spas and health resorts using highly mineralised spring water can potentially face extended downtimes due to plumbing systems gradually getting blocked with mineral deposits. This involves expensive repairs, frequent replacement of parts and represents an additional burden for the general upkeep of the facilities.

3.4.2 Ore Mineral Deposits

Ore deposits can be an important product of hydrothermal mineralisation. These are accumulations of ore minerals and associated rock forming minerals that can be exploited for valuable metals. Ore deposits can have different origins and are categorised either as hydrothermal, igneous, metamorphic or sedimentary (Allaby 2013). Not all ore deposits are produced by hydrothermal activity, but where mineralisation is related to hydrothermal solutions, the resulting sediments are classed as primary deposits, consisting mainly of metallic components such as cobalt (Co), copper (Cu), gold (Au), iron (Fe), lead (Pb), mercury (Hg), molybdenum (Mo), silver (Ag), tin (Sn), tungsten (W), and zinc (Zn) with the potential to be commercially extracted (Shanks 2012; Wilkinson et al. 2009).

Fig. 3.14 Mineralisation is covering parts of the Geyser Flat area of the Whakarewarewa Thermal Valley in Rotorua, New Zealand

Hydrothermal ore deposits are formed when ore minerals are precipitated from hot aqueous fluids moving through permeable rocks and structures. These fluids may contain up to 50% of dissolved solids. The hydrothermal fluids act as both a transporting and depositing agent for the ore bodies or mineral veins, which develop in suitable openings, pore spaces or by replacement of existing rocks. Known hydrothermal deposits include the gold-quartz veins and lodes of the Kalgoorlie's gold field in Western Australia, the silver-nickel-uranium veins of the Erzgebirge (*Ore Mountains*) in Germany and the tin-copper-lead–zinc veins of Cornwall in the south of England (Allaby 2013; Encyclopædia Britannica 2019; Klein 2002; Shanks 2012; Wilkinson et al. 2009). Apart from veins, hydrothermal solutions also form disseminated masses of fine-grained ore minerals such as copper together with smaller amounts of other metals (McGeary et al. 2001).

A related and important style of hydrothermal ore deposits are the volcanogenic massive sulphide ores (VMS), which are rich in metallic sulphide minerals, e.g. copper–nickel sulphides and copper–zinc–lead sulphides. VMS deposits have been discovered on every continent, where they are typically associated with hydrothermal activity, and are a highly sought-after source of Cu, Pb, Zn including precious metals such as Au and Ag (McClenaghan and Peter 2015; Piercey et al. 2015). Resulting from hydrothermal fluids circulating in previous submarine environments or near the seafloor, where hydrothermal vents precipitate ore minerals, VMS developed into large layers of mineral deposits.

3.4.3 Hydrothermal Alteration

Hydrothermal alteration is a different geochemical process associated with hydrothermal activity and is caused by convecting hot solutions entering and passing through rock layers. During this process, mineral components are redistributed leading to chemical and mineralogical changes in the host rock. The formation of alteration minerals is

Fig. 3.15 Communal hot spring water supply is common in areas rich in hydrothermal resources. Due to extreme mineralisation, new installations are added to deteriorated and weakened older hot water tanks (Kyushu, Japan)

not only based on the initial composition and permeability of the parent rock, but also on the duration of the hydrothermal activity, the fluid composition as well as the temperature and pressure; all factors that can influence the volume and the level of hydrothermal alteration (Carlino et al. 2016; Lagat 2009; Mayer et al. 2016). Depending on the conditions in the hydrothermal system, the composition of these fluids regarding their mineral, gas and metallic element content, also varies considerably. Most importantly, hydrothermal alteration can greatly reduce the strength and density of the original rock structure by increasing the porosity and permeability of the rock (Robb 2005; Wyering et al. 2014). Such structural weakness has the potential to cause destabilisation and deformation, making the rock mass more susceptible to slope failure and disintegration (Vallance and Iverson 2015; Van Wyk de Vries and Davies 2015).

The alteration process creates different textures and colours in the altered rock (Fig. 3.16; Appendix 3.3). Extreme cases of alteration are often exhibited in extinct or dormant eroding high-temperature hot spring regions (Branney and Acocella 2015; Franzson et al. 2008; Lagat 2009; Mayer et al. 2016; Marks et al. 2010; Stix 2015). In some alteration areas (around 200–300 °C), the rock colour takes on various shades of green based on the occurrence of epidote and chlorite (Fig. 3.17; Appendix 3.3, Image 3). At high-temperature systems (>300 °C) other minerals can form, including actinolite, garnet, prehnite and wollastonite (Sæmundsson and Gunnlaugsson 2002). At temperatures ranging between 50 and 400 °C the process of hydrothermal alteration is frequently referred to as metasomatism, a mass transfer process that is a common occurrence in hydrothermal areas (Stefánsson and Kleine 2018).

Stages of alteration can range from minor changes affecting only some minerals in the host rock to the complete replacement of the original minerals with alteration minerals.

3.5 Extremophiles and Hot Springs

3.5.1 Heat-Loving Microorganisms

Hot springs, geysers and fumaroles are some of the remarkable features that demonstrate the visible activity of

Fig. 3.16 Hydrothermal alteration of a crater wall in an active volcanic environment (Mount Garandake, Kyushu, Japan)

hydrothermal systems; often accompanied by a display of bright colours that surround erupting vents and hot pools. Many of these colourful manifestations are caused by microbial growth and are ideal research sites for scientists who use them to study the geomicrobiology of extreme life forms residing in and around hot springs (Renaut and Jones 2003; Te Ara 2019). To exist in high temperature environments organisms must be thermotolerant, an ability that allows certain microorganisms to adapt to life under extreme conditions, e.g. hot springs with high acidity and/or toxic concentrations of metallic elements. For these microorganisms, known as *extremophiles*, hydrothermal systems and their discharge areas (Fig. 3.18) are a perfect breeding ground (Rzonca and Schulze-Makuch 2003).

Extremophiles are single-celled microorganisms or microbes that occur in all three domains of life: *archaea*, *bacteria* and *eukaryotes* and have the capacity to adapt to life in extreme environments. The collective term extremophile includes *thermophiles* and *hyperthermophiles*, which all have an extraordinary tolerance to extreme hydrothermal conditions. As far as hydrothermal systems are concerned, extremophiles survive and even thrive at high temperatures, extreme pH levels (acidophiles <5.0; alkaliphiles >9.0) and in highly saline hydrospheres (halophiles). Extremophiles are also tolerant to seafloor environments, withstanding high pressure while living near hydrothermal vents where hydrogen sulphide in high concentrations is supporting unusual life forms such as sulphide-oxidising bacteria (Amin et al. 2018; GNS Science 2019; Sigurdsson 2015) .

Note
The term extremophile also covers microbes surviving in extreme cold environments.

3.5.2 Thermophiles and Hyperthermophiles

Thermophiles are microbes equipped with specialised enzymes that enable them to withstand high temperatures with an optimum growth temperature between 55 and 65 °C. Regardless of the hostile surroundings, often close to active volcanoes, hydrothermal environments provide ideal conditions for thermophiles to maintain their existence by feeding

Fig. 3.17 Extreme hydrothermal alteration of entire hillsides at Landmannalaugar, Iceland

on inorganic compounds (Kochetkova et al. 2011). For **hyperthermophiles** the optimum growth temperature is even higher at above 80 °C up to over 110 °C, while temperatures below 90 °C seem to prevent them from growing and reproducing (Stetter 2005).

Microorganisms in hot spring areas can grow into large colonies, which establish microbial mats around hydrothermal vents (Fig. 3.19). These microbial formations of thermophilic bacteria and algae are often brightly coloured, representing different temperature ranges. They also develop in hot streams and pools, often close to the surface (Fig. 3.20; Appendix 3.6).

Thermophilic organisms are a relatively recent discovery; however, decades of research and studies of their geomicrobiology have increased the knowledge and understanding of life forms in extreme environments (Oliverio et al. 2018). In the mid-1960s Thomas Brock, a microbiologist, visited Yellowstone National Park, where he discovered new thermophilic bacteria and their microbial formations. Based on existing knowledge about extremophiles, Brock analysed and described in detail the organisms his research team found in Yellowstone around hot springs of varying temperatures (Brock 1967, 1972, 1978; Brock and Freeze 1969). Since then, hydrothermal environments all over the world have attracted microbiologists (Example 3.5), who are carrying out studies into the diversity of bacterial communities that inhabit hydrothermal systems including hot springs, geysers, streams and lakes (Renaut and Jones 2003; Song et al. 2009).

Example 3.5 Kamchatka, Russia

The Kamchatka Peninsula has many areas with volcanic hot springs and other hydrothermal features, which favours the ongoing study of complex thermophilic microbial communities by national and international research teams. In a journal article from 2005, scientists reported that over twenty thermophilic microorganisms were identified at several hot springs studied in Kamchatka. Research projects are further

Fig. 3.18 Overflow area next to erupting geysers with microbial communities and evidence of mineralisation (Haukadalur, Iceland)

exploring the potential of extreme life forms as well as the microbial silicification of hot springs in the Peninsula.

Source Kochetkova et al. (2011), Kyle et al. (2007), Merkel et al. (2017) and Zhao et al. (2005, 2011)

A further interesting aspect of the research into thermotolerant microbes are investigations that focus on unique thermophilic properties such as the thermal stability of their enzymes. This is an essential characteristic of the diversity of thermophiles, which is becoming increasingly important because of their biotechnological potential that includes practical applications in agriculture, commercial industry and in medicine (Brock 1967; Ghilamicael et al. 2018; Sahoo et al. 2015; Song et al. 2009; Tang et al. 2018). In fact, the discovery of thermotolerant bacteria in hot spring environments has made a significant difference in the study of DNA (deoxyribonucleic acid) by helping to speed up the copying process of genetic material, a procedure known as polymerase chain reaction (PCR). At the time of writing the COVID-19 pandemic was sweeping the globe and it is interesting to note that the necessary testing for this virus is based on the same copying process, although with some extra steps (Wei-Haas 2020). Who knows, what other benefits natural hot springs have to offer in the future while science is further investigating this unique resource?

3.5.3 Cyanobacteria and Other Microorganisms

The growth of unicellular organisms such as cyanobacteria (blue-green algae), a bacteria group with the ability to use photosynthesis to produce oxygen, is another form of microbial activity in warm and hot springs (Kalender et al. 2015; Zhang et al. 2007). Existing and thriving in thermal waters, cyanobacteria develop into microbial communities, with the potential to influence the accumulation of mineral deposits (e.g. travertine) by colonising runoff channels close to hot springs and geysers (Renaut and Jones 2003;

Fig. 3.19 This iron-rich hot spring is causing algae growth while depositing minerals over a cone originally formed with rocks (Karahayit, Turkey)

Fig. 3.21). Here they grow either in resistant mats or as loose filaments, trapping precipitates around vents and flow paths (Bonny and Jones 2003; Pentecost 2003).

Research from as far back as the 1930s established the occurrence of a diversity of microalgal and cyanobacterial communities in the **Dead Sea area** between Israel and Jordan (Elazari-Volcani 1940; Oren et al. 2008), especially at the hot springs of Zarka Ma'in and the Zara springs in **Jordan**. Other bacterial groups, including proteobacteria and acidobacteria members, have been reported from hydrothermal environments associated with volcanic areas in the **Canary Islands** (Portillo and Gonzalez 2008), as well as in western **Thailand** (Kanokratana et al. 2004; Portillo et al. 2009). In **Australia**, the Paralana Hot Springs, heated to 60 °C by radioactive decay, are also hosting a wide diversity of thriving microbial communities including cyanobacteria and proteobacteria (Anitori et al. 2004; Greene et al. 2016; Wright et al. 2012). In the **Philippines**, on the island of Luzon, phototrophic green non-sulphur bacteria have been observed in the tropical hot springs (Lacap et al. 2007), which confirms that thermotolerant microorganisms exist in hot spring areas all over the world.

Some thermophilic microbes have the potential to cause serious health issues. These include *Legionella bacteria* and *Naegleria fowleri*, which have been identified at a number of hot springs used for recreational purposes (Erfurt-Cooper 2018), a topic that is discussed in detail in Chap. 9.

Fig. 3.20 Microbial communities have formed in a small hydrothermal stream in Furnas, São Miguel (Azores)

3.6 Final Comments

The geochemistry and geomicrobiology of hot springs are complex in detail and still not fully understood. This chapter has presented an overview of the main geochemical aspects, particularly related to the fundamental chemical components, minerals, gases and microbiota present in hot springs. Geochemical processes like hydrothermal alteration and mineralisation, as well as the existence of unique life forms (e.g. thermotolerant microorganisms) are also associated with hydrothermal environments and were briefly discussed as well. To assist the interested reader with additional information, a considerable amount of suitable and useful source material was reviewed and cited to provide relevant reference points for further reading.

Fig. 3.21 White silica layers are visible beneath well-developed bacterial mats, which are mainly composed of thermophilic cyanobacteria (blue-green algae) and inhabit the edge of a pool fed by a boiling spring (Kyushu, Japan)

Hydrothermal field in the Krafla volcano area, Iceland

3.7 Appendices

Appendix 3.1 Selection of Travertine Deposits of Hydrothermal Origin

Country	Geographic location	Source temperature	Type of formation	Composition of deposit	Site status utilisation
Algeria	Hammam Meskhoutine[a]	Several natural hot springs Temperatures between 70 and 98 °C Artesian origin	Travertine formations Mounds, domes, cascades, terraces, petrified water channels (Fig. 3.9)	Calcium carbonate deposition up to 6 m high, predominantly magnesium bicarbonate and magnesium sulphate, colour from microbial communities	Tourist attraction Spa resort Greenhouse heating
Hungary Heves County	Egerszalók[b]	Hot spring resulted from exploration drilling in 1961 65–68 °C Artesian origin—limestone aquifer	Slope-deposited travertine mound with waterfalls, stepped pools and terraces	Carbonate precipitation almost pure calcite at point of discharge	Open-air spa Tourist attraction
Iran Mazandaran Province	Badab-e Surt[c]	Water from two different hot mineral springs—one spring has a high salt content	Red and orange stepped travertine terraces, cascades and mound formations	Calcium carbonate deposits containing large concentrations of iron oxide (Appendix 7.4)	Tourist attraction under visitor pressure—in danger of degradation
Italy	Tivoli (and Guidonia)[d] Lapis Tiburtinus Travertine	Acque Albule basin, an area with abundant hot springs, 23–25 °C Low-temperature hydrothermal activity based on volcanic origin	Travertine ridges, compact strata Ancient succession of travertine deposits	White to grey calcium-carbonate deposits up to 90 m thick Calcite and sulphate precipitates	Active quarries mined for over 2000 years Travertine is used as building material
Italy	Saturnia Travertine Deposits[e]	Sulphur springs 37.5 °C Hydrothermal water associated with volcanic origin	Travertine rock pools Waterfalls, cascades	Travertine deposits consist of several carbonate banks separated by erosion up to 25 m thick	Active quarries and tourist attraction—hot springs have been used since Etruscan times
Mexico	Hierve el Agua, Oaxaca[f]	Several warm springs at temperatures between 22 and 27 °C Artesian origin	Terraced travertine pools, petrified waterfalls up to 30 m high, travertine irrigation channels	Calcium carbonate deposits containing other minerals Accumulated layers up to 5 m thick	Natural tourist attraction under visitor pressure—in danger of degradation
New Zealand North Island	Pink and White Terraces of Rotomahana[g]	Original location of both terraces beside Lake Rotomahana, Current research indicates remnants of both terraces still exist at lake bottom	Stepped travertine formations, terraced pools with scalloped buttresses	Silica formations White and salmon coloured siliceous sinter deposits (Appendices 7.1 and 7.2)	Deemed the 8th World Wonder until the eruption of Mount Tarawera in 1886, which destroyed the terraces
New Zealand North Island	Whakarewarewa Thermal Valley, Rotorua[h]	Geyser Flat geothermal area, Pōhutu Geyser, Te Tohu Geyser Other features >70 °C and up to and above boiling point	Precipitation takes place mainly around erupting geyser vents, hot springs	Geyserite, siliceous sinter, pale grey and white silica layers (Appendix 3.4, Image 7)	Whakarewarewa Living Māori Village—Tourist attraction GNS Earth Science Facility
Turkey Denizli Region	Pamukkale[i] Clusters of hot springs	Four hot springs at the top of the cliff at 35–57 °C Reports of discharge capacity vary between 20 and >40 million l/d	Large-scale terraced pools, petrified channels, mounds, silicified vegetation, banded travertine, microterracettes	White to cream coloured calcium bicarbonate deposits, cyanobacteria mats and diatoms, past algae growth due to anthropogenic actions (Figs. 7.5, 7.6 and 7.7)	UNESCO World Heritage Area Tourist attraction (Appendix 3.2)

(continued)

Country	Geographic location	Source temperature	Type of formation	Composition of deposit	Site status utilisation
Turkey Denizli Region	Karahayıt[j]	33–61.5 °C Due to overuse many hot springs have stopped flowing	Terraced pools, mounds and domes, petrified outflow channels	Red travertine deposits coloured by metal oxides (e.g. iron), microbial mats and algae communities growing on terraces (Figs. 2.8, 3.19 and 8.13)	Close to Pamukkale (5 km) Tourist attraction Open-air spa
Turkey Elazığ Province and Tunceli Province	Karakoçan-Yoğunağaç Mazgirt-Dedebağ[k]	Several hot springs 24.5–44.4 °C	Travertine ridges, terraced pools, banded travertines Travertine deposition controlled by tectonic activity	White calcite deposits, coloured deposits (light grey, ivory and ochre), inhabited by algal species	Active quarries mined for building material
Uganda Rwenzori	Buranga Hot Springs[l] East African Rift	Cluster of 37 hot springs 60 and >98 °C Hydrothermal system (150–160 °C) related to volcanic origin	Travertine mounds, cones and pools Most springs are associated with travertine deposits	Sulphur deposits Carbonate deposits High salinity	Tourist attraction
Wyoming, USA	Mammoth Hot Springs[m] Yellowstone National Park	Large hot springs complex 73–80 °C Volcanic origin Meteoric water (Norris Geyser Basin)	Travertine mounds and slopes, channels and aprons, terraced pools, scalloped buttresses, microterracettes	Calcium carbonate deposits Sequence of five different travertine facies Diversity of microbial communities	UNESCO World Heritage Tourist attraction
Wyoming, USA	Thermopolis[n] Big Horn Hot Springs at Hot Springs State Park	Three active hot springs including one sulphur spring—up to 57 °C, springs may have moved over time, artesian origin, no obvious volcanic association	Extensive travertine deposits known as Rainbow Terraces travertine ledges, rock pools, manmade domes (e.g. Teepee Fountain; Appendix 3.8)	Calcium carbonate deposits, coloured by varies types of algae, concentrations of calcium, magnesium, and bicarbonate combined >50% of TDS, Rainbow Terrace layers approximately 12 m thick	Natural Monument IUCN category III Tourist attraction

Sources
[a]Benamara et al. (2017), Himri et al. (2009), Office de Tourisme de Guelma (2019) and Saibi (2015)
[b]Benkhard (2015) and Kele et al. (2006, 2008)
[c]Khansha et al. (2018) and Sotohian and Ranjbaran (2015)
[d]De Filippis et al. (2013), Faccenna et al. (2008) and Giggenbach et al. (1989)
[e]Ronchi and Cruciani (2015)
[f]Castañeda Camacho and Reyes Perez (2015), Reyes Pérez and Sanchez Crispín (2005), Wass (2007) and Winsborough et al. (1996)
[g]Cole (1970), De Ronde et al. (2016, 2018) and Keam (2016)
[h]Jones and Renaut (2003)
[i]Açikel and Ekmekçi (2007), Altunel and Hancock (1993), De Boever et al. (2016), Dilsiz (2006), Özkul et al. (2013), Scoon (2015), Simsek et al. (2000) and Taşeli (2016)
[j]Gokgoz et al. (2010), Ökçesiz (2014) and Taşeli (2016)
[k]Kalender et al. (2015)
[l]Bahati (2003), Kato and Kraml (2010) and Li et al. (2012)
[m]Chafetz and Guidry (2011), De Boever et al. (2016), Fouke et al. (2003) and Fouke (2011)
[n]Darton (1906), Geology of Wyoming (2019), Kaszuba et al. (2014) and Lund (1993)

Note Deposits classified as travertine can include various characteristics, both from a geochemical as well as from a morphological perspective. Travertine deposits are further influenced by climate, terrain, temperature, discharge rates, bacteria and anthropogenic activities (Pentecost and Viles 1994).

Appendix 3.2 Pamukkale Travertine

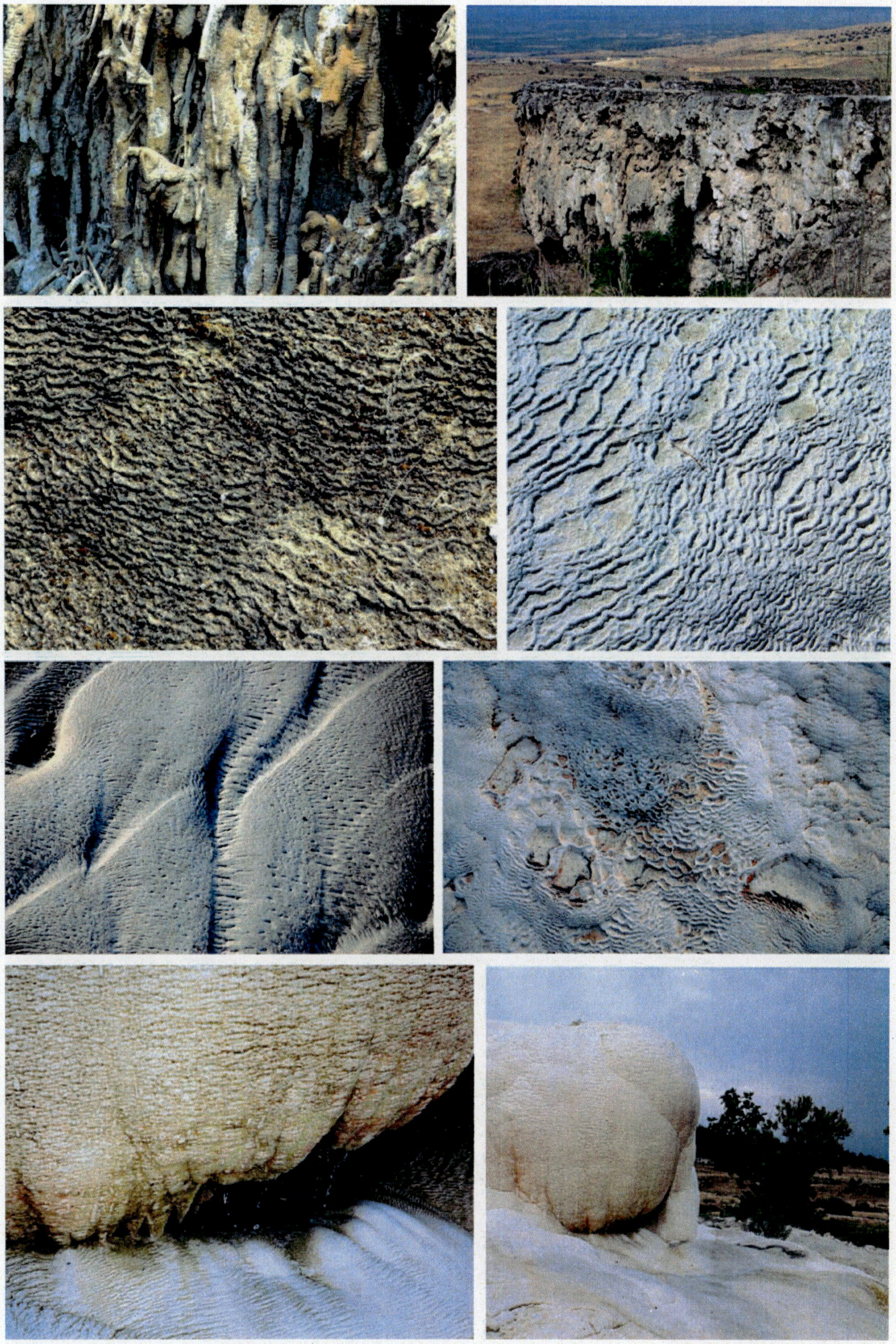

1, 2—Mineral deposits covering vegetation and exposed root systems; 3, 4—Contrasting travertine layers, on the left an older and discoloured section of the terraces, on the right fresh travertine deposits with more clearly defined terracettes; 5, 6—Different surface structures, e.g. microterracettes, at different sites of the terraces; 7, 8—Dome-shaped mineral deposits with vermicular surface texture and clearly defined rills

Appendix 3.3 Hydrothermal Alteration/Metasomatism

1, 2—Hydrothermally altered sections of the crater walls inside the Stefanos Crater, Nisyros (Greece); 3, 4—Exposed hydrothermal alteration at Landmannalaugar and Krýsuvík (Iceland); 5, 6—Hydrothermal activity inside the crater of Mt. Garandake in Kyushu and on the flanks of Showa Shinzan in Hokkaido (Japan); 7, 8—Metasomatism caused by hydrothermal volatiles/steam, Terceira Island (Azores), and White Island (New Zealand).

Appendix 3.4 Mineralisation Effects

1, 2—Mineral deposits forming wave-shaped encrustations on the volcanic rocks at the Blue Lagoon (Iceland); 3, 4—Mineral encrusted hot water tanks fed by local hot springs for domestic use (Japan); 5, 6—Hot spring ponds are emptied to remove the build-up of mineral deposits, Kamado Jigoku, Kyushu (Japan—See Fig. 3.12a, b); 7, 8—Geyserite layers at Whakarewarewa (Rotorua, New Zealand) and at the Caldeira das Furnas, São Miguel (Azores)

Appendix 3.5 Fumaroles, Sulphur Deposits

1, 2—Sulphur crystals are forming at small fumaroles in Krýsuvík (Iceland) and on the crater floor of Nisyros volcano (Greece); 3, 4—Sulphur encrustations in pipes and gutters in Myoban, Beppu (Japan); 5, 6—Featherlike sulphur crystals at a degassing fumarole (Greece) and thick layers of sulphur around steam vents in the crater of White Island volcano (New Zealand); 7, 8—Degassing fumaroles at Whakarewarewa (Rotorua, New Zealand) and spear-like sulphur crystals, Nisyros (Greece)

Appendix 3.6 Thermophiles

1, 2—Thermotolerant microbial growth at high-temperature degassing fumaroles (Japan); 3, 4—Microbial mats covering silica deposits in a hot spring fed pond (Japan); 5, 6—Algae communities in a creek fed by hot springs at Landmannalaugar (Iceland) (left) and at Kerosene Creek (New Zealand); 7, 8—Algal growth blanketing parts of the crystal clear water of a hot spring pond. The image on the right shows the bottom of the pond (Northern Territory, Australia)

Appendix 3.7 Thermopolis

The 'Big Spring' in Thermopolis, Wyoming, is over four metres deep and has a flow rate of more than eight million litres a day. The water emerges at a temperature of 57 °C and is used for the state bath house and several other facilities. The Big Spring also supplies the water for the Teepee Fountain (Appendix 3.8). *Source* Steele and Fisher (2016). *Photo credit* Mark Fisher (2017) (with permission)

Appendix 3.8 Thermopolis

The Teepee Fountain in Thermopolis is a manmade structure dating back to 1906. Hot spring water from the Big Spring is piped to the Teepee Fountain during spring and summer. The growth rate of the travertine cone is estimated between two and five centimetres a year. *Source* Steele and Fisher (2016). *Photo credit* Mark Fisher (2017) (with permission)

References

Açikel Ş, Ekmekçi M (2007) Assessment of sorption characteristics for enhancement of remediation of Pamukkale travertines, Turkey. In: L Candela et al (eds) Water pollution in natural porous media at different scales. Assessment of Fate, Impact and Indicators. WAPO^2. Madrid: Instituto Geológico y Minero de España. ISBN:978–84–7840–676–0

Ajlouni A-W, Abdelsalam M, Abu-Haija O, Almasa'efah YS (2010) Radiation doses due to natural radioactivity in the Afra Hot Springs, Jordan. Int J Low Radiat 7(1):48–52. https://doi.org/10.1504/IJLR.2010.03277

Allaby M (2013) A Dictionary of Geology and Earth Sciences, 4th edn. Oxford University Press, Oxford

Altunel E, Hancock PL (1993) Morphology and structural setting of quaternary travertines at Pamukkale, Turkey. Geol J 28:335–346. https://doi.org/10.1002/gj.3350280312

Amin A, Ahmed I, Khalid N, Zhang Y, Xiao M, Li W-J (2018) Insights into the thermophile diversity in hot springs of Pakistan. In: Egamberdieva D et al (eds) Extremophiles in Eurasian ecosystems: ecology, diversity, and applications, microorganisms for sustainability, vol 8. https://doi.org/10.1007/978-981-13-0329-6_1

Andrews JE (2006) Palaeoclimatic records from stable isotopes in riverine tufas: synthesis and review. Earth Sci Rev 75(1–4):85–104. https://doi.org/10.1016/j.earscirev.2005.08.002

Anitori RP, Trott C, Saul DJ, Bergquist PL, Walter MR (2004) The microbial community of a radon hot spring. In: Norris R, Stootman F (eds) Bioastronomy 2002: life among the stars, proceedings of IAU symposium #213. Astronomical Society of the Pacific, San Francisco, p 374

Arnórsson S, Thórhallsson S, Stefánsson A (2015) Utilization of geothermal resources. In: Sigurdsson H (ed) The encyclopedia of volcanoes, 2nd edn. Academic Press, London, pp 851–864

Bahati G (2003) Geothermal energy in Uganda, country update. In: International geothermal conference. Reykjavík, Sept 2003

Benamara A, Kherici-Bousnoubra H, Bouabdallah F (2017) Thermo-mineral waters of Hammam Meskoutine (north-east Algeria): composition and origin of mineralization. J Water Land Dev 34:47–57. https://doi.org/10.1515/jwld-2017-0037

Benkhard B (2015) Travertine mound of Egerszalók: the Hungarian Pamukkale. In: Lóczy D (ed) Landscapes and landforms of Hungary. World geomorphological landscapes series. Springer International Publishing AG, Switzerland, pp 201–207 (Chapter 23)

Bonny S, Jones B (2003) Microbes and mineral precipitation, Miette Hot Springs, Jasper National Park, Alberta, Canada. Can J Earth Sci 40(11):1483–1500

Bonotto DM, Santos TO (2007) Health risk related to radon in a thermal spa utilizing waters from Guarani aquifer. WIT Trans Biomed Health 11:137–146. https://doi.org/10.2495/EHR070151

Boudreau AE, Lynne BY (2012) The growth of siliceous sinter deposits around high-temperature eruptive hot springs. J Volcanol

Geoth Res 247–248:1–8. https://doi.org/10.1016/j.jvolgeores.2012.07.008

Branney M, Acocella V (2015) Calderas. In: Sigurdsson H (ed) The encyclopedia of volcanoes, 2nd edn. Academic Press, London, pp 299–315

Breckenridge RM, Hinckley BS (1978) Thermal springs of Wyoming. Geol Surv Wyom Bull 60:104

Brock TD (1967) Life at high temperatures. Science 158:1012–1019

Brock TD (1972) Ecology of a hot spring. Rediscovering yellowstone—Tom Brock. Encyclopaedia Britannica Films. Retrieved from https://www.youtube.com/watch?v=YdkTW30Gv64

Brock TD (1978) Thermophilic microorganisms and life at high temperatures. Springer, New York

Brock TD, Freeze H (1969) Thermus aquaticus gen. n. and sp. N., a non-sporulating extreme thermophile. J Bacteriol 98(1):289–297

Brugger J, Long N, McPhail DC, Plimer I (2005) An active amagmatic hydrothermal system: the Paralana Hot Springs, Northern Flinders Ranges, South Australia. Chem Geol 222:35–64. https://doi.org/10.1016/j.chemgeo.2005.06.007

Capezzuoli E, Gandin A, Pedley M (2014) Decoding tufa and travertine (freshwater carbonates) in the sedimentary record: the state of the art. Sedimentology 61:1–21. https://doi.org/10.1111/sed.12075

Carlino S, Troiano A, Di Giuseppe MG, Tramelli A, Troise C, Somma R, De Natale G (2016) Exploitation of geothermal energy in active volcanic areas: a numerical modelling applied to high temperature Mofete geothermal field, at Campi Flegrei caldera (Southern Italy). Renew Energy 87(1):54–66. https://doi.org/10.1016/j.renene.2015.10.007

Castañeda Camacho X, Reyes Pérez Ó (2015) Ecoturismo y Problemas Ambientales Subsecuentes: El Caso de Cascadas de Hierve el Agua, Oaxaca. México, Coordinacion de Ciencias Sociales y Humanidades

Chafetz HS, Guidry SA (2011) Deposition and diagenesis of Mammoth Hot Springs travertine, Yellowstone National Park, Wyoming, USA. Can J Earth Sci 40(11):1515–1529. https://doi.org/10.1139/e03-051

Cole JW (1970) Structure and eruptive history of the Tarawera volcanic complex. NZ J Geol Geophys 13(4):879–902. https://doi.org/10.1080/00288306.1970.10418208

Cowan D, Tuffin M, Mulako I, Cass J (2012) Terrestrial hydrothermal environments. In: Bell EM (ed) Life at extremes: environments, organisms and strategies for survival. In: CAB international, pp 219–241 (Chapter 12)

Daintith J (ed) (2000) A dictionary of chemistry, 4th edn. Oxford University Press, New York

Darton NH (1906) The hot springs at Thermopolis, Wyoming. J Geol 14:194–200. https://archive.org/metadata/jstor-30055838

Daubeny C (1848) Thermal waters, their gaseous impregnations, etc. In: Daubeny C (ed) A description of active and extinct volcanoes, of earthquakes and of thermal springs; with remarks on the causes of these Phænomena, the character of their respective products, and their influence on the past and present conditions of the globe. 2nd edn. Richard and John E. Taylor, London

De Boever E, Foubert A, Lopez B, Swennen R, Jaworowski C, Özkul M, Virgone A (2016) Comparative study of the Pleistocene Cakmak quarry (Denizli Basin, Turkey) and modern Mammoth Hot Springs deposits (Yellowstone National Park, USA). Quat Int 437 (Part A):129–146. https://doi.org/10.1016/j.quaint.2016.09.011

De Filippis L, Anzalone E, Billi A, Faccenna C, Poncia PP, Sella P (2013) The origin and growth of a recently-active fissure ridge travertine over a seismic fault. Geomorphology, Tivoli. https://doi.org/10.1016/j.geomorph.2013.04.019

De Ronde CEJ, Fornari DJ, Ferrini VL, Walker SL, Davy BW, LeBlanc C, Caratori Tontini F, Kukulya AL, Littlefield RH (2016) The pink and white terraces of Lake Rotomahana: what was their fate after the 1886 Tarawera Rift eruption? J Volcanol Geoth Res 314:126–141. https://doi.org/10.1016/j.jvolgeores.2016.02.003

De Ronde CEJ, Caratori Tontini F, Keam RF (2018) Where are the pink and white terraces of Lake Rotomahana? J R Soc N Z. https://doi.org/10.1080/03036758.2018.1474479

Delmelle P, Henley RW, Bernard A (2015) Volcano-related lakes. In: Sigurdsson H (ed) The encyclopedia of volcanoes, 2nd edn. Academic Press, London, pp 851–864

Dilsiz C (2006) Conceptual hydrodynamic model of the Pamukkale hydrothermal field, southwestern Turkey, based on hydrochemical and isotopic data. Hydrogeol J 14:562–572. https://doi.org/10.1007/s10040-005-0001-4

Elazari-Volcani B (1940) Studies on the microflora of the Dead Sea. Ph.D. thesis. Hebrew University, Jerusalem

Encyclopædia Britannica (2019) Hydrothermal mineral deposit. Retrieved from https://www.britannica.com/science/hydrothermal-mineral-deposit

Erfurt P (2012) An assessment of the role of natural hot and mineral springs in health, wellness and recreational tourism. Unpublished doctoral dissertation, School of Business, James Cook University, Cairns, Queensland, Australia

Erfurt-Cooper P (2018) Active hydrothermal features as tourist attractions. In: Fearnley CJ, Bird DK, Haynes K, McGuire WJ, Jolly G (eds) Observing the volcano world: volcano crisis communication. Advances in volcanology series. Springer, Berlin (Chapter 27). https://doi.org/10.1007/11157_2016_33

Erfurt-Cooper P, Cooper M (2009) Health and wellness tourism: spas and hot springs. Channel View Publications, Bristol

Faccenna C, Soligo M, Billi A, De Filippis L, Funiciello R, Rossetti C, Tuccimei P (2008) Late Pleistocene depositional cycles of the Lapis Tiburtinus travertine (Tivoli, Central Italy): possible influence of climate and fault activity. Glob Planet Change 63:299–308. https://doi.org/10.1016/j.gloplacha.2008.06.006

Fischer TP, Chiodini G (2015). Volcanic, magmatic and hydrothermal gases. In: Sigurdsson H et al (eds) The encyclopedia of volcanoes, 2nd edn. Elsevier, Academic Press, London, pp 779–797

Fisher M (2017) Photo credit: The Teepee Fountain in Thermopolis. With permission. Retrieved from https://www.geowyo.com/thermopolis-hot-springs.html

Fouke BW (2011) Hot-spring systems geobiology: abiotic and biotic influences on travertine formation at Mammoth Hot Springs, Yellowstone National Park, USA. Sedimentology 58:170–219. https://doi.org/10.1111/j.1365-3091.2010.01209.x

Fouke B, Bonheyo GT, Sanzenbacher B, Frias-Lopez J (2003) Partitioning of bacterial communities between travertine depositional facies at Mammoth Hot Springs, Yellowstone National Park, USA. Can J Earth Sci 40(11):1531–1548. https://doi.org/10.1139/E03-067

Franzson H, Zierenberg R, Schiffman P (2008) Chemical transport in geothermal systems in Iceland—evidence from hydrothermal alteration. J Volcanol Geoth Res 173(3–4):217–229. https://doi.org/10.1016/j.jvolgeores.2008.01.027

Fridleifsson IB, Bertani R, Huenges E, Lund JW, Rybach L (2008) The possible role and contribution of geothermal energy to the mitigation of climate change. In: Hohmeyer O, Trittin T (eds) IPCC scoping meeting on renewable energy sources—proceedings. IPCC Special Report

Gallois RW (2006) The geology of the hot springs at Bath Spa, Somerset. Geosci South-West Engl 11:168–173

Geology of Wyoming (2019) Thermopolis hot springs—the smoking waters. Retrieved from https://www.geowyo.com/thermopolis-hot-springs.html

Ghersetich I, Lotti TM (1996) Immunologic aspects: immunology of mineral water spas. Clin Dermatol 14(6):563–566. https://doi.org/10.1016/S0738-081X(96)00085-5

Ghilamicael AM, Budambula NLM, Anami SE, Mehari T, Boga HI (2018) Thermotolerant bacteria of biotechnological potential from hot springs in Eritrea. Afr J Microbiol Res 12(22):512–524. https://doi.org/10.5897/AJMR2017.8626

Giggenbach WF (1987) Redox processes governing the chemistry of fumarolic gas discharges from White Island, New Zealand. Appl Geochem 2:143–161

Giggenbach WF, Minissale AA, Scandiffio G (1989) Isotopic and chemical assessment of geothermal potential of the Coili Albani area, Latium region, Italy. Appl Geochem 3:475–486

Girault F, Perrier F, Przylibski TA (2016) Radon-222 and radium-226 occurrence in water: a review, vol 451. Geological Society, London. Special Publications. https://doi.org/10.1144/SP451.3

GNS Science (2019) Microorganisms inhabiting submarine volcanoes. Retrieved from https://tinyurl.com/yahrur8k

Gokgoz A, Yilmazli IE, Gungor I, Yavuzer I (2010) Hydrogeology and environmental study at the Karahayit geothermal field (Western Turkey). In: Proceedings world geothermal congress 2010, Bali, Indonesia, 25–29 Apr 2010

Goldscheider N, Mádl-Szönyi J, Eröss A, Schill E (2010) Review: thermal water resources in carbonate rock aquifers. Hydrogeol J 18 (6):1303–1318

Gostin V (2020) Paralana Hot Springs—photo. With permission by Professor Victor Gostin from the University of Adelaide

Greene AC, Wright MH, Aldosary HA (2016) Bacterial diversity and metal reducing bacteria in Australian thermal environments. In: Méndez-Vilas A (ed) Microbes in the spotlight: recent progress in the understanding of beneficial and harmful microorganisms. BrownWalker Press, Boca Raton

Guo Q, Wang Y (2012) Geochemistry of hot springs in the Tengchong hydrothermal areas, Southwestern China. J Volcanol Geoth Res 215–216:61–73. https://doi.org/10.1016/j.jvolgeores.2011.12.003

Guo Q, Planer-Friedrich B, Liu M, Li J, Zhou C, Wang Y (2017) Arsenic and thioarsenic species in the hot springs of the Rehai magmatic geothermal system, Tengchong volcanic region, China. Chem Geol 453:12–20

Himri Y, Malik AS, Boudghene-Stambouli A, Himri S, Draoui B (2009) Review and use of the Algerian renewable energy for sustainable development. Renew Sustain Energy Rev 13(6–7):1584–1591. https://doi.org/10.1016/j.rser.2008.09.007

Hodgson S (2004) A beautiful spa—thermal waters at San Bartolo Agua Caliente, Mexico. In: GHC bulletin. Geo-Heat Center. Oregon, June 2004

Horvath Á, Bohus LO, Urbani F, Marx G, Piróth A, Greaves ED (2000) Radon concentrations in hot spring waters in northern Venezuela. J Environ Radioact 47:127–133

Horwell CJ, Patterson JE, Gamble JA, Allen AG (2005) Monitoring and mapping of hydrogen sulphide emissions across an active geothermal field: Rotorua, New Zealand. J Volcanol Geoth Res 139:259–269. https://doi.org/10.1016/j.jvolgeores.2004.08.010

Hunt TM (1998) Geothermal resources in New Zealand—an overview. GHC Bull, Sept 1998. Retrieved from https://tinyurl.com/yagq4ln7

Ilani S, Minster T, Kronfeld J, Even O (2006) The source of anomalous radioactivity in the springs bordering the sea of Galilee, Israel. J Environ Radioact 85:137–146. https://doi.org/10.1016/j.jenvrad.2005.07.003

IMA (International Mineralogical Association) (2021) The new IMA list of minerals—a work in progress—updated. Jan 2021. Retrieved from http://cnmnc.main.jp/

Italiano F, Nuccio PM (1992) Volcanic steam output directly measured in fumaroles: the observed variations at Vulcano Island, Italy, between 1983 and 1987. Bull Volcanol 54:623–630

Jones B, Renaut RW (2003) Hot spring and geyser sinters: the integrated product of precipitation, replacement, and deposition. Can J Earth Sci 40(11):1549–1569. https://doi.org/10.1139/e03-078

Kaki H (2016) Travertine mound of the Hammam Mesqoutine area in Guelma, Algeria. Public Domain. Retrieved from https://tinyurl.com/rwfvkgy

Kalender L, Öztekin Ökan O, İnceöz M, Çetindağ B, Yildirim V (2015) Geochemistry of travertine deposits in the Eastern Anatolia District: an example of the Karakoçan-Yoğunağaç (Elazığ) and Mazgirt-Dedebağ (Tunceli) travertines, Turkey. Turk J Earth Sci 24:607–626. https://doi.org/10.3906/yer-1504-27

Kano A, Okumura T, Takashima C, Shiraishi S (2019a) Geochemical model for rapid carbonate precipitation of travertines: with a focus on Japanese sites. In: Geomicrobiological properties and processes of travertine. Springer, Berlin, pp 113–131. https://doi.org/10.1007/978-981-13-1337-0_6

Kano A, Okumura T, Takashima C, Shiraishi S (2019b) Travertines in Japan: with a focus on Japanese sites. In: Geomicrobiological properties and processes of travertine. Springer, Berlin, pp 133–173. https://doi.org/10.1007/978-981-13-1337-0_6

Kanokratana P, Chanapan S, Pootanakit K, Eurwilaichitr L (2004) Diversity and abundance of bacteria and archaea in the Bor Khlueng Hot Spring in Thailand. J Basic Microbiol 44(6):430–444. https://doi.org/10.1002/jobm.200410388

Karlstrom KE, Crossey LJ, Hilton DR, Barry PH (2013) Mantle ^3He and CO_2 degassing in carbonic and geothermal springs of Colorado and implications for neotectonics of the Rocky Mountains. Geology 41(4):495–498. https://doi.org/10.1130/G34007.1

Kaszuba JP, Sims KWW, Pluda AR (2014) Aqueous geochemistry of the Thermopolis hydrothermal system, southern Bighorn Basin, Wyoming, USA. Rocky Mount Geol 49(1):1–16. https://doi.org/10.2113/gsrocky.49.1.1

Kato V, Kraml M (2010) Geochemistry of Rwenzori Hot Spring waters. Proc Afr Rift Geotherm (ARGeo) 3:172–183

Keam RF (2016) The Tarawera eruption, Lake Rotomahana, and the origin of the Pink and White Terraces. J Volcanol Geoth Res 314:10–38. https://doi.org/10.1016/j.jvolgeores.2015.11.009

Kebir H, Boucenna A (2017) Natural radionuclide concentrations in thermal springs of east Algeria. Environ Earth Sci 76. https://doi.org/10.1007/s12665-016-6373-z

Kele S, Ramaley R, Siklósy Z, Kulja A, Demény A (2006) Microbial studies of an isolated active travertine hot spring (Egerszalók) located in Hungary. Presented at Extremophiles 2006, the 6 International Congress on Extremophiles Quartz, Brest France, 17–21 Sept 2006

Kele S, Ramaley R, Siklósy Z, Kulja A, Demény A (2008) Chemical and stable isotope composition of recent hot-water travertines and associated thermal waters, from Egerszalók, Hungary: depositional facies and non-equilibrium fractionation. Sediment Geol 211(3–4):53–72. https://doi.org/10.1016/j.sedgeo.2008.08.004

Khansha J, Ranjbaran M, Amoozegar MA (2018) Isolation and identification of halophilic and halotolerant bacteria from Badab-e Surt travertine spring, Kiasar, Iran, and investigation of calcite biomineralization induction. Geomicrobiol J 35(1):64–73. https://doi.org/10.1080/01490451.2017.1321065

Kiyosu Y, Okamoto Y (1998) Variation in fumarolic H, gas and volcanic activity at Nasudake in Japan. J Volcanol Geoth Res 80:27–37. https://doi.org/10.1016/S0377-0273(97)00039-5

Klein C (2002) The Manual of Mineral Science, 22nd edn. Wiley, USA

Kochetkova TV, Rusanov II, Pimenov NV, Kolganova TV, Lebedinsky AV, Bonch-Osmolovskaya EA, Sokolova TG (2011) Anaerobic transformation of carbon monoxide by microbial communities of Kamchatka Hot Springs. Extremophiles 15:319–325. https://doi.org/10.1007/s00792-011-0362-7

Konhauser KO, Jones B, Reysenbach A-L, Renaut RW (2003) Hot spring sinters: keys to understanding Earth's earliest life forms. Can J Earth Sci 40(11):1713–1724. https://doi.org/10.1139/E03-059

Kyle JE, Schroeder PA, Wiegel J (2007) Microbial Silicification in Sinters from Two Terrestrial Hot Springs in the Uzon Caldera, Kamchatka, Russia. Geomicrobiol J 24(7–8):627–641. https://doi.org/10.1080/01490450701672158

Lacap DC, Barraquio W, Pointing SB (2007) Thermophilic microbial mats in a tropical geothermal location display pronounced seasonal changes but appear resilient to stochastic disturbance. Environ Microbiol 9:3065–3076. https://doi.org/10.1111/j.1462-2920.2007.01417.x

Lagat J (2009) Hydrothermal alteration mineralogy in geothermal fields with case examples from Olkaria domes geothermal field, Kenya. Presented at short course IV on exploration for geothermal resources, organized by UNU-GTP, KenGen and GDC, at Lake Naivasha, Kenya, 1–22 Nov 2009

Li QF, Dong FQ, Dai QW, Huo TT, An DJ, Tang S (2012) The microbial factor of travertine deposition between Yellowstone National Park (YNP), USA and Huanglong Scenic, Sichuan. Adv Mater Res 518–523:136–139. https://doi.org/10.4028/www.scientific.net/AMR.518-523.136

Liu Y, Zhou X, Fang B, Zhou H, Yamanaka T (2012) A preliminary analysis of the formation of travertine and travertine 1 Cones in the Jifei Hot Spring, Yunnan, China. Environ Earth Sci 66(7):1887–1896. https://doi.org/10.1007/s12665-011-1413-1

Lund JW (1993) Thermopolis, Wyoming. Geo-Heat Center (GHC) Bulletin, Mar 1993

Lynne BY (2007) Diagenesis of siliceous sinter deposits in the U.S.A. and New Zealand. Doctoral dissertation. Department of Geology, University of Auckland, New Zealand

Lynne B, Smith G, Heasler H, Jaworowski C, Smith I, Foley D, Sahdarani D (2017) Post-depositional alteration of siliceous sinter near old faithful geyser, Yellowstone National Park, USA. In: Proceedings 39th New Zealand Geothermal Workshop. Rotorua, New Zealand, 22–24 Nov 2017

Marks N, Schiffman P, Zierenberg RA, Franzson H, Fridleifsson GÓ (2010) Hydrothermal alteration in the Reykjanes geothermal system: insights from Iceland deep drilling program well RN-17. J Volcanol Geoth Res 189(1–2):172–190. https://doi.org/10.1016/j.jvolgeores.2009.10.018

Mayer K, Scheu B, Montanaro C, Yilmaz TI, Isaia R, Aßbichler D, Dingwell DB (2016) Hydrothermal alteration of surficial rocks at Solfatara (Campi Flegrei): petrophysical properties and implications for phreatic eruption processes. J Volcanol Geoth Res 320:128–143. https://doi.org/10.1016/j.jvolgeores.2016.04.020

McClenaghan MB, Peter JM (2015) Till geochemical signatures of volcanogenic massive sulphide deposits: an overview of Canadian examples. Geochemistry: exploration, environment, analysis. The Geological Society, London. https://doi.org/10.1144/geochem2014-315

McGeary D, Plummer C, Carlson D (2001) Physical Geology—Earth revealed, 4th edn. McGraw Hill, USA

McPhie J, Cas R (2015) Volcanic successions associated with ore deposits: facies characteristics and ore-host relationships. In: Sigurdsson H (ed) The encyclopedia of volcanoes, 2nd edn. Academic Press, London, pp 865–879

Merkel AY, Pimenov NV, Rusanov II, Slobodkin AI, Slobodkina GB, Tarnovetckii IY, Frolov EN, Dubin AV, Perelova AA, Bonch-Osmolovskaya EA (2017) Microbial diversity and autotrophic activity in Kamchatka Hot Springs. Extremophiles 21 (2):307–317. https://doi.org/10.1007/s00792-016-0903-1

Montazeri H, Abbasnejad A, Negarestani A (2011) Continuous radon monitoring in the Jowshan Hot Spring as an earthquake precursor, SE Iran. Geochem J 24:463–472. https://doi.org/10.2343/geochemj.1.0148

Office de Tourisme de Guelma (2019) Hammam meskhoutine. Retrieved from https://tinyurl.com/ye7lb9sj

Ökçesiz M (2014) Karahayit thermal spa center. Retrieved from https://www.tourmakerturkey.com/karahayit.html

Oliverio AM, Power JF, Washburne A, Cary SC, Stott MB, Fierer N (2018) The ecology and diversity of microbial eukaryotes in geothermal springs. ISME J. https://doi.org/10.1038/s41396-018-0104-2

Oren A, Ionescu D, Hindiyeh M, Malkawi H (2008) Microalgae and cyanobacteria of the Dead Sea and its surrounding springs. Israel J Plant Sci 56(1–2):1–13. https://doi.org/10.1560/IJPS.56.1-2.1

Özkul M, Kele S, Gökgöz A, Shen C-C, Jones B, Baykara MO, Fórizs I, Németh T, Chang Y-W, Alçiçek MC (2013) Comparison of the quaternary travertine sites in the Denizli extensional basin based on their depositional and geochemical data. Sed Geol 294:179–204. https://doi.org/10.1016/j.sedgeo.2013.05.018

Park C, Allaby M (2013) A Dictionary of Environment and Conservation, 2nd edn. Oxford University Press, Oxford

Pentecost A (2003) Cyanobacteria associated with hot spring travertines. Can J Earth Sci 40(11):1447–1457. https://doi.org/10.1139/e03-075

Pentecost A, Viles H (1994) A review and reassessment of travertine classification. Géograph Phys Quat 48(3):305–314. https://doi.org/10.7202/033011ar

Piercey SJ, Peter JM, Herrington RJ (2015) Zn-rich volcanogenic massive sulphide (VMS) deposits. In: Archibald SM, Piercey SJ (eds) Current perspectives on zinc deposits. Published by the Irish Association for Economic Geology

Portillo MC, Gonzalez JM (2008) Microbial communities and immigration in volcanic environments of Canary Islands (Spain). Naturwissenschaften 95:307–315. https://doi.org/10.1007/s00114-007-0330-3

Portillo MC, Sririn V, Kanoksilapatham W, Gonzalez JM (2009) Differential microbial communities in hot spring mats from Western Thailand. Extremophiles 13:321–331. https://doi.org/10.1007/s00792-008-0219-x

Renaut RW (2004) Mineral precipitation at thermal springs in the Kenya Rift Valley. Geol Soc Am Abstr Program 36(5):471

Reyes Pérez Ó, Sanchez Crispín Á (2005) Metodología para determinar el potencial de los recursos turísticos naturales en el estado de Oaxaca, México. Cuad Turismo16:153–173. ISSN: 1139-7861

Robb L (2005) Introduction to ore-forming processes. Blackwell Publishing, Oxford

Rockel W, Hoth P, Seibt P (1997) Charakteristik und Aufschluß hydrogeothermaler Speicher. Geowissenschaften: Organ der Alfred-Wegener-Stiftung 15(8):244–252

Rodriguez A, Eggenkamp HGM, Martinez-Cruz M, van Bergen MJ (2016) Chlorine isotope and Cl–Br fractionation in fluids of Poás volcano (Costa Rica): insight into an active volcanic–hydrothermal system. J Volcanol Geoth Res 325:70–85. https://doi.org/10.1016/j.jvolgeores.2016.05.020

Ronchi P, Cruciani F (2015) Continental carbonates as a hydrocarbon reservoir, an analog case study from the travertine of Saturnia, Italy. AAPG Bull Am Assoc Petrol Geol 99(4):711–734. https://doi.org/10.1306/10021414026

Rouwet D, Hidalgo S, Joseph EP, González-Ilama G (2017) Fluid geochemistry and volcanic unrest: dissolving the haze in time and space. In: Gottsmann J, Neuberg J, Scheu B (eds) Volcanic unrest: from science to society, advances in volcanology book series. Springer, Berlin, pp 221–239. https://doi.org/10.1007/11157_2017_12

Rzonca B, Schulze-Makuch D (2003) Correlation between microbiological and chemical parameters of some hydrothermal springs in New Mexico, USA. J Hydrol 280(1–4):272–284. https://doi.org/10.1016/S0022-1694(03)00231-2

Sæmundsson K, Gunnlaugsson E (2002) Icelandic rocks and minerals. Mal og mennig, Reykjavík, Iceland

Sahoo RK, Subudhi E, Kumar M (2015) Investigation of bacterial diversity of hot springs of Odisha, India. Genom Data 6:188–190. https://doi.org/10.1016/j.gdata.2015.09.018

Saibi H (2015) Geothermal resources in Algeria. In: Proceedings world geothermal congress. Melbourne, Australia, 19–25 Apr 2015

Saqan SA, Kullab MK, Ismail AM (2001) Radionuclides in hot mineral spring waters in Jordan. J Environ Radioact 52(1):99–107. https://doi.org/10.1016/S0265-931X(00)00096-5

Schellschmidt R, Sanner B, Pester S, Schulz R (2010) Geothermal energy use in Germany. In: Proceedings world geothermal congress 2010. Bali, Indonesia. 25–29 Apr 2010

Scoon R (2015) Pamukkale and Hierapolis, Turkey: spectacular travertine terraces and an ancient city. Geotraveller technical report. 10.13140/RG.2.1.1761.4567

Seibt P, Kabus F, Hoth P (2005) The Neustadt-Glewe geothermal power plant—practical experience in the reinjection of cooled thermal waters into sandstone aquifers. In: Proceedings world geothermal congress 2005. Antalya, Turkey, 24–29 Apr 2005

Shanks WCP III (2012) Hydrothermal alteration. scientific investigations report 2010–5070–C. U.S. Geological Survey

Shinohara H, Hirabayashi J, Nogami K, Iguchi M (2011) Evolution of volcanic gas composition during repeated culmination of volcanic activity at Kuchinoerabujima volcano, Japan. J Volcanol Geoth Res 202:107–116. https://doi.org/10.1016/j.jvolgeores.2011.01.011

Sigurdsson H, Houghton B, McNutt S, Rymer H, Stix J (eds) (2015) The encyclopedia of volcanoes, 2nd edn. Academic Press, London

Simsek S, Günay G, Elhatip H, Ekmeçi M (2000) Environmental protection of geothermal waters and travertines at Pamukkale, Turkey. Geothermics 29(4–5):557–572. https://doi.org/10.1016/S0375-6505(00)00022-5

Sotohian F, Ranjbaran M (2015) Depositional system and facies analysis of travertine deposits: Badab-e Surt Spring Mazandaran, Iran. Arab J Geosci 8(7):4939–4947. https://doi.org/10.1007/s12517-014-1544-2

Song Z, Zhi X, Li W, Jiang H, Zhang C, Dong H (2009) Actinobacterial diversity in hot springs in Tengchong (China), Kamchatka (Russia), and Nevada (USA). Geomicrobiol J 26:256–263. https://doi.org/10.1080/01490450902892373

State Library of South Australia (2019) Paralana Hot Springs. Old photo taken by the Rev. Robert Mitchell, circa 1898

Steele K, Fisher M (2016) Thermopolis hot springs—the smoking waters. Geol Wyoming. Retrieved from https://www.geowyo.com/thermopolis-hot-springs.html

Stefánsson A, Kleine BI (2018) Hydrothermal alteration. In: White WM (ed) Encyclopedia of Geochemistry. Encyclopedia of Earth Sciences Series. Springer International Publishing

Stefánsson A, Arnórsson S, Gunnarsson I, Kaasalainen H, Gunnlaugsson E (2011) The geochemistry and sequestration of H_2S into the geothermal system at Hellisheidi, Iceland. J Volcanol Geoth Res 202:179–188. https://doi.org/10.1016/j.jvolgeores.2010.12.014

Stetter KO (2005) Volcanoes, hydrothermal venting and the origin of life. In: Marti J, Ernst GG (eds) Volcanoes and the environment. Cambridge University Press, pp 175–206 (Chapter 6)

Stimac J, Goff F, Goff CJ (2015) Intrusion-related geothermal systems. In: Sigurdsson H et al (eds) The encyclopedia of volcanoes, 2nd edn. Academic Press, London, pp 799–822

Stix J (2015) Volcanic interactions. In: Sigurdsson H et al (eds) The encyclopedia of volcanoes, 2nd edn. Academic Press, London, pp 777–778

Sun S, Dong F, Ehrlich H, Zhao X, Liu M, Dai Q, Li Q, An D, Dong H (2014) Metabolic influence of psychrophilic diatoms on travertines at the Huanglong natural scenic district of China. Int J Environ Res Publ Health 11:13084–13096. https://doi.org/10.3390/ijerph111213084

Symonds RB, Gerlach TM, Reed MH (2001) Magmatic gas scrubbing: implications for volcano monitoring. J Volcanol Geoth Res 108(1–4):303–341. https://doi.org/10.1016/S0377-0273(00)00292-4

Tang J, Liang Y, Jiang D, Li L, Luo Y, Shah MdMR, Daroch M (2018) Temperature-controlled thermophilic bacterial communities in hot springs of western Sichuan, China. BMC Microbiol 18:134. https://doi.org/10.1186/s12866-018-1271-z

Taşeli BK (2016) Achievements in the use of renewable energy in Turkey's special environmental protection areas (SEPA). J Multi Eng Sci Technol (JMEST) 3(3):4282–4285. ISSN: 2458–9403

Te Ara (2019) Life in hot springs: extremophiles. The encyclopedia of New Zealand. Retrieved from https://teara.govt.nz/en/life-in-hot-springs/page-1

Tsunogai U, Cheng L, Ito M, Komatsu DD, Nakagawa F, Shinohara H (2016) Remote determinations on fumarole outlet temperatures in an eruptive volcano. In: Geophysical research letters. AGU Publications, pp 11,620–11,627. https://doi.org/10.1002/2016GL070838

UNESCO (2019) Huanglong scenic and historic interest area. Retrieved from https://whc.unesco.org/en/list/638

Vallance JW, Iverson RM (2015) Lahars and their deposits. In: Sigurdsson H et al (eds) The encyclopedia of volcanoes, 2nd edn. Academic Press, London, pp 649–664

Van Wyk de Vries B, Davies T (2015) Landslides, debris avalanches, and volcanic gravitational deformation. In: Sigurdsson H et al (eds) The encyclopedia of volcanoes, 2nd edn. Academic Press, London, pp 665–685

Voronov AN (2003) Radon-rich waters in Russia. Environ Geol 46:630–634. https://doi.org/10.1007/s00254-003-0857-3

Wang H, Liu Z, Zhang J, Sun H, An D, Fu R, Wang X (2010) Spatial and temporal hydrochemical variations of the spring-fed travertine-depositing stream in the Huanglong Ravine, Sichuan, SW China. Acta Carsol 39(2):247–259. https://doi.org/10.3986/ac.v39i2.97

Wass EF (2007) El espacio indígena. Los pueblos de Oaxaca y la lucha por la autonomía. In: Araucaria. Revista Iberoamericana de Filosofía, Política y Humanidades, vol 18, pp 130–149

Wei-Haas M (2020) Key ingredient in coronavirus tests comes from Yellowstone's lakes—a curious life-form that lives in the park's thermal pools makes a protein that changed the course of biomedical history. National Geographic. Retrieved from https://tinyurl.com/qsuyzup

Werner C, Hurwitz S, Evans WC, Lowenstern JB, Bergfeld D, Heasler H, Jaworowski C, Hunt A (2008) Volatile emissions and gas geochemistry of Hot Spring Basin, Yellowstone National Park, USA. J Volcanol Geoth Res 178:751–762. https://doi.org/10.1016/j.jvolgeores.2008.09.016

White DE, Hutchinson RA, Keith TEC (1988) The geology and remarkable thermal activity of Norris Geyser Basin, Yellowstone National Park, Wyoming. U.S. geological survey professional paper 1456. USGS, Washington DC Wikimedia Commons (2009a) Map Wyoming

Wilkinson JJ, Stoffel B, Wilkinson CC, Jeffries TE, Appold MS (2009) Anomalously metal-rich fluids form hydrothermal ore deposits. Science 323:764–767. https://doi.org/10.1126/science.1164436

Winsborough BM, Caran SC, Neely JA, Valastro S Jr (1996) Calcified microbial mats date prehistoric canals-radiocarbon assay of organic extracts from travertine. Geoarchaeol Int J 11(1):37–50. CCC 0883-6353/96/010037-14

Woith H, Barbosa S, Gajewsli C, Steinitz G, Piatibratove O, Malik U, Zschau J (2011) Periodic and transient radon variations at the Tiberias hot spring, Israel during 2000–2005. Geochem J 45 (6):473–482. https://doi.org/10.2343/geochemj.1.0147

Wright MH, Patel BKC, Greene AC (2012) Thermophilic bacteria from Paralana Hot Springs in the northern flinders ranges of South Australia. In: Conference paper. https://doi.org/10.13140/RG.2.1.2765.5525

Wyering LD, Villeneuve MC, Wallis IC, Siratovich PA, Kennedy BM, Gravley DM, Cant JL (2014) Mechanical and physical properties of hydrothermally altered rocks, Taupo volcanic zone, New Zealand. J Volcanol Geoth Res 288:76–93. https://doi.org/10.1016/j.jvolgeores.2014.10.008

Yan R, Woith H, Wang R, Wang G (2017) Decadal radon cycles in a hot spring. Sci Rep 7:12120. https://doi.org/10.1038/s41598-017-12441-0

Zhang CL, Huang Z, Li Y-L, Romanek CS, Mills GL, Gibson RA, Talbot HM, Wiegel J, Noakes J, Culp R, White DC (2007) Lipid biomarkers, carbon isotopes, and phylogenetic characterization of bacteria in California and nevada hot springs. Geomicrobiol J 24 (6):519–534. https://doi.org/10.1080/01490450701572515

Zhao W, Romanek CS, Mills G, Wiegel J, Zhang C (2005) Geochemistry and microbiology of hot springs in Kamchatka, Russia. Acta Metall Sin 11(2):217–223. ISSN: 1006-7493 CN: 32-1440/P

Zhao W, Song Z, Jiang H, Li W, Mou X, Romanek CS, Wiegel J, Dong H, Zhang CL (2011) Ammonia-oxidizing archaea in Kamchatka Hot Springs. Geomicrobiol J 28(2):149–159. https://doi.org/10.1080/01490451003753076

Zuniga A, Su M, Sanchez M (2003) Thermal manifestations in Nicaragua. GHC Bull 24(4):23–25. ISSN 0276-1084

Vigorously boiling mud pot at Námafjall (Hverir), North–East Iceland

4 The Conservation of Hot Springs

Contents

P. Erfurt, *The Geoheritage of Hot Springs*, Geoheritage, Geoparks and Geotourism,
https://doi.org/10.1007/978-3-030-60463-9_4

Thermal stream flowing through the city of Beitou (Taiwan)

Conservation of the natural environment must include the responsible management of all hot springs worldwide to prevent them from depletion and to protect them from pollution.

4.1 Introduction

All hot springs, regardless of their temperature and flow rate, their size and location, or their economic usage, face threats to their sustainability. These threats are to a large degree based on human activities, which have led to the contamination and over-exploitation of a vast number of natural water resources on a global scale. This applies to groundwater reservoirs as well as hydrothermal systems. Researchers have analysed satellite observations spanning 14 years and were able to directly associate undesirable changes in freshwater storage areas with several key factors: human impact, climate change and natural variability (NASA 2018). While perhaps not much can be done about natural variability such as changing rainfall patterns influenced by climatic phenomena like the El Niño Southern Oscillation (ENSO), the threat of climate change however should be taken more seriously and counteracted in every possible way; especially, where human interference might accelerate a looming water crisis.

To sustain the continuous flow of endangered natural hot springs, human activities need to be adjusted according to the availability of water resources, at the very least during prolonged droughts. This is not the case in many regions, where groundwater aquifers lack sufficient recharge, while at the same time industry and agriculture have essentially ignored the fact that over-drafting of a diminishing water supply leads to environmental problems that will have long-term consequences. With special consideration of hydrothermal systems as the source of hot spring water, excessive draw-down of subterranean reservoirs can result in the reduction of pressure within the reservoir, which subsequently can lead to the decreased flow of hot springs in the affected area; a problem, which has been reported from several countries, where over-exploitation has led to the depletion of hydrothermal resources.

In **New Zealand** for example, hydrothermal systems have been historically exploited for direct use and energy generation. In Rotorua, the demand for geothermal resources started to grow in the 1950s and led to the largely uncontrolled extraction of more than 1000 thermal bores two decades later. To exacerbate the situation, much of the extracted hot water was discharged unused and the drawdown gradually started to affect the surrounding hydrothermal features to the point that their performance was declining. In the case of the *Waikite Geyser* at the Whakarewarewa Māori Village, it resulted in the total loss of this feature (Doorman and Barber 2017; NZ Geothermal Association 2018a). This is one of many cases where human activities have been responsible for negatively affecting a natural resource. For this reason, it is important that legislation relating to groundwater protection includes effective methods to protect all hot spring systems.

This is possible by preventing the depletion of hydrothermal reservoirs, thereby sustaining them as a renewable resource. This chapter provides a general overview of the potential threats to available water resources and some points in relation to mitigating measures.

4.2 Conservation of Hot Springs

Conservation is a concept that can be applied to all natural as well as cultural environments. In the case of hot springs, conservation needs to focus not only on the ethics of preservation and stewardship of a natural resource but also on the socio-cultural aspects (Chap. 6) and their responsible and best practice management to preserve these resources for future generations. Many hot spring areas are already protected in national parks, geoparks, world heritage sites, biospheres and nature reserves (Chap. 7). However, these designated sites do not cover the vast amount of hot springs that are exploited for domestic, industrial, agricultural and tourism purposes. All these are areas that should be addressed without delay to avoid irreversible harm of an otherwise renewable resource. The protection of natural hot spring environments can be achieved in different ways, and with long-term sustainability as the main objective. This requires strategic planning for future developments and a responsible management approach, based on present-day water acts and legislations that include natural hot springs, which is not the case in every country. Without strict water laws it is not possible to enforce co-operation at all stakeholder levels.

In many countries natural hot springs are also an important part of the national health system and play a significant role in health and wellness tourism as well as for recreational use (Erfurt 2012). The hot spring tourism industry is built on strong partnerships for sustainable destination development; although, for hot spring tourism to be sustainable requires adherence to the actual carrying capacity of a region. This includes the responsible use of all natural resources by avoiding over-development and aiming for low impact land use to support the conservation of natural environments.

In the case of geothermal energy developments, which are generally considered as 'safe' and not necessarily affect groundwater availability, there have nevertheless been suggestions that their infrastructure could be viewed as having a negative impact on 'the aesthetic quality of the landscape' and the 'cultural, historical or spiritual significance'. A concern that probably arises because such developments are frequently located either in or close to protected natural areas (Shortall et al. 2015). Other impacts related to geothermal developments could include land subsidence and the potential for increased micro-seismic activity as well as adverse effects on ecosystems located within the geothermal field (NZ Geothermal Association 2018b). In addition, there is a possibility for land-use conflicts between geothermal resource extraction and conservation of the natural environment. Because the development of geothermal power stations requires sizeable areas that potentially affect the flow of natural hot springs and other hydrothermal features, it is paramount that appropriate policies and strategies for sustainable planning are in place (Sæþórsdóttir and Ólafsson 2010; Shortall et al. 2015).

4.2.1 Destruction of Hydrothermal Features

Over time, hydrothermal features have been destroyed during natural events, either through earthquakes, landslides or volcanic eruptions. In these cases, it is too late for conservation of the area, unless some of the features can be recovered. Sometimes nature itself overcomes the destruction caused by tectonic events and hot springs start flowing and geysers start spouting again, although perhaps not in the same place where they emerged before. In the year 2007 the Valley of the Geysers on the Kamchatka Peninsula in far eastern Russia, one of the largest geyser fields worldwide, was buried under a large mudslide. Since then, many hydrothermal features have recovered, and adventure tourism has developed into a lucrative business for tour operators who fly visitors in and out of the area by helicopter. The main reason for visiting the Kamchatka are the many volcanoes with the hydrothermal attractions as an additional bonus.

Another major disappearance of a natural hydrothermal phenomenon was caused by the eruption of Mount Tarawera (New Zealand) in 1886 during which the famous *Pink and White Terraces* vanished from sight (Appendices 7.1 and 7.2). Despite much speculation in the past it is generally assumed that they are completely submerged, and their remnants are located at the bottom of Lake Rotomahana. Recent research indicates that after the eruption of Mount Tarawera the lake level rose significantly, which means that despite underwater surveys the Pink and White Terraces are gone forever (De Ronde et al. 2016, 2018; GNS Science 2018). In other cases, hot springs are flooded during unprecedented rainfall and overflow as is happening occasionally at the thermal pools in *Mataranka*, Australia. During storm surges in coastal areas, sand or silt can contribute to blocking hot springs while rocks and other debris are left behind after the stormwater recedes. The same happens when rivers, carrying high water, are flooding hot springs located on their banks, in which case the damage from natural events may only be temporary.

Sometimes it is environmental vandalism that damages a hot spring to the point that it stops flowing, or in the case of geysers, ceases to erupt. In attempts to create a visual attraction on demand, some geysers are prompted to spout by various means. In the past people tried to induce a geyser eruption by throwing rocks and other foreign objects into the

Fig. 4.1 Early environmental vandalism was described and illustrated in a book about Iceland following a visit by the author in 1858. Several old drawings exist that depict this form of destruction of geysers to trigger an eruption. *Source* Winkler (1861)

open pipe (Fig. 4.1). This environmental destruction consequently leads to the deterioration of the geyser by blocking the conduit (Hróarsson and Jónsson 1992). In more recent times geysers are tempted to erupt by less harmful methods, such as adding a surfactant (e.g. soap, detergent) to the conduit, which seems to have the desired effect of causing the spring to spout. To date, there is no evidence whether this method is affecting the hydrothermal system in the long term, although there are no reliable reports to confirm that this interference is safe either.

Elsewhere, hot springs in public areas are used to dump empty bottles and other waste, a thoughtless act that spoils the sight and leaves visitors with an unpleasant impression (Fig. 4.2). Again, these are human activities and can be avoided.

4.3 Depletion of Water Levels in Hydrothermal Systems

4.3.1 Protection of Groundwater from Over-Exploitation

When groundwater is extracted at a faster rate than it can be replenished, the natural water level decreases and the reservoir faces depletion. In fact, this does not just affect groundwater aquifers that are used to supply water for domestic, industrial and agricultural purposes, but hydrothermal systems as well. Once groundwater resources are over-exploited, this can lead to the deterioration of the water quality and a reduced supply of water feeding into surface water systems such as lakes and rivers. If hydrothermal systems are exhausted, hot springs stop flowing and geysers stop erupting to the point of permanent damage. For example, the exploitation of the *Wairakei* geothermal field in New Zealand contributed to ground subsidence and the decline of hydrothermal features in the Geyser Valley and the Waimangu Volcanic Valley due to the depletion of hot groundwater (White and Hunt 2005). To counteract this imbalance, wastewater from geothermal power generation is now re-injected into the ground to recharge the reservoir.

Another area of concern is the Great Australian Basin (GAB). Based on the excessive draw down of artesian groundwater in the past the impact on the reservoir health has resulted in a greatly diminished discharge rate over time. In the year 1915 the GAB's combined discharge from over 1,500 artesian bores that were sunk since 1878 was estimated at more than 2,000 mega litres per day (ML/d). As the pressure declined, the total discharge in the year 2000 was estimated at 1,500 ML/d, an amount above the calculated recharge, which also declined when the natural equilibrium was no longer sustained (Chavasse 2006; Ponder 2002; Wilmott 2017). The protection of groundwater sources from over-exploitation is the main focus of conservation programs

Fig. 4.2 One of many hot springs all over the world that have been neglected and used as a rubbish disposal. The bottom of this 100 °C spring is covered in bottles and other refuse

in areas, where the problems have been identified and are actively confronted (Example 4.1). While human activities can greatly influence the volume of hot spring discharge, there are ways to halt or even reverse excessive groundwater withdrawals and the correlated land degradation.

Example 4.1 Great Artesian Basin Hot Springs, Australia

Thousands of bores sunk into the Great Artesian Basin resulted in vast amounts of water wasted through uncontrolled discharge from uncapped bores and contributed to the pressure reduction in the reservoir, as well as causing considerable damage to the ecosystem. Draining away in open channels artesian water at temperatures between 30 and 100 °C caused the increase of feral animals, which in turn damaged the native habitats around the hot springs. The ecosystems of the artesian hot springs were further compromised by the introduction of foreign animals and plants causing weed invasion as well as the disturbance of native vegetation around the springs by cattle and sheep. This led to a reduction in the diversity of native plants, while overgrowth by invasive species caused the loss of open hot spring pools. In addition to cattle and sheep farmers accessing GAB water, mining companies in the area also claimed a large share of the precious resource. As the native flora and fauna were displaced, the natural equilibrium of the fragile systems of the springs was disturbed and habitats were destroyed. However, interest in the conservation of the springs and addressing the main threats to the integrity of the GAB springs has led to some signs of rehabilitation. Large areas have been designated as protected sites (e.g. Wabma Kadarbu Mound Springs Conservation Park) or national parks (Witjira [Dalhousie Springs] National Park and Kati

Thanda—Lake Eyre National Park), which are closely monitored by the traditional owners such as the *Lower Southern Arrente* and the *Irrwanyere Aboriginal Corporation.* Groups interested in the rehabilitation of the endangered artesian springs such as the *Friends of Mound Springs* (FOMS) in South Australia have been actively engaged in a *Springs Protection and Monitoring Program* for many years, which includes weed control and protective fencing to exclude cattle and other animals, as well as measuring, monitoring and recording any changes and improvements. Government sponsored sustainability initiatives and programs going back several decades have gradually led to the capping of many open bores, which has helped to restore pressure in the GAB. But the artesian hot springs in Australia's desert are still exploited for irrigation and as a water supply for cattle and sheep on private land, although many landholders now control the water flow to minimise the waste of bore water.

Source Ah Chee 2014; Burt 2019; Chavasse 2006; Davidson 2020; FOMS 2020; Habermehl 2000, 2001; Ponder 2002; Wilmott 2017; (see also Fig. 2.20, Fig. 6.7; Appendices 4.1, 4.2).

4.4 Pollution of Aquifers

4.4.1 Urban, Agricultural and Industrial Runoff

Human activities are to a large degree responsible for environmental degradation, which affects the sustainability of water as a renewable resource. One of the worst impacts is caused by the contamination of water, which can affect all reservoirs, including hydrothermal systems. To compound the problem, quite often different forms of pollution are taking place at the same time in the same area. The following types of anthropogenic pollution enter the natural environment from point sources and non-point sources:

- Effluent spills and sewage leaks
- Landfill seepage and oil spills
- Hazardous substances
- Wastewater leaks
- Agricultural runoff and industrial pollution
- Stormwater and urban runoff
- Mining activities and spills
- Thermal pollution
- Oxygen depletion.

The list does not include the additional dangers of microbial pollution such as pathogens or microorganisms that cause water-borne diseases; this would lead too far away from the overall scope of this chapter. However, it is worth keeping in mind, that many other factors can also affect water quality as well as reservoir capacity.

Pollution and contamination with hazardous substances are the worst threat to all groundwater resources and include seepage from mining operations and landfills, oil and chemical spills, leaking septic systems and mismanaged wastewater facilities. These are all contributing factors that are causing not just environmental degradation on the Earth's surface but affect hydrothermal systems and other groundwater reservoirs forever.

Other factors causing pollution of natural hot springs are agricultural, industrial and urban run-off, which infiltrates the ground and finds a way into nearby hot and cold-water reservoirs. Thermal pollution can occur, when geothermal wastewater is re-introduced into the natural environment after temperatures have been reduced during the process of power generation (Figs. 4.3, 4.4), although the water can still be hot enough to be averse to the environment into which it is released. While there may not be any obvious environmental damage at present, the eventual consequences are uncertain. This can be met with the argument, that in hot spring areas hot water is constantly released into the environment but it is often the human intervention, which is causing the greatest damage to nature.

4.5 Groundwater Systems: The Potential for Contamination

4.5.1 Hydraulic Fracturing—Fossil Fuel Extraction

One of the major threats to groundwater reservoirs is pollution on a large scale and to a point where the damage cannot be reversed. Here we need to critically reflect on the process of hydraulic fracturing, commonly known as fracking or hydrofracking, which is used by an increasing number of countries to access trapped natural gas and oil. This technology is used extensively for the unconventional extraction of resources and has become a highly controversial issue with one of the main concerns being the potential for groundwater contamination. The initial preparations require the drilling of wells to reach the oil and gas bearing rock formations. During this process, a special water-based fracturing fluid is injected under high pressure into the rock formation to open up additional cracks and fissures (Fig. 4.5). These particular fluids contain apart from water and sand, toxic chemicals such as arsenic, barium sulphate, benzene, ethylene glycol, formaldehyde, hydrocarbons, hydrochloric acid, lead, mercury, methanol, naphthalene, toluene and xylene. Studies are suggesting that over 200 different toxic chemicals can be used in fracking fluids, with

Fig. 4.3 Hot wastewater from a geothermal power plant is discharged into a creek. Clearly visible is the point of entry, where the cold water turns into a steaming water course

at least a dozen of the more commonly used chemicals (some sources claim even higher numbers) identified not only as hazardous and/or as extremely toxic, but also as carcinogenic, with reports that groundwater reservoirs have already been contaminated in some areas (DiGiulio and Jackson 2016; Example 4.2).

Promotional online videos insist that the percentage of the potentially toxic chemicals used is extremely low (<1%), but the picture changes when considering that millions of gallons of fracking fluid are required during each individual extraction process. Even less than one percent adds up to more than enough toxic substances to contaminate entire groundwater reservoirs beyond remediation in case of an equipment failure underground or spills on the fracturing pad above ground. The toxicity of the used fracking fluids is so severe, that it cannot be removed in wastewater treatment plants, and at this time there are no long-term studies on the subject of how water contaminated with fracking fluid chemicals will behave in years to come. Many thousand fracking wells are already established with critics arguing that the accumulated impact of chemical spills from well pads and leaks underground may not be evident for perhaps another decade, although it is not just anecdotal evidence showing that contamination has already happened in several areas with an environmental disaster in the making (DiGiulio and Jackson 2016; Earth Focus 2011).

Example 4.2 Farmers concerned about Coal Seam Gas Extraction, Australia

Back in 2011, a concerned Queensland cattle farmer started to investigate a fracking incident near Dalby, which was caused by the Queensland Gas Company (QGC) in 2009. Fears were raised that chemicals used in the process found their way into the aquifer and would affect other water reservoirs as well. While QGC maintained there was no breach of regulations, the company did not alert authorities or nearby landholders about the problem until over a year later. Further investigations discovered that safety data for an Environment Authority application supplied by the company was years out of date and incomplete. As a

Fig. 4.4 The wastewater is released into the natural environment, where it generates near tropical temperatures with lush vegetation along the stream as the water cools down gradually

result, affected cattle farmers and environmentalists are frustrated that there does not seem to be any accountability regarding what happens underground.

Source ABC News 2011.

Once the fractures have been created in the rock formations at depths of up to and over three kilometres, in theory the fracturing fluids are meant to return to the surface, either to be reused or to be stored until they can be disposed of safely. However, according to some research reports only 25 to 50% of the used fracking fluid are recovered with the flowback stored in plastic lined holding dams on the surface, which may not be a safe enough practice for toxic wastewater (Trial Exhibits Inc 2012). On account of the growth of this industry, the disposal of fracking fluids is becoming an increasing problem due to the amount of harmful chemicals, metals and salts, which makes recycling extremely difficult.

Another concern related to hydraulic fracturing is the emission of methane. Leaks can occur through poorly constructed well casings made from low quality concrete. As a flammable gas, methane leaking from wells also has the potential to burn uncontrolled unless the leaks are sealed or capped (Example 4.3). Methane rising up in lakes and rivers, if set alight, continues to burn (Example 4.4) and has been the cause of much controversy. Once escaped, on its way to the surface, methane has the potential to dissolve into overlying aquifers, where it can enter residential water wells, which act as a conduit for this gas to reach living areas inside homes. As an explosion hazard as well as an asphyxiant, methane can be one of the dangerous 'side effects' of fracking (CSIRO 2020; Trial Exhibits Inc 2012).

Example 4.3 Methane Gas Fire in Queensland, Australia

Escaping methane gas burning at an old mine site near Dalby was blamed on the coal seam gas company Arrow Energy. Environmentalists say that the process of de-watering and de-pressurising the coal seams on the western Darling Downs is the cause for the leaking methane gas. The gas is escaping from an exploration well, which Arrow Energy insists has nothing to do with their current operations, a claim that is rejected by

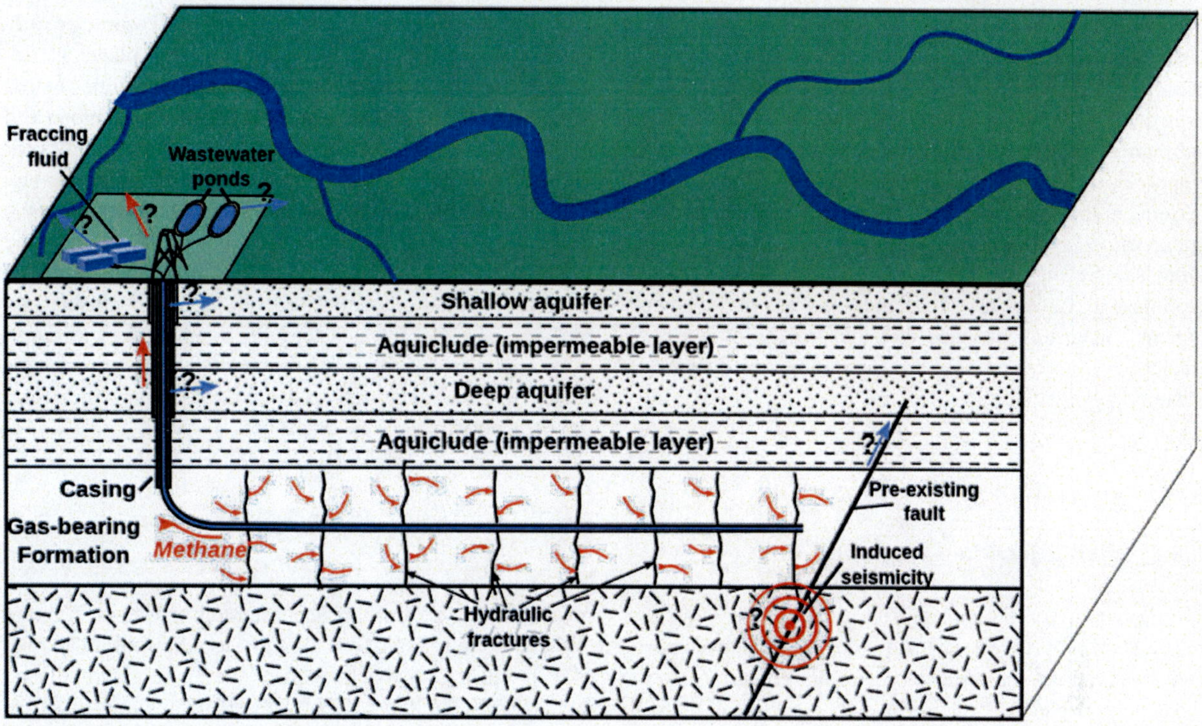

Fig. 4.5 Hydraulic Fracturing diagram. *Source* Mike Norton 2013 (Public Domain)

the president of the 'Lock the Gate Alliance' as defying common sense.

Source Benny-Morrison 2012.

Example 4.4 Burning River, Australia

In 2016 the Condamine River in south western Queensland made the news, as the vigorously bubbling methane gas escaping from the water was set on fire by a politician, who tried to raise awareness about the environmental degradation of the river since the coal seam gas (CSG) exploitation nearby started. The gas rising up in the river was reported in 2012 with many locals and landholders convinced that hundreds of CSG wells owned by energy companies (Origin Energy, QGC and Arrow Energy) are responsible for the methane seeps, which reportedly had intensified over the previous 12 months. While this increase in gas emissions has been acknowledged by the *CSIRO* (Commonwealth Scientific and Industrial Research Organisation), they do not agree that it is connected to the fracking industry nearby, which unsurprisingly is echoed by the energy companies, who deny that the methane seeps in the river are caused by CSG activity. This however is rejected by environmentalists, whose view is that the CSIRO is receiving funding from the energy companies and is therefore compromised. The *Queensland's Department of Natural Resources and Mines* argues that there is not enough information to pinpoint the cause of the methane seeps and requires a long-term investigation to gather more data. The Queensland Government maintains that it does not have sufficient information either to identify the cause of the gas seeps. In essence the official view is that the methane leaks 'are the result of natural geological faults' and rise 'through natural cracks and fissures to the surface' into the river. This is in contradiction to the time frame when the gas leaks started, as they were not observed before the CSG industry moved into the area.

Source Brisbane Times 2016; Willacy 2016.

In addition to the above-mentioned concerns, the water-use intensity of hydraulic fracturing in production areas has steadily increased over time, requiring ever larger volumes of water and thereby placing a burden on already overused aquifers (Kondash et al. 2018). In recognition of the controversy surrounding the practice of fracking, its potential environmental impacts have become the subject of scientific research to assess the effects on wildlife habitats, land degradation and water contamination. While currently the long-term consequences are unforeseeable, the risk of lasting contamination of groundwater reservoirs in favour of economically viable fossil fuel extraction cannot be underestimated. There are reasonable grounds to assume that the subsequent impact of hydraulic fracturing on natural water resources equally applies to any geothermal/hydrothermal systems in the same area. It is therefore paramount to take a conservative approach and err on the side of caution to protect reservoir health because contaminated water is worthless for human use ad infinitum. At this point the only logical conclusion indicates that the process of hydraulic fracturing is not as safe for the environment as claimed by industry representatives.

4.6 Re-Injection of Wastewater

4.6.1 Well-Injection—Pros and Cons

To dispose of wastewater, utilising drilled wells is another method whereby excess hot and cold fluids from domestic and industrial use are re-injected into the ground to aid the recharge of depleted aquifers. However, water cannot just be pumped back into the ground; especially, if it is still at elevated temperatures after being used for the generation of geothermal power, as overflow from industrial applications or from hot spring resorts and thermal waterparks. Hence, the process of re-injection requires a thorough assessment of all positive and negative aspects and the potential impact for all related hydrologic systems.

In **Alaska** (USA), discharge permits are required before used geothermal water can be re-introduced into either the ecosystem or via pressurised injection into the ground. The Department of Environmental Conservation (DEC) has imposed strict rules regarding the disposal of wastewater via injection wells and discharge permit applications are reviewed based on thermal, chemical, ecological as well as anthropogenic assessments before a permit is issued (Roe et al. 2015). To get a discharge permit, a Notice of Intent (NOI) must be submitted and a Class 1 injection permit must be obtained from the Environmental Protection Agency (EPA). Conditions are strict and can include discharge limitations and adherence to monitoring requirements, which must be met to maintain the injection process without causing any degradation to subsurface water reservoirs (Alaska Department of Environmental Conservation 2018).

Re-injection is an important factor for geothermal energy production, and surface water and treated waste water are generally re-injected to recharge geothermal systems (e.g. Japan, Turkey, Iceland), although there appears to be a correlation between re-injection and micro-earthquakes (seismicity) in some geothermal areas similar to seismic events in relation to hydraulic fracturing. To avoid negative impacts on the environment and wherever commercially viable, wastewater from geothermal power plants is frequently used for district heating and bathing facilities before re-injection (Rivera Diaz et al. 2016). The Blue Lagoon and the Mývatn Nature Baths in Iceland are prime examples for the use of clean excess water from geothermal power stations. In Australia, in the state of Victoria for example, the extraction of groundwater (hot and cold) is controlled by the Water Act (1989) with conditions relating to the subsurface disposal of wastewater. For this reason, the Peninsula Hot Springs spa resort has installed a re-injection well to return excess geothermal water to the aquifer, a process that is subject to close monitoring to avoid any long-term impact on the geothermal source reservoir.

4.6.2 Geologic Carbon Sequestration

Not just wastewater and fracking fluids are injected into the ground. Geologic Carbon Sequestration or Carbon Capture and Sequestration (CCS) is regarded as a key option and the most promising approach for the reduction of anthropogenic carbon dioxide (Global CCS Institute 2015; Kim et al. 2018; Massachusetts Institute of Technology 2003; Newmark et al. 2010; Takaya et al. 2018; Wilkin and DiGuilio 2010; Zhang and Song 2014). Carbon dioxide (CO_2), a gas that is generated during various industrial processes including natural gas extraction and processing, as well as at power stations, can be captured and sequestered in sedimentary basins at a (minimum) depth of one kilometre or more (Feitz et al. 2018; Nico et al. 2017; Massachusetts Institute of Technology 2003; Virginia.gov 2015). As the storage conditions in underground rock formations vary, research articles rarely include more accurate definitions of what constitutes an acceptable safe depth for the long-term storage of CO_2. Instead, authors frequently refer to 'deep sedimentary basins', 'deep saline aquifers', 'deep unmineable coal seams', 'a given depth' or even less specific 'deeper formations'; references that only vaguely suggest a recommended safe distance from the surface or more importantly, from groundwater reservoirs.

While the storage of CO_2 underground technically is supposed to have a mitigating effect on the accumulation of anthropogenic greenhouse gas (GHG) emissions by reducing their amount released into the atmosphere, to achieve a substantial reduction this requires storage on a large scale, which is not without risk to water quality in underground reservoirs (Carroll et al. 2014; Eldardiry and Habib 2018; Jacobs and Stump 2010; Little and Jackson 2010; Kim et al. 2018; Massachusetts Institute of Technology 2003; Newmark et al. 2010; Takaya et al. 2018; Wilkin and DiGuilio 2010).

Depending on the storage depth, the sequestered CO_2 is compressed into a supercritical liquid, which could escape through fault lines caused by earthquakes, through natural fissures or compromised well integrity on its way to the surface. These are potential migration pathways for fluid leakage with the possibility of groundwater contamination including hydrothermal systems (Carroll et al. 2014, 2016; Feitz et al. 2018; Kim et al. 2018; Nico et al. 2017). The growing concerns regarding impacts on groundwater quality could be significant enough to act as a barrier to large-scale geosequestration (Nico et al. 2017).

In 2010 the Harvard Law School published a liability framework with a legal focus on the uncertainty regarding the responsibility for geosequestration, in which some of the main risks were identified. For instance, *human health concerns* are based on the escape of sequestered CO_2 to the surface through unpredicted and uncontrolled migration, potentially causing harm to the general public as well as employees. *Geological hazards* could include pressure changes underground caused by the injection of CO_2, and which could lead to induced seismicity (Ndzelen and Reinhard 2016). *Ecological problems* consist of the potential for groundwater contamination though saltwater intrusion, an irreversible impact for potable water. A change of pH levels in groundwater systems can lead to the movement of organic compounds and metallic elements and can *contaminate groundwater reservoirs* (Jacobs and Stump 2010; Little and Jackson 2010; Wilkin and DiGuilio 2010). Nevertheless, the Harvard liability framework supports the sequestration of CO_2, and aims to *provide certainty, assuage public concerns,* and *remove barriers* (Jacobs and Stump 2010:i). While CO_2 sequestration is considered by many researchers and scientists as the most promising option to reduce greenhouse gas emissions, uncertainties associated with the subterranean storage of CO_2 remain (Feitz et al. 2018; Jacobs and Stump 2010; Nico et al. 2017; Takaya et al. 2018; Wilkin and DiGuilio 2010), with research studies indicating significant concerns about the potential contamination of hydrologic systems. A review of the current literature covering geosequestration reveals that not all studies investigating the safety of CCS are in agreement, with the concern that sequestered CO_2 does not necessarily stay trapped underground indefinitely as a common denominator. As is frequently the case with controversial environmental issues, one side (e.g. Global CCS Institute 2015) claims that the injection of CO_2 into the ground is completely safe, has been carried out for decades by the oil and natural gas industry without adverse effects, and that there are no reasons why liquid CO_2 could not be stored underground safely and secure indefinitely if following correct procedures.

On the other side, critics of CCS raise concerns that dangerous levels of contaminants could be released in increased concentrations at rates currently neither foreseeable, nor fully understood. In fact, there is no guarantee that CO_2 leaks cannot occur, even if the groups in favour of geosequestration insist that all precautions are taken to prevent any potentially negative impacts on the environment such as compromising the quality of groundwater. Indeed, most research articles acknowledge that there are areas requiring further study; for instance, *on the effect of inadvertent releases of CO_2 on the rock between the storage reservoir and the shallow subsurface* (Lawter et al. 2017:3053). Other key issues were raised in a publication by the Massachusetts Institute of Technology (2003:43), questioning whether *we understand storage mechanisms well enough* and *once injected, can we monitor and verify the movement of subsurface CO_2* or *what is the probability of CO_2 escaping from injection sites?* (Massachusetts Institute of Technology 2003).

> **Note** Geologic Carbon Sequestration or geosequestration is different from Biologic or Terrestrial Carbon Sequestration.

4.7 Hot Spring Laws and Regulations

4.7.1 Examples from Different Countries

Because hot spring environments include such remarkable features, they are often protected by environmental laws or have been designated as natural or cultural geoheritage to prevent unsustainable development in their proximity or they have been declared World Heritage sites (e.g. Pamukkale, Turkey; Yellowstone, USA). There are, however, a variety of special laws and water acts that apply to the management of hot springs as a natural resource. Globally speaking, the efficient and sustainable use of water resources can normally be regulated by laws that concentrate on water rights, the control of ownership and environmental sustainability. Apart from governments, there are also national and international non-government organisations and non-profit associations advocating for the sustainable management and pollution control of all freshwater resources. While there are water laws in most countries, government policies and legislations not always include a provision for the protection of hot spring water from over-exploitation or pollution.

Countries such as **Japan, Taiwan, Korea, Germany**, and **Austria** for instance take these issues very seriously with special legislations to control the appropriate use of hot springs as an important natural and economic resource. Laws in these countries are quite specific regarding water temperature and dissolved elements, which are required to meet certain standards to be classed as a natural hot spring suitable for health and recreational purposes. The *Japanese Hot Spring Law from 1948* requires a natural hot spring, commonly referred to as *onsen*, to meet certain criteria to be classed as a hot spring; (a) hot water emerging from the ground at a temperature in excess of 25 °C and/or (b) containing a certain amount of designated substances such as naturally occurring mineral elements. This Hot Spring Law (Law No. 125 of 1948) and subsequent amendments also provides for the protection of Japan's hot springs by aiming for the proper utilisation of onsen facilities using on-site inspections to monitor the onsen water quality. Another important issue is the prevention of disasters following the accumulation of combustible or toxic gases such as hydrogen sulphide (Ministry of the Environment 2018a, b).

Measures to protect hot springs therefore incorporate certain requirements. To open a new onsen facility, the business owner and/or manager must first obtain the permission of the prefectural governor or from the city where the onsen, hot spring spa or health clinic is established. Any excavations or drilling related to hot spring water extraction require the permission of the prefectural governor and must comply with technical standards to prevent accidents caused by combustible gas. The Ministry of the Environment has further determined that every authorised onsen facility must display the mineral elements contained in their hot spring water together with possible contraindications. The water must be periodically analysed (every ten years) as per the Enforcement Regulation of the Hot Spring Act by a registered laboratory that is authorised to conduct the required analyses (Government of Japan, 2018a, b). Other conditions focus on national hot spring health resorts, which are designated by the Ministry of the Environment and based on the Hot Spring Law with the purpose to promote the public use of hot springs (Government of Japan, 2018a). Conditions related to the environment of the hot spring area are applied to correctly implement measures to protect hot spring resources, the management of hygiene and sanitation of hot spring facilities, the public use of hot springs, and the care for elderly people and persons with disabilities (Ministry of the Environment 2018b).

In **Germany** hot spring spas and health resorts play a central role in the health and recreational tourism industry as well as in the rehabilitation process of patients who want to reclaim their health (Chap. 8). Important criteria are first and foremost the exceptional quality of the hot spring water together with public access, a mild climate and attractive landscapes. Hot spring spas and health resorts must be officially recognised by the German government and abide by the laws and legal decrees of individual states as well as follow guidelines provided by organisations such as the German Spa Association (*Deutscher Heilbäderverband*) and the German Tourism Association (*Deutscher Fremdenverkehrsverband*). Government legislations further define quality standards such as the characteristics of natural hot spring water (>20 °C, discharge rate, certain amount of mineral elements), which requires official recognition for its healing properties based on regulations that apply to all hot spring destinations with the objective to contribute to the conservation of hydrothermal resources for the future (Erfurt 2012).

In **Austria** the water laws state that landowners hold the rights to the groundwater below their land, which has restricted the development of geothermal energy in favour of hot spring spas, resorts and health clinics and their importance for the country's tourism industry. Most of the hot water is derived from deep bores, which supply health and wellness centres, skiing resorts and other recreational facilities such as thermal pools (Goldbrunner and Goetzl 2016).

Protective restrictions are in place for the hydrothermal system of the famous hot spring spa town *Karlovy Vary* (Karlsbad, **Czech Republic**), which have been repeatedly upgraded since 1861. Protection zones extend in all directions from the hydrothermal system and its source and were established to restrict or exclude activities such as building, mining and the transport of hazardous waste materials (Vylita and Žák 2009). This is a clear indication of the importance of natural hot springs for Karlovy Vary and shows the efforts and the commitment of the science community of the Czech Republic to protect natural resources for future use.

In **Iceland** one of the main objectives of the Ministry for the Environment and Natural Resources is the conservation of geological formations including hydrothermal features and geothermal areas as sources of renewable energy. This means, uninhabited natural wilderness areas are supposed to be excluded from development and should remain untouched (Ministry for the Environment 2002; Sæþórsdóttir and Ólafsson 2010). Legislations such as the Act on Water Management and the Nature Conservation Act are intended to provide universal protection for all water- and ecosystems and coordinate the management of natural resources to ensure their sustainability. Article 37 of the Nature Conservation Act identifies the type of landscapes in Iceland that require special protection and should not be disturbed: *waterfalls, hot springs and other thermal sources, as well as surficial geothermal deposits (sinter and travertine), 100 m^2 or more in area.* Nature conservation councils and committees provide advice on nature conservation to the Minister for the Environment and to local authorities (Ministry for the Environment 2018).

In **South America** the transboundary Guarani Aquifer System (GAS) is shared by four countries, Argentina, Brazil Paraguay and Uruguay (Fig. 2.2), which has potential implications in terms of defining efficient and sustainable strategies for resource management and development as well as protection measures for groundwater resources (Foster et al. 2009). Some areas extract vast amounts of water from wells drilled into the thermal aquifers, which has led to signs of depletion after only a few years, and which is putting the future sustainability of this important reservoir at risk (Pesce 2002). Without binding legal agreements to administer the Guarani Aquifer, only non-binding arrangements such as a *Declaration of Basic Principles* to protect the shared water resource from contamination and guidelines for sustainable aquifer management are in place. A *Strategic Action Program* (SAP) had the purpose of promoting the coordination of sustainable groundwater management while at the same time respecting the national sovereignty of each of the four countries over the Guarani Aquifer System (International Waters Governance 2018). However, in 2018, after years of stagnation, the *Guarani Aquifer Agreement* (initially signed by Argentina, Brazil, Uruguay and Paraguay in 2010) was finally ratified, although the question remains when and how the Agreement will be implemented (Sindico and Hawkins 2015; Villar 2018). At the time of writing the agreement still was not entered into force. The Guarani Aquifer Agreement is based on the principles of the *United Nations Law of Transboundary Aquifers*, which include the equitable use of water resources, a commitment not to cause harm, to cooperate, to exchange data and information as well as the sovereignty of each of the member countries, with the latter principle causing extensive criticism, as sovereignty can be used as a defence *for imprudent, inadequate or illegal actions* (Villar and Ribeiro 2011). The Guarani Aquifer is a complex example for the time it takes to get an agreement about the management of an important natural resource such as water in place, without even knowing when the agreement will be finally implemented.

The laws and regulations from different countries illustrate the diversity of legislative procedures required to address the range of hydrothermal resources and their diverse purposes. Returning to the subject of hydraulic fracturing and the potential impact on hot or cold groundwater systems, suggestions are that the fracking industry is not as strictly regulated as would be preferable with loopholes that allow the bypassing of safer exploitation practices (Trial Exhibits Inc 2012). In **Western Australia** environmental campaigners claimed that fracking regulations failed to address potential environmental issues caused by shale gas extraction and were therefore giving the industry special treatment. The state government on the other hand has granted the Department of Mines and Petroleum (DMP) greater power to enforce regulations that cover water monitoring and well management and assured the public that companies will face harsh penalties breaching regulations (News.com.au 2014).

It seems therefore that regulations associated with hydraulic fracking are debatable, and in some cases outdated or not in place at all. Or else, in cases where guidelines are available, these may not be enforced together with safe groundwater management. Recommendations, that require applicants for drilling permits to apply best management practices *whenever reasonably practicable* to avoid landscape and groundwater conservation problems such as chemical contamination during the fracking process (Macey 2018) sound hollow and are subject to interpretation.

4.8 Hot Spring Destinations Under Threat

As long as hydrothermal reservoirs are not over-exploited, they can be classed as a renewable and sustainable resource. Regrettably, this is not always the case when it comes to the tourism industry in some countries, although in recent years a more responsible approach is being adopted when developing new facilities such as hot spring spas, resorts, health clinics and public pools. But equally important is the protection of existing hot spring resources. For decades managers and owners of hot spring resorts relied on a seemingly endless supply of natural hot spring water and invested in large-scale hotel developments based on a freely available resource without sufficient consideration for the future.

4.8.1 Visitor Pressure—Some Examples

Pamukkale in **Turkey** is a classic example of environmental degradation caused by unrestricted development in the past. The spectacular travertine terraces, created by the continuous flow of hot spring water, together with the ancient ruins of Hierapolis located above the petrified waterfalls, have a long history as a combined natural and cultural attraction and in 1988 were designated as a UNESCO World Heritage site (Erfurt-Cooper 2010a, b; Erfurt 2012; UNESCO 2020). However, due to extreme overuse by visiting tourists and extensive anthropogenic pollution from large hotel developments at the top of the travertine terraces the site increasingly deteriorated and required protective measures. In 1995, after carrying out scientific studies, the Turkish Ministry of Culture took steps to prevent further contamination of the travertine terraces by placing them under environmental protection (Açikel and Ekmekçi 2007; Simsek et al. 2000; Taşeli 2016; United Nations 2011). As a consequence, the hotel buildings at the top of the travertine terraces were demolished and access to the historic site of Hierapolis was restricted by closing public road access. Swimming in the natural pools and walking on the terraces was prohibited with rangers patrolling the area. Part of the protective measures included the removal of leaking septic systems and the covering of channels carrying hot spring water to and from the thermal pools, which had caused uncontrolled and unsightly algae growth, and subsequently led to the discolouration of the previously white terraced pools (Açikel and Ekmekçi 2007).

Once the unrestrained commercial tourism developments and activities as the main source of pollution were removed, the effective remediation of the terraces was not immediately obvious. The travertine had absorbed such large amounts of contaminants that these continued to find their way through the porous limestone layers (Açikel and Ekmekçi 2007; Fig. 3.13). Other protective measures involved the controlled alternating release of hot spring water over the travertine pools to combat the algae communities that continued to colonise large parts of the travertine (Kalender et al. 2015). In addition, to help keep the terraces looking as natural and pristine as possible, special regulations for visitor access in protected zones were created and are enforced by rangers in attendance on site (Fig. 4.6).

What remains today, is a rehabilitated and managed natural attraction in accordance with conditions imposed on the area to retain its protected site status. Based on conversations with tour guides and rangers at Hierapolis the impressive and exceptional blend of natural hot springs and ancient archaeological remains attracts approximately 2.5 million people every year with the majority of visitors travelling on Russian cruise ships, who visit the area by bus as part of their trip agenda (Fig. 4.7). Together with other tourists the travertine terraces of Pamukkale are visited on average by between 5000 and 7000 people every day. Needless to say, the visitor pressure remains, and environmental concerns are ongoing, even though there are still moments of solitude early in the morning before the first tour buses arrive.

Undeveloped natural hot spring locations such as *Hakereteke Stream* (commonly known as Kerosene Creek) in **New Zealand** generally receive enthusiastic reviews from visitors (Fig. 4.8). This small watercourse in the *Waiotapu* geothermal field is fed by natural hot spring water around 33 °C and flows over rocky terraces through a native bush setting (Duggan et al. 2007). With shallow pools of pleasantly warm water, Kerosene Creek certainly sounds like a

Fig. 4.6 Rangers guarding the travertine terraces and observing visitor behaviour at Pamukkale to protect the fragile limestone formations from damage. Shoes for example are not allowed on the terraces

Fig. 4.7 Day visitors are arriving by tour bus at the World Heritage site Pamukkale/Hierapolis (Turkey) early in the morning. The vast majority are cruise ship tourists from Russia and eastern European countries

piece of paradise but is a popular destination for ever increasing numbers of tourists. Once only known and appreciated by local visitors, word-of-mouth quickly spread the information about this unique hot stream. Over the years the site has become known for being overrun with tourists at times, which can be a disappointing experience. The lack of facilities constantly leads to rubbish left behind, including broken glass and other unpleasant remnants, indicating that Kerosene Creek is increasingly under threat from growing visitor numbers. A sign that a certain amount of improvement together with some protective management of this unusual hot spring setting may be a positive development.

In comparison with the two hot spring destinations discussed above, the Japanese *Spa Resort Hawaiians* presents a very different case with a focus on mass tourism (Fig. 4.9, see also Chap. 7). Established in 1966 as a hot spring resort this unique oversized theme park is blessed with an abundant supply of hydrothermal resources, which allows for aquatic entertainment on a very large scale with five different theme parks including water slides, pools at different temperatures and with distinctive features. This facility is located in the Fukushima prefecture, an area that suffered not just from a large earthquake and a devastating tsunami in March 2011, but also from the potential danger of being located in the close vicinity (50 km) of a nuclear power station that was destroyed at the same time. Even though all operations of the theme park and the resort had to be closed following the disastrous events, the Spa Resort Hawaiians was back in business early the following year. This is a remarkable example of responding to the ongoing demand for hot spring tourism without delay, while at the same time dealing with the aftereffects of a major natural and manmade catastrophe.

The multi-facility resort has created an artificial tropical indoor environment and borrowed period themes from

Fig. 4.8 One of Kerosene Creek's shallow pools in a quiet bushland setting before tourists arrive. The creek is fed by hot spring water, which contributes to a pleasant temperature for a soak surrounded by nature (New Zealand)

different countries to attract visitors, of which a large percentage is made up of young families, couples and groups of friends and colleagues, who do not mind noise and large crowds (Erfurt 2012). Spa Resort Hawaiians has continuously attracted large visitor numbers of mainly domestic tourists with an average of 3200 visitors per day and up to 10,000 on weekends (Figs. 4.10, 7.16 and 7.17).

The *Blue Lagoon* in **Iceland** is another popular hot spring destination (Figs. 4.11 and 4.12), that has seen a vast increase in visitor numbers over recent years. From 1992, when the Blue Lagoon officially opened to the public, more and more domestic and international tourists made sure that this unique hot spring facility was on their trip agenda and by 2018 the numbers had reached 1.2 million visitors per year. To successfully manage an average of over 3000 people every day pre-bookings are required and the amount of bathing guests in the water is limited per hour. A personal communication in 2018 with the Marketing & Communication Director of the Blue Lagoon A.S. Kristjánsson revealed that long opening times during summer (18 h from 7.00 to 1.00 am) allow for welcoming more guests. Visitor demand is controlled based on data collected in relation to the average length of stay in the water, and the Blue Lagoon has become a benchmark and a role model for the control of visitor pressure and the protection of natural resources. While the Blue Lagoon has long been famous based on its own special attraction, it is also located within the Reykjanes UNESCO Global Geopark (designated in 2015), which adds even more incentive to visit this region.

Visitor management is also a growing problem in the *Yellowstone National Park,* famous for its more than 10,000 hydrothermal features (Figs. 4.13, 9.5 and 7.12). With the park's visitation continuing to grow (4,020,288 in 2019; NPS Stats 2020), conservation of the environment has become

Fig. 4.9 The hot spring Spa Resort Hawaiians in Fukushima was closed after the Great East Japan Earthquake in 2011. Following extensive renovations, upgrades and restructuring the resort was back in business in less than a year. This photo was taken just prior to the earthquake

more challenging and the upkeep and repairs of infrastructure increasingly costly. Besides the visitor pressure, the maintenance of nine visitor centres, hundreds of kilometres of paved and unpaved roads, boardwalks between hot springs and geysers, hiking trails, 900 historic buildings and salaries for 4000 employees are a massive burden for the park management.

Overcrowding of roads at peak times has led to traffic congestions resulting in an increase in accidents and emergencies within the park boundaries, and as a result of millions of visitors, over a billion litres of water and waste water are treated in Yellowstone's 23 water treatment and wastewater treatment facilities every year. To help improving traffic flow and offering visitors a more positive experience, visitor surveys have been carried out and are used to increase staffing during peak periods, and to understand visitor use and expectations as well as the environmental impact of visitors on the natural resources (Landers 2016; NPS Stats 2020; Chap. 9; Fig. 9.5).

All the above examples demonstrate different issues with extreme visitor demand and overcrowding, which need to be responsibly managed to avoid the deterioration of a natural resource.

4.9 What Can Be Done?

4.9.1 Recommendations for the Conservation of Natural Hot Springs

In fact, the heading should read 'What Must be Done?' to protect the sustainability of all natural hot and mineral springs. The subject of controversial issues varies from one country to another, based on whether adequate environmental regulations and water laws are in place or not. To avoid the pollution and depletion of groundwater and to safeguard hot spring environments, some basic measures include:

- Avoiding reservoir depletion through excessive draw-down
- Correct treatment and storage of all types of wastewater
- Avoiding pollution through safe storage of hazardous substances
- Sustainable land use to avoid agricultural, urban and industrial runoff.

Fig. 4.10 Spa Resort Hawaiians on a weekday. At weekends and school holidays the pools are facing much larger crowds

Nonetheless, this alone is not enough to protect hydrothermal systems from the effects of anthropogenic pollution that leads to the contamination of freshwater resources. Even though we may not have much influence on natural events such as earthquakes, tsunamis, volcanic eruptions, landslides, floods or droughts, threats to the sustainability of all water sources from human activities can and must be prevented (Fig. 4.14). To reduce direct and indirect impacts on the environment and preferably avoid them altogether, conservation programs for hot spring areas all should include upper limits of the extraction volume for commercial purposes. Monitoring and regular evaluation of reservoir health are other precautionary measures to be combined with procedures to remediate any environmental degradation in hydrothermal areas (Cernik et al. 2010).

Industrial pollution and urban developments are human activities and main causes of groundwater pollution, affecting not just drinking water resources, but also contribute to the deterioration of hydrothermal systems, resulting in over-exploitation and degradation, which leads to:

- Destruction of ecosystems
- Decreased water quality
- Loss of biodiversity and geodiversity
- Loss of economic value
- Loss of tourist attractions and recreational value.

Combined with insufficient aquifer recharge caused by natural variability and climate change effects, deforestation and wetland drainage due to human action, it is obvious that conservation of natural resources must be taken more seriously. In order to protect all catchment areas and groundwater systems the surrounding natural environment and its ecosystem must be preserved as buffer zones together with low-impact land use to minimise harmful effects on the ecology. To effectively prevent long-term problems for hydrothermal systems, responsible use of hot springs, reduced visitor pressure and close monitoring of exploitation and recharge are part of a responsible and sustainable management approach. Strategic planning for the future of endangered hot spring destinations further involves strong partnerships and co-operation of all stakeholders from the tourism industry as well as from relevant government bodies. Without raising more awareness about the effect of encroachment by industry, agriculture and urban development on hydrothermal features, conservation efforts lack sufficient support.

As studies have shown, conservation programs rely heavily on community participation, with local communities

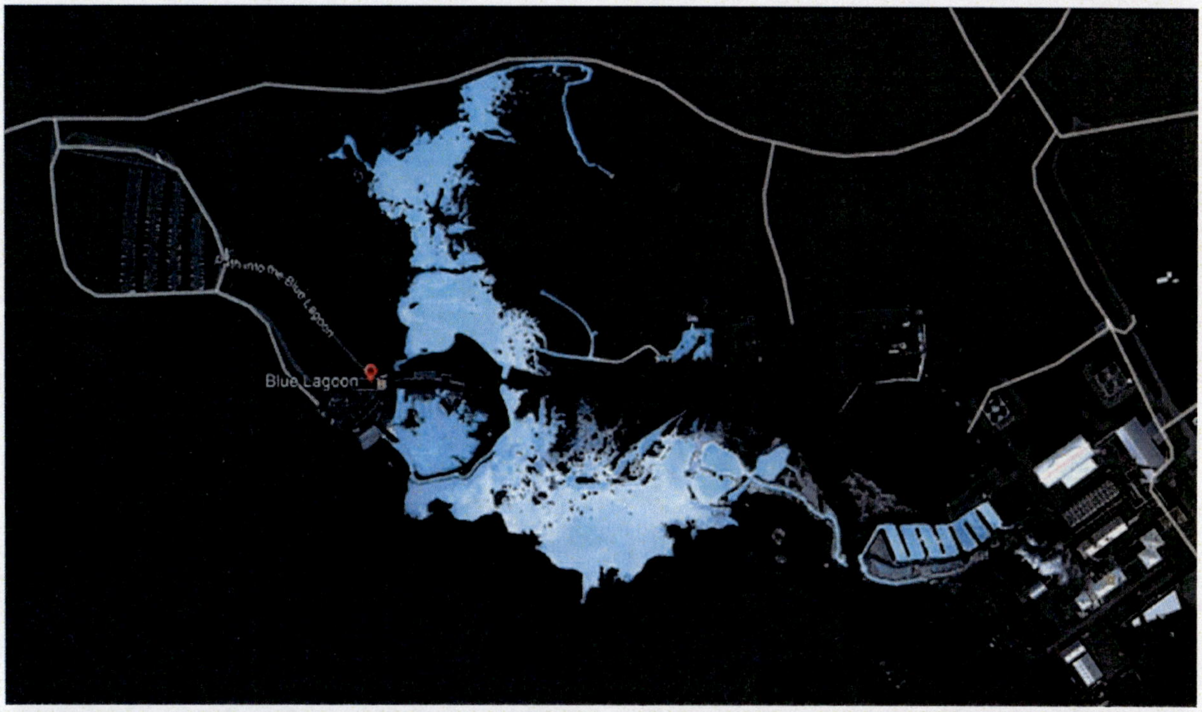

Fig. 4.11 This aerial view of the Blue Lagoon shows the extent of water that has created ponds in the surrounding lava fields. Only a comparatively small part is used as the actual bathing area for visitors. In the right bottom corner, the Svartsengi geothermal power station peacefully co-exists with the facilities of the Blue Lagoon and accommodation buildings (Map data: Google Image © 2018 DigitalGlobe)

contributing significantly to the sustainable management of natural environments and resources. Through their direct involvement, residents have an opportunity to improve their appreciation and their sense of belonging, while at the same time they are encouraged and motivated to contribute to the decision making process (Gursoy et al. 2002; Rasoolimanesh et al. 2017; Zhang et al. 2013).

4.9.2 Best Practice Management

The integration of nature conservation with a special focus not just on biodiversity but also on geodiversity is essential to counteract the many challenges to the environment caused by human activities (Matthews 2014). Recreational activities in national parks as well as health resorts and hot spring spas depend on natural resources that remain sustainable and renewable. However, the natural environment is under increasing pressure because hot spring destinations are a major tourist attraction. To reduce the impact of hot spring tourism in its various forms, the oversized ecological footprint in many recharge areas must be considerably reduced through best practice management. With the aim to protect natural resources such as hot springs, limits to growth and carrying capacity at tourist destinations must be re-assessed and taken into consideration. To avoid capacity constraints and undesirable environmental impacts at popular locations due to extensive visitor demand, overcrowding during peak seasons can be regulated by restricting visitor numbers using a permit system and/or promoting alternative attractions (Navarro Jurado et al. 2012; Prats and Bassols 2014; Schwartz et al. 2012).

Development of hot spring destinations looks very different from a conservation and sustainability angle compared to stakeholder concerns about the economic sustainability of the hot spring tourism industry. Hot spring environments with their distinctive features and unique natural surroundings are of substantial commercial and recreational value and therefore, responsible management is one of the key factors for the protection of renewable resources to ensure the long-term sustainability of natural hot springs as tourist attractions (Buckley 2007; Dowling and Newsome 2006; Erfurt-Cooper 2009). Without question, further developments in the hot spring tourism sector require environmental impact assessments (EIA) and a firm commitment to the principles of sustainability to prove that natural resources will not be depleted or contaminated.

Apart from national parks and world heritage sites, geoparks at national and global levels have been developed in recent times as another option to preserve the geoheritage

Fig. 4.12 The Blue Lagoon's expanse of geothermal water on a winter's day in Iceland. *Source* Rita Erfurt 2019

of a region including hydrothermal features. These recreational areas are further explored in Chap. 7, although from a different perspective.

4.10 Conclusion

At this point in time many conservation programs are focusing primarily on protecting the biodiversity of natural habitats and endangered species of flora and fauna. However, the development of geoconservation strategies with a focus on protecting the geodiversity of the natural environment is a growing activity with an increasing profile worldwide (Burek and Prosser 2008; Farsani et al. 2014; Henriques et al. 2011; Hose 2012). In **Madagaskar**, a country rich in geoheritage, conservation of natural resources such as hot springs and volcanic lakes includes the biodiversity as well as the geodiversity as both are considered of equal importance (Raharimahefa 2012). In many countries hydrothermal features such as hot springs, geysers and travertine terraces are already protected in national parks, geoparks and world heritage sites, where conservation is coordinated with promoting best practice approaches and policies covering land use planning and sustainable resource management (Bentivenga and Geremia 2015; Erfurt 2012).

When managing hot spring resources, it is essential to take a long-term view, which is not always the case and has resulted in the deterioration of many hydrothermal features. Successful conservation requires understanding the value of a feature or site to appreciate its need to be preserved as geoheritage for future generations (Burek and Prosser 2008; Prosser 2011). This chapter has identified a number of key issues potentially affecting the sustainability of natural hot springs as a renewable resource, which clearly indicates that the deterioration of countless hot springs worldwide is based on human activities. It is therefore our responsibility to ameliorate the damage that is already done and to ensure that the future use of natural hydrothermal systems is strictly regulated, and all extraction processes are monitored on a constant basis.

Fig. 4.13 A partial bird's-eye view of Yellowstone National Park from over 100 years ago. *Source* Wellge 1904 (Public Domain)

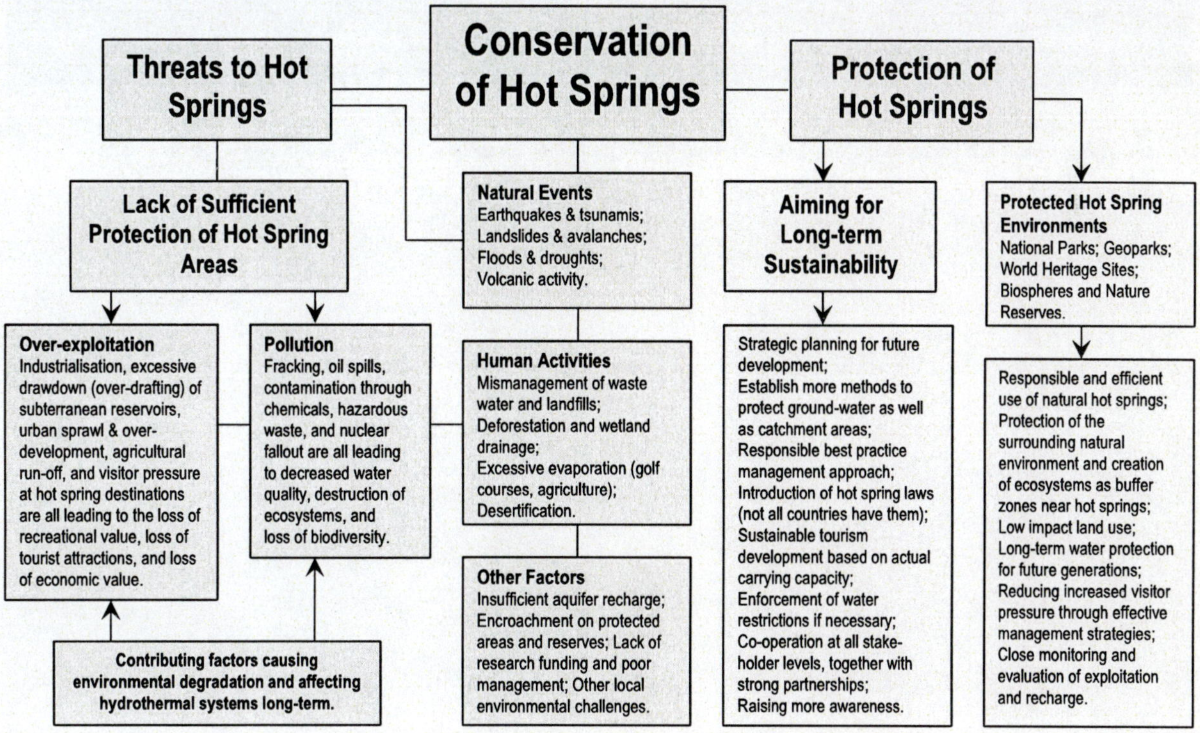

Fig. 4.14 A framework outlining key issues related to the conservation of natural hot springs. *Source* Author

The most photographed egg basket in the world is submerged in a small boiling crater lake at Umi Jigoku in Beppu, Japan

4.11 Appendices

Appendix 4.1 Great Australian Basin History

This poster tells the Story of the Great Artesian Basin (Australia) and was produced by the Great Artesian Basin Coordinating Committee (GABCC) to illustrate the historical significance and to raise awareness for the need of conservation of endangered species and landforms

Appendices 4.2 Desert Mound Springs, Australia

1—Aerial View of the Dalhousie Springs Complex, which is part of the Witjira National Park in the Australian Desert. *Source* Charles Davidson, 2020 (with permission); 2—Diagram of mound springs emerging from the Great Artesian Basin (GAB), the largest subterranean water reservoir worldwide, which underlies 22% of permission); 3—North Freeling Spring on the Peake Pastoral Lease. *Source* Simon Lewis, 2018 (with permission); 4—Aerial view of the largest Dalhousie thermal spring, which is open to visitors and bathing is permitted by the traditional owners, the Lower Arrente people, who are managing the Witjira National Park. *Source* Charles Davidson

References

ABC News (2011) Farmers count cost of coal seam gas rush. Four Corners. Retrieved from https://tinyurl.com/vge32sx

Açikel Ş, Ekmekçi M (2007) Assessment of sorption characteristics for enhancement of remediation of pamukkale travertines, Turkey. In: L Candela et al (eds) WAter pollution in natural porous media at different scales. Assessment of Fate, Impact and Indicators. WAPO^2. Madrid: Instituto Geológico y Minero de España. ISBN: 978–84–7840–676–0

Alaska Department of Environmental Conservation (DEC) (2018) Industrial or nondomestic discharges—applications. Wastewater discharge permits—wastewater discharge authorization program. Retrieved from https://dec.alaska.gov/water/wastewater/permitentry/industrial-or-nondomestic

Benny-Morrison A (2012) Work begins to cap burning well. Retrieved from https://tinyurl.com/sa56efg

Bentivenga M, Geremia F (2015) VII International Symposium ProGEO on the Conservation of the Geological Heritage 'Geoheritage: Protecting and Sharing', Bari (Apulia, Italy), 24th to 28th September 2012. Geoheritage 7:1–3. https://doi.org/10.1007/s12371-015-0146-z

Brisbane Times (2016) Condamine River on fire in CSG mining area. Retrieved from https://tinyurl.com/szup9zu

Buckley R (2007) Adventure tourism. CABI, London, UK

Burek CV, Prosser CD (eds) (2008) The history of geoconservation. Geological Society, London, Special Publications 300, 1–5. https://doi.org/10.1144/SP300.1

Burt M (2019) How we can preserve our Great Artesian Basin. NSW Farmers. Retrieved from https://tinyurl.com/t2o4n93

Carroll SA, Keating E, Mansoor K, Dai Z, Sun Y, Trainor-Guitton W, Brown C, Bacon D (2014) Key factors for determining groundwater impacts due to leakage from geologic carbon sequestration reservoirs. Int J Greenhouse Gas Control 29:153–168. https://doi.org/10.1016/j.ijggc.2014.07.007

Carroll S, Carey JW, Dzombak D, Huerta NJ, Li L, Richard T, Um W, Walsh SDC, Zhang L (2016) Review: role of chemistry, mechanics, and transport on well integrity in CO_2 storage environments. Int J Greenhouse Gas Control 49:149–160. https://doi.org/10.1016/j.ijggc.2016.01.010

Cernik M, Sima J, Plesinger V (2010) Development and protection of geothermal sources in Peru. In: Proceedings World Geothermal Congress Bali, Indonesia, 25–29 April 2010

Chavasse L (2006) The great Artesian Basin. Queensland Government Natural Resources, Mines and Water. Facts—Water Series W68 QNRM03322. Produced by: Central West Region

CSIRO (2020) Hydraulic fracturing—what is unconventional gas? Retrieved from https://www.csiro.au/en/Research/Energy/Hydraulic-fracturing/What-is-unconventional-gas

Davidson C (2020) Great Artesian Basin Hot Springs. Personal communication

De Ronde CEJ, Fornari DJ, Ferrini VL, Walker SL, Davy BW, LeBlanc C, Caratori Tontini F, Kukulya AL, Littlefield RH (2016) The Pink and White Terraces of Lake Rotomahana: what was their fate after the 1886 Tarawera Rift eruption? J Volcanol Geoth Res 314:126–141. https://doi.org/10.1016/j.jvolgeores.2016.02.003

De Ronde CEJ, Tontini FC, Keam RF (2018) Where are the Pink and White Terraces of Lake Rotomahana? J Royal Soc New Zealand. https://doi.org/10.1080/03036758.2018.1474479

DiGiulio D, Jackson RB (2016) Impact to underground sources of drinking water and domestic wells from production well stimulation and completion practices in the Pavillion. Environmental Science & Technology, Wyoming, Field. https://doi.org/10.1021/acs.est.5b04970

Doormann P, Barber J (2017) Rotorua geothermal system: current and future management. In: Proceedings 39th New Zealand Geothermal Workshop, 22–24 Nov 2017 Rotorua, New Zealand

Dowling R, Newsome D (2006) Geotourism. Elsevier, London, UK

Duggan IC, Boothroyd IKG, Speirs DA (2007) Factors affecting the distribution of stream macroinvertebrates in geothermal areas: Taupo Volcanic Zone, New Zealand. Hydrobiologia 592:235–247. https://doi.org/10.1007/s10750-007-0748-9

Earth Focus (2011) Fracking hell: the untold story. Retrieved from https://www.youtube.com/watch?v=dEB_Wwe-uBM

Eldardiry H, Habib E (2018) Carbon capture and sequestration in power generation: review of impacts and opportunities for water sustainability. Energy Sustain Soc 8(6). https://doi.org/10.1186/s13705-018-0146-3

Erfurt-Cooper P, Cooper MJ (2009) Health and wellness tourism: spas and hot springs. Channel View Publications, Bristol, UK

Erfurt-Cooper P (2010a) The importance of natural geothermal resources in tourism. In: Proceedings World Geothermal Congress, Bali, Indonesia, 25–29 April 2010

Erfurt-Cooper P (2010b) Volcano and Geothermal Tourism: Sustainable Geo-Resources for Leisure and Recreation. Earthscan, London

Erfurt P (2012) An assessment of the role of natural hot and mineral springs in health, wellness and recreational tourism. Unpublished Doctoral Dissertation, School of Business, James Cook University, Cairns, Queensland, Australia

Erfurt R (2019) Blue Lagoon, Iceland. Photo credit

Farsani NT, Coelho COA, Costa CMM, Amrikazemi A (2014) Geo-knowledge management and geoconservation via geoparks and geotourism. Geoheritage 6:185–192. https://doi.org/10.1007/s12371-014-0099-7

Feitz AJ, Boreham C, de Caritat P, Langford R, Causebrook R, Preda M, Draper J (2018) Groundwater and geological storage of carbon dioxide. Geoscience Australia. Retrieved from https://d28rz98at9flks.cloudfront.net/70671/70671.pdf

FOMS (Friends of Mound Springs) (2020) Great Artesian Basin (GAB) Springs Protection and Monitoring Program, 1984–2005. Retrieved from https://tinyurl.com/rdsdu7o

Foster S, Hirata R, Vidal A, Schmidt G, Garduño H (2009) The Guarani Aquifer initiative—towards realistic groundwater management in a transboundary context. Worldbank—Water Partnership Program

Global CCS Institute (2015) Secure geologic storage of CO_2. Retrieved from https://www.globalccsinstitute.com/publications/secure-geologic-storage-co2

GNS Science (2018) Scientists reaffirm earlier findings on the fate of Pink and White Terraces–03 Aug 2018. Retrieved from https://gns.cri.nz/Home/News-and-Events/Media-Releases/earlier-findings-reaffirmed

Goldbrunner J, Goetzl G (2016) Geothermal energy use, country update for Austria. European Geothermal Congress, Strasbourg, France, 19–24 Sept 2016

Government of Japan (2018a) National hot spring health resorts. Ministry of the Environment. Retrieved from https://www.env.go.jp/en/nature/nps/hshr.html

Government of Japan (2018b) The Criteria for the Notification, etc. on Contraindications and Cautions for Bathing and Drinking as specified in Article 18, Paragraph 1 of the Hot Springs Law. Ministry of the Environment. Retrieved from https://www.env.go.jp/en/nature/nps/hshr.html

Gursoy D, Jurowski C, Uysal M (2002) Resident attitudes: a structural modeling approach. Annals of Tourism Research 29(1):79–105. https://doi.org/10.1016/S0160-7383(01)00028-7

Habermehl MA (2000) Great Artesian Basin springs, inventory and assessment. Pages 14–16. In: Proceedings of the 3rd Mound Spring Researchers Forum, Adelaide February 9th, 2000. Department for Environment, Heritage and Aboriginal Affairs: Adelaide

Habermehl MA (2001) Hydrogeology and environmental geology of the Great Artesian Basin, Australia. In V A Gostin (ed), Gondwana to Greenhouse—Australian Environmental Geoscience. Geological Society of Australia Inc. Special Publication 21: 127–143, 344–346
Henriques MH, Pena dos Reis R, Brilha J, Mota T (2011) Geoconservation as an emerging geoscience. Geoheritage 3:117–128. https://doi.org/10.1007/s12371-011-0039-8
Hose TA (2012) Editorial: geotourism and geoconservation. Geoheritage 4:1–5
Hróarsson B, Jónsson SS (1992) Geysers and Hot Springs in Iceland. Mál og Menning, Reykjavik, Iceland
International Waters Governance (2018) Guaraní Aquifer System. Retrieved from https://www.internationalwatersgovernance.com/guaraniacute-aquifer-system.html
Jacobs WB, Stump D (2010) Proposed liability framework for geological sequestration of Carbon Dioxide. Harvard Law School, Emmett Environmental Law and Policy Clinic. Cambridge, Mass
Kalender L, Öztekin Ökan O, İnceöz M, Çetindağ B, Yildirim V (2015) Geochemistry of travertine deposits in the Eastern Anatolia District: an example of the Karakoçan-Yoğunağaç (Elazığ) and Mazgirt-Dedebağ (Tunceli) travertines, Turkey. Turkish Journal of Earth Sciences 24:607–626. https://doi.org/10.3906/yer-1504-27
Kim CY, Han WS, Park E, Jeong J, Xu T (2018) CO_2 Leakage-induced contamination in shallow potable aquifer and associated health risk assessment. Geofluids, Article ID 4834601, p 19. https://doi.org/10.1155/2018/4834601
Kondash AJ, Lauer NE, Vengosh A (2018) The intensification of the water footprint of hydraulic fracturing. Sci Adv 4(8):eaar5982. https://doi.org/10.1126/sciadv.aar5982
Landers R (2016) Yellowstone Park under pressure from crowds, budget. The Spokesman-Review. Retrieved from https://www.spokesman.com/stories/2016/dec/11/yellowstone-park-under-pressure-from-crowds-budget/
Lawter AR, Qafoku NP, Asmussen RM, Bacon DH, Zheng L, Brown CF (2017) Risk of geologic sequestration of CO_2 to groundwater aquifers: current knowledge and remaining questions. Energy Procedia 114:3052–3059
Lewis S (2018) North Freeling Spring on the Peake Pastoral Lease. Friends of Mound Springs (FOMS). Photo credit
Little MG, Jackson RB (2010) Potential impacts of leakage from deep CO_2 geosequestration on overlying freshwater aquifers. Environ Sci Technol 44(23):9225–9232. https://doi.org/10.1021/es102235w
Macey GP (2018) The incomplete ecology of hydraulic fracturing governance. Arizona State Law J 581–614
Massachusetts Institute of Technology (2003) Geological Carbon Sequestration. The future of Coal. An Interdisciplinary MIT Study. ISBN 978-0-615-14092-6
Matthews TJ (2014) Integrating geoconservation and biodiversity conservation: theoretical foundations and conservation recommendations in a european union context. Geoheritage 6:57–70. https://doi.org/10.1007/s12371-013-0092-6
Ministry for the Environment (Iceland) (2002) Welfare for the future—Iceland's national strategy for sustainable development 2002–2020. ISBN 9979–839–02–3
Ministry for the Environment (Iceland) (2018) The nature conservation Act No. 44/1999. Ministry for the Environment and Natural Resources. Retrieved from https://www.government.is/Publications/Legislation/Lex/?newsid=4dfffdb5-fb1d-11e7-9423-005056bc4d74
Ministry of the Environment 環境省 (2018a) Overview of Hot Spring Law (温泉法の概要). Retrieved from https://www.env.go.jp/nature/onsen/outline/
Ministry of the Environment 環境省 (2018b) 温泉熱の有効活用について (On the effective utilization of hot spring heat). Retrieved from https://www.env.go.jp/nature/onsen/spa/spa_utilizing.html
NASA (2018) NASA satellites reveal major shifts in global freshwater. Retrieved from https://www.nasa.gov/press-release/nasa-satellites-reveal-major-shifts-in-global-freshwater
Navarro Jurado E, Tejada Tejada M, Almeida García F, Caballo González J, Cortés Macías R, Delgado Peña J, …Solís Becerra, F. (2012) Carrying capacity assessment for tourist destinations. Methodology for the creation of synthetic indicators applied in a coastal area. Tourism Manage 33:1337–1346. https://doi.org/10.1016/j.tourman.2011.12.017
Ndzelen R, Reinhard N (2016) Hydraulic fracturing and seismicity. Working Paper. https://doi.org/10.13140/RG.2.1.3891.6082
Newmark RL, Friedmann SJ, Carroll SA (2010) Water challenges for geologic carbon capture and sequestration. Environ Manage 45:651–661. https://doi.org/10.1007/s00267-010-9434-1
News.com.au. (2014) Western Australia—new fracking rules give industry special treatment: environmentalists. Retrieved from https://tinyurl.com/y9ys9z7j
Nico PS, Fox P, Davis J, Birkholzer J, Spycher N, Zheng L (2017) Potential geochemical impacts of carbon capture and storage operations on groundwater quality. Report prepared by Lawrence Berkeley National Laboratory for California Energy Commission
Norton M (2013) Hydraulic fracture diagram. Public Domain. Retrieved from https://commons.wikimedia.org/wiki/File:HydroFrac2.svg
NPS Stats (2020) Recreation visitors (1904—Last Calendar Year). Retrieved from https://tinyurl.com/kgk5xh8
NZ Geothermal Association (2018a) NZ geothermal fields. Retrieved from https://nzgeothermal.org.nz/nz_geo_fields/
NZ Geothermal Association (2018b) Surface environmental effects. Retrieved from https://nzgeothermal.org.nz/nz_geo_fields/
Pesce A (2002) Thermal Spas: an economic development alternative along both sides of the Uruguay River. In: GHC Bulletin. September 2002. Oregon: Geo–Heat Center
Ponder WF (2002) Desert springs of the Australian Great Artesian Basin. In Conference proceedings. Spring-fed wetlands: important scientific and cultural resources of the intermountain region, 2002
Prats and Bassols (2014) La Garrotxa Volcanic Zone, Spain. In P Erfurt-Cooper (Ed), Volcanic Tourist Destinations. Chap. 10, pp 139–143. Berlin, Heidelberg: Springer
Prosser CD (2011) Principles and practice of geoconservation: lessons and case law arising from a legal challenge to site-based conservation on an Eroding Coast in Eastern England, UK. Geoheritage 3:277–287. https://doi.org/10.1007/s12371-011-0042-0
Raharimahefa T (2012) Geoconservation and geodiversity for sustainable development in Madagascar. Madagascar Conservation & Development 7(3):126–134. https://doi.org/10.4314/mcd.v7i3.5
Rasoolimanesh SM, Jaafar M, Ahmad AG, Barghi R (2017) Community participation in world heritage site conservation and tourism development. Tourism Manage 58:142–153. https://doi.org/10.1016/j.tourman.2016.10.016
Rivera Diaz A, Kay E, Zarrouk SJ (2016) Reinjection in geothermal fields—a worldwide review update. Renew Sustain Energy Rev 53:105–162. https://doi.org/10.1016/j.rser.2015.07.151
Roe G, Benoit D, Pike C (2015) Study of direct use options for hot Springs in Adak, Akutan and Atka. Prepared for: The Alaska Energy Authority and The Aleutian Pribilof Islands Association
Sæþórsdóttir AD, Ólafsson R (2010) Nature tourism assessment in the Icelandic Master Plan for geothermal and hydropower development. Part II: assessing the impact of proposed power plants on tourism and recreation. J Herit Tourism 5(4):333–349. https://doi.org/10.1080/1743873X.2010.517840
Schwartz Z, Stewart W, Backlund EA (2012) Visitation at capacity-constrained tourism destinations: exploring revenue management at a national park. Tourism Management 33:500–508. https://doi.org/10.1016/j.tourman.2011.05.008

Shortall R, Davidsdottir B, Axelsson G (2015) Geothermal energy for sustainable development: a review of sustainability impacts and assessment frameworks. Renew Sustain Energy Rev 44:391–406. https://doi.org/10.1016/j.rser.2014.12.020

Simsek S, Gunay G, Elhatip H, Ekmeçi M (2000) Environmental protection of geothermal waters and travertines at Pamukkale. Turkey. Geothermics 29(4–5):557–572. https://doi.org/10.1016/S0375-6505(00)00022-5

Sindico F, Hawkins S (2015) The Guarani Aquifer Agreement and Transboundary Aquifer Law in the SADC: comparing Apples and Oranges? Review of European Community and International Environmental Law 24(3):318–329. https://doi.org/10.1111/reel.12139

Takaya Y, Nakamura K, Kato Y (2018) Long-term reaction characteristics of CO_2–Water–Rock interaction: insight into the potential groundwater contamination risk from underground CO_2 storage. Resource Geology 68(1):93–100. https://doi.org/10.1111/rge.12147

Taşeli BK (2016) Achievements in the use of renewable energy in Turkey's special environmental protection areas (SEPA). J Multidisciplinary Eng Sci Tech (JMEST) 3(3):4282–4285. ISSN: 2458–9403

Trial Exhibits, Inc. (2012) Hydraulic fracturing 3D animation. Retrieved from https://www.youtube.com/watch?v=fFUxq9UolN4

UNESCO (2020) Hierapolis-Pamukkale. Retrieved from https://whc.unesco.org/en/list/485

United Nations (2011) Turkey—Hierapolis–Pamukkale Rescue Plan. World Conservation Monitoring Centre

Villar P C (2018) Countdown to the Guarani Aquifer agreement coming into force: will it be effective in promoting transboundary groundwater governance? International Water Law Project Blog. Retrieved from https://tinyurl.com/y7gmndaw

Villar PC, Ribeiro WC (2011) The Agreement on the Guarani Aquifer: a new paradigm for transboundary groundwater management? Water International 36(5):646–660. https://doi.org/10.1080/02508060.2011.603671

Virginia.gov. (2015) Carbon sequestration. Division of geology and mineral resources. Retrieved from https://www.dmme.virginia.gov/DGMR/carbonstorage.shtml

Vylita T, Žák K (2009) Travertine deposits of the Karlovy Vary thermal water system. Environ Geol 58:1639–1644

Wellge H (1904) Yellowstone National Park. Public Domain. Retrieved from https://commons.wikimedia.org/wiki/File:Yellowstone_National_Park_by_Wellge,_1904.jpg

White P, Hunt T (2005) Simple modelling of the effects of exploitation on hot springs, Geyser Valley, Wairakei, New Zealand. Geothermics 34:187–207

Wilkin RT, DiGuilio DC (2010) Geochemical impacts to groundwater from geologic carbon sequestration: controls on pH and inorganic carbon concentrations from reaction path and kinetic modeling. Environ Sci Technol 44(12):4821–4827. https://doi.org/10.1021/es100559j

Willacy M (2016) Condamine River's mysterious bubbling 'intensifying'. Retrieved from https://tinyurl.com/sc39s86

Wilmott W (2017) Underground water—lifeblood of the West. In: Wilmott W, Cook A, Neville B (eds) Rocks, landscapes & resources of the great artesian basin—a handbook for travellers. Geological Society of Australia, Queensland Division, pp 41–50

Winkler GG (1861) Island: Seine Bewohner, Landesbildung und Vulkanische Natur. Georg Westermann Verlag, Braunschweig, De

Zhang Y, Cole ST, Chancellor HC (2013) Residents' preferences for involvement in tourism development and influences from individual profiles. Tourism Plan Dev 10(3):267–284. https://doi.org/10.1080/21568316.2012.747984

Zhang D, Song J (2014) Mechanisms for geological carbon sequestration. Procedia IUTAM 10:319–327. https://doi.org/10.1016/j.piutam.2014.01.027

The ‘Great Bath’ at the historical Roman Complex in Bath, UK

Hot Springs Throughout History

5

Contents

P. Erfurt, *The Geoheritage of Hot Springs*, Geoheritage, Geoparks and Geotourism,
https://doi.org/10.1007/978-3-030-60463-9_5

The ruins of Hierapolis, also called Pambouk Kalesi [Pamukkale], originally painted by Thomas Allom on the way from Laodicea circa 1838. *Source* Allom and Walsh (1838) (Appendix 5.2)

5.1 Introduction

The use of hot springs as places of relaxation, healing and worship goes back to the dawn of history and the earliest civilisations worldwide. Throughout time, natural hot springs were a reason for establishing settlements where communities had access to the benefits of the thermal waters for health and recreation. The aim of this chapter is to explore the origins and the different stages of development of hot spring use to provide reference points for a realistic timeframe that is based on historical records. The objective of exploring the global hot spring heritage includes the identification of historical settings and to determine how, when and where hot springs became an integral part of the socio-cultural context and what led to their development as destinations for health, wellness and recreation. Extensive literature reviews and the evaluation of existing records indicate that the history of natural hot spring use has global roots with only a few countries without direct access to hydrothermal resources.

Note At this point I would like to add a short disclaimer to address what may seem perhaps as a somewhat disproportionate representation of individual countries. While the hot spring heritage of some regions is covered in more detail than others, this is not based on personal preferences, but on the understandable limitation of carrying out literature reviews of source material from every country and in every language. Nevertheless, many sources were translated during the research for this book and used to enrich the chapter content. The sequence of countries discussed in this chapter does not follow any specific order, other than the quest to identify the first nation that used hot springs by reaching back into time as far as possible (Appendix 5.1).

5.1.1 Historical Development of Hot Springs—Taking the Waters

This historical overview contains examples from various countries that were selected to determine a maximum historical timeframe of hot spring use, at times a challenging task due to inconsistent sources or conflicting dating systems. The analysis of existing accounts nevertheless revealed that natural hot springs have been used since times immemorial and that their usage has worldwide origins. Supported by acceptable chronicles the maximum historical timeframe of hot spring use may reach back to at least around 3000 BC (Erfurt-Cooper and Cooper 2010; Erfurt 2012). Earlier human interactions with hydrothermal phenomena are placed in the realms of prehistory and cannot be accurately documented owing to the absence of written records. However, most of the knowledge about the use of hot springs for domestic purposes, healing practices and various cultural traditions from ancient times were passed on through generations using traditional stories and narratives (Erfurt-Cooper and Cooper 2009). Very often the existence of natural hot springs was associated with certain divinities and spirits of the Earth; for example, in Greece, temples with thermal baths were frequently dedicated to *Asclepios*, the god of medicine in the Greek mythology (Melillo 1995; Fig. 5.1; Chap. 6).

A review of the more reliable historical information has led to a few early accounts by writers of the antiquity, who described hydrothermal manifestations including hot springs, their occurrence and their characteristics (Table 5.1). A detailed chronological order of hot spring developments is included as a separate timeline at the end of this chapter (Appendix 5.1).

The early attention directed at hydrothermal manifestations confirms not only that such places motivated people to settle in their vicinity, but also demonstrates the enduring trust in the curative value of natural hot spring water. The strong belief in these therapeutic benefits throughout history has led to the establishment of thermal bathing facilities where various medical conditions were treated with external and internal applications of hot and cold spring water (Erfurt-Cooper and Cooper 2009; Erfurt 2012).

Almost certainly the use of hot springs for bathing and healing purposes started in the Neolithic age (Stone age—c. 12,000 BC) or even earlier. This is based on the reasonable assumption that humans are naturally drawn to thermal water sources given the opportunity. But even so, tracing back the use of hot springs to prehistorical times is largely subject to speculation and guesswork, frequently supported by indirect evidence found at locations close to natural hot springs. This gives room to the argument that ancient peoples in many countries used natural resources such as hot springs for recreation, cooking, heating and their curative effect (Calderón 1999; Erfurt 2012), as well as various geothermal and volcanic byproducts where these were available (Özgüler and Kasap 1999).

For example, the native peoples of the **Americas** have long considered hot springs as sacred places and believed in the healing powers of hot spring water (Lund 2005). Similarly, this belief was held by the Indigenous population of **Australia**, who have historically used natural hot springs for thousands of years, also regarding them as sacred sites (Ah Chee 2014; Erfurt-Cooper and Erfurt-Cooper 2010; Pearn

Fig. 5.1 Asclepios, the god of medicine in Greek mythology. *Source* Mehnert (2008) (Public Domain). Image modified by Author

and Little 1998). And while the explorer *Henry Morton Stanley* travelled various countries on the **African** continent in the 1870s, he came across hot springs used by local tribes as a meeting place and for their healing power (Fig. 5.2). It is understood that these springs were used long before the explorer Stanley observed them on his visit.

On the **European** continent, between the eighth and sixth century BC, a systematic spread of hot spring use took place under the rule of the Etruscans, who inhabited the region Etruria in central Italy (Fig. 5.3). During the fifth and fourth century BC the development of hot springs increased, spreading not only to the territories occupied by the Etruscans, but caused initial interest in other Mediterranean areas, followed by moderate developments. At this point the Roman military became aware of some locations in Italy with thermal springs and started to use them for their own purposes (Cataldi and Burgassi 1999).

Following the annihilation of the Etruscans in the third century BC and as Rome extended its power over Italy (De Cazanove 2007), the use of natural hot springs appears to have temporarily declined in the former Etruscan territories until the second century BC. Still, other Mediterranean areas continued to enjoy thermal bathing with the Roman elite starting to visit natural hot spring destinations. Without any written records left behind by the Etruscans, it was left to the Romans to provide historical accounts for posterity, which were perhaps not completely objective and impartial due to Rome's relationship with Etruria (De Cazanove 2007). From the first century BC to the third century AD the use of hot springs for balneotherapy peaked under the Romans and multipurpose facilities offering thermal baths using water, steam and mud for health and recreation were developed throughout the Roman Empire (Erfurt 2012).

During the fourth and fifth century AD however, the use of hot springs went into a general decline in Europe and thermal bathing took place at significantly reduced levels. Although the main function of hot spring bathing was supposed to be for healing purposes and hygiene, their use became increasingly regarded with disapproval. This was

Table 5.1 Between the fifth century BC and the third century AD, several Greek and Latin writers explored natural phenomena and wrote about various hydrothermal manifestations

Historic authors and their recorded interest in natural hot springs	
5th century BC—Herodotus, Greek historian and author of *The Nine History Books*	Described many Greek hydrothermal manifestations in his publications
c. 350—c. 290 BC—Megasthenes, Greek historian, diplomat, explorer of India and author of the book *Indica*	Mentioned in his writings the medicinal value of thermal springs in India
1st century BC—Strabo, Greek geographer, philosopher, historian and author of two book series titled *Geography* and *History*	Described the main hydrothermal manifestations in the Mediterranean area
1st century BC—Vitruvius, Roman architect, engineer and the author of *The Ten Books on Architecture*	Interpreted the hydrothermal manifestations of the Phlegraean Fields
1st century AD—Pliny the Elder, Roman naturalist, philosopher, military commander of the early Roman Empire and author of the encyclopaedia *Natural History*	Described many of the major hydrothermal manifestations of Southern Europe and the Mediterranean area

Source Cataldi et al. (1999)

Fig. 5.2 The Mtagata Hot Springs (Central Africa) with the explorer Henry M. Stanley observing the therapeutic powers of the hot springs. Depicted on a wood engraving from 1878. *Source* Wellcome Library (2018)

largely due to the condemnation by the Christian church of public thermal baths as facilities where immorality was encouraged and therefore considered as dangerous breeding grounds for sexually transmitted diseases (Erfurt 2012; Maier and Fiedler 2004). Later, during the fifteenth century AD, and despite the attitude of the church, hot spring destinations such as Bad Wildbad (Germany) became popular places, where aristocracy and members of the clergy met for health and recreational purposes, regardless of the potential risk to their morals (Rumpf and Sollner 2006).

The history of hot spring use is much more extensive than commonly acknowledged, as societies on every continent made use of this natural resource for various purposes including personal hygiene, healing and relaxation. While the Romans may be the most historically renowned society who developed hot spring facilities for human use at many locations throughout Europe, the Middle East and North Africa (Verdel and Pittler 2003), they certainly were not the only ones, let alone the first (Erfurt 2012). To prove this, we start with the Asian continent.

5.2 The Hot Spring Heritage of Asia

5.2.1 The Indus Valley Civilisations (IVC)

From an archaeological and anthropological point of view the Indus Valley, located in today's Pakistan, is considered as one of the cradles of ancient civilisation. Discoveries made during the excavations of *Harappa* and *Mohenjo Daro* are placing these cities as well as other settlements in the Indus Valley at circa 3000 BC (Alam et al. 2007; Dasgupta and Chattopadhyay 2004; Kenoyer 2019; Robbins Schug et al. 2013; Robinson 2016; Staubwasser et al. 2003; Wheeler 1968). But to this day the question remains why these IVC settlements were abandoned within what is generally considered a rather short time in the historical context.

Based on paleoclimate modelling researchers have suggested that weather extremes may have caused hydro-climatic stress, which could have resulted in negative

Fig. 5.3 The extent of the region Etruria and Etruscan territories inhabited by the Etruscan civilisation between 750 and 500 BC. *Source* Norman Einstein (2005) (Public Domain). Many of these areas are still popular hot spring destinations

impacts on the environment inhabited by the IVC and affected sustainable water management practices, probably toward the end of the third millennium BC (Robbins Schug et al. 2013). Several contributing factors are thought to have led to the termination of the IVC:

- The result of climate change, e.g. Holocene Monsoon change, possibly with devastating floods alternating with long seasons of drought
- Change of direction of the Indus river (e.g. ponding, tectonic uplift) leaving the land barren and infected with germs
- Inadequate maintenance of dams and irrigation channels as well as silt deposits at the river mouth
- Lowering of agricultural standards, e.g. unrestrained grazing leading to soil erosion and desertification
- Large scale deforestation resulting in uncontrolled salinity of irrigated areas
- Lack of planning and administrative disintegration
- People falling victim to diseases
- Increasing indifference towards maintenance, law and order (Alam et al. 2007; Dasgupta and Chattopadhyay 2004; Robbins Schug et al. 2013; Staubwasser et al. 2003; Wheeler 1968).

The disintegration of the IVC after 1900 BC could be therefore attributed to economic and social change, exacerbated by climate deterioration as well as by disruptions of regional interaction. This would have led to the vulnerability of the IVC, who failed to protect themselves against a changing climate and the effects of wide-ranging cultural exchange. Together with the potential of spreading mycobacterial diseases such as tuberculosis, leprosy, and other infectious diseases through interacting with other countries the impact on human health would have contributed to the downfall of the IVC (Robbins Schug et al. 2013).

Origin of Water Supply

All major cities and other important sites of the Indus Valley Civilisations (IVC) were in the vicinity of either the ancient course of the Indus river or its tributaries (Fig. 5.4). This alone however may not have been the principal reason for the establishment of some of these settlements. According to data collected at the excavated sites of *Harappa* and *Mohenjo Daro*, the foundations and other remnants of public and domestic buildings revealed advanced drainage systems connected to internal bathrooms and other conveniences that required an uninterrupted supply of fresh cold and hot water. It appears that many buildings had access to freshwater wells, fountains and cisterns, which points to the availability of a constant adequate water source for such sophisticated amenities (Erfurt 2012; Pakistan Water Gateway 2019). This raises the question about the origin of the water that was required for these complex facilities in a climate that largely relied on seasonal rainfall to replenish groundwater aquifers.

Extensive reviews of hydrogeological data and geological maps of Pakistan for this book have led to the theory that the confirmed presence of clusters of natural hot springs in the Indus Valley ought to be considered seriously in solving the mystery of the abandonment of settlements like Mohenjo Daro and Harappa (Erfurt-Cooper and Cooper 2009, 2010; Erfurt 2012). This theory lends support to the explanation for the existence of a substantial sanitation and plumbing infrastructure, which points to the dependence on reliable water sources that may be derived from several origins such as (a) the nearby river, (b) freshwater wells connected to underground aquifers, or (c) hot and cold springs. Natural springs would have allowed for the constant flow of fresh and unpolluted water required to keep pools and large baths clean and to avoid the growth of algae. The residents of the Indus Valley cities and other smaller settlements almost certainly would have valued the access to natural hot springs and would have used them like many other cultures for domestic purposes as well as for health and recreation, although possibly at a more advanced level (Erfurt 2012). Further supporting the theory of extensive hot spring use by the IVC is the fact that underneath the Indus Basin large unconfined aquifers, covering circa 15 million acres in surface area, are a known source for natural springs (Pakistan Water Gateway 2019). Underlying active hydrothermal systems, supported by high geothermal gradients, are a source for an abundance of natural hot springs to this day (Meltzer et al. 2001). In northern Pakistan for example, numerous hydrothermal manifestations exist, including boiling hot springs with maximum temperatures reported as reaching up to 98 °C (Amin et al. 2017; Meltzer et al. 2001).

Natural hot springs with a sufficient discharge rate could even solve the question surrounding the *Great Bath* at Mohenjo Daro (Fig. 5.5), a structure that archaeologists believe to be the ruins of either a public bath or a large cistern for water storage (Danino 2018; Kenoyer 2019). However, compared to the more recent (200–400 AD) stepped wells and stepped ponds of India (Livingston and Beach 2002), Mohenjo Daro's water basin seems quite small, shallow and ordinary from an architectural standpoint. Hence, the size and depth (12 m long × 7 m wide × 2.4 m deep) would point more towards a pool for either ritual bathing, recreational use and/or healing purposes (if supplied by hot springs). It is possible though that this water basin may have been a forerunner of the much later constructed stepwells. To a certain degree the *Great Bath* of Mohenjo Daro resembles the *Great Bath* of the Roman hot spring complex in Bath (England). This basin however is still filled with hot spring water and although being shallower, is nearly double the size (25 m × 10 m × 1.6 m).

Based on the hydrogeology of Pakistan major groups of currently active natural hot springs exist along the Indus Basin and throughout other areas (Bakht 2000). Their presence strongly indicates that hot springs existed throughout history, and therefore could have supplied Mohenjo Daro, Harappa and other Indus Valley settlements with thermal water for domestic purposes including bathing, washing and cooking. However, it is important to note that the prehistoric environment of the Indus Valley was frequently affected by seismic activity, which could be a logical and realistic explanation for the existence, as well as the vanishing of active hot springs throughout the region (compare Example 5.1). It is a fact that earthquakes of greater magnitude have the destructive power to reposition topographic features, including the location of riverbeds and natural springs. Even if the damage is only minor to moderate, this is often enough to trigger additional natural events such as landslides or mudslides that bury everything beneath them, including flowing hot and cold springs. Consequently,

Fig. 5.4 This map of Pakistan shows the location of two major settlements of the IVC, Mohenjo Daro and Harappa. The main water source of the Indus Valley is the river itself, although large aquifers, replenished by seasonal rainfall, are feeding into subterranean reservoirs and hydrothermal systems from where natural hot springs rise to the surface. *Source* Public Domain. Map modified by Author

seismic activity associated with hydrothermal systems may have caused some hot springs to change their location, dry up completely or turn saline.

Example 5.1 Great Aachen Earthquake in 1756, Germany

Repeatedly occurring earthquakes in the vicinity of Aachen's hot springs have in the past affected the flow rate and the spring temperature, although usually only lasting for a short time. Numerous writings reported that some smaller hot springs in a certain area of Aachen's hot spring clusters stopped flowing after the great earthquake in February 1756 (assumed magnitude approximately 6.1 or 6.4 on the Richter scale), and that the local government had to defend itself against rumours that the springs had completely dried up. However, while some of Aachen's hot springs ran dry, others increased in strength.

Source Leydecker (2011), Zitterland (1836).

Because natural catastrophes like floods and earthquakes are not uncommon in the Indus Valley, such events could have restricted the access to natural water sources either by redirecting rivers (Alam et al. 2007; Dasgupta and Chattopadhyay 2004; Wheeler 1968), submerging freshwater wells and/or by terminating the flow of hot springs that were reliable for centuries, and subsequently contributing to the abandonment of settlements (Erfurt-Cooper and Cooper 2009; Erfurt 2012).

Note Since 2011 eight (8) earthquakes at magnitudes between 6.1 and 7.7 have occurred in Pakistan (at the time of publication, 2021).

5.2.2 China

Throughout the Asian continent traditions of hot spring use developed at locations with a plentiful supply of thermal water. In China for example, natural hot springs are widely

Fig. 5.5 The excavated remnants of a structure commonly described as the 'Great Bath' of Mohenjo Daro. Compared to other hot spring sites where water continually flows through basins or pools to keep the water fresh and the temperature constant, the Great Bath has indeed the appearance of a facility for thermal bathing if the water was sourced from natural hot springs. *Source* Qayyum (2014) (Public Domain)

distributed throughout the country including offshore islands such as Hainan (PRC) and Taiwan (ROC) (Erfurt 2012). Speculation by historians and archaeologists estimates the geoheritage of China's hot springs is dating back several thousand years to circa 3000 or 4000 BC, which would place the people of China among the earliest potential users of hot springs in history and leaders in the development of hot spring bathing (Erfurt-Cooper and Cooper 2010; Erfurt 2012; Talmadge 2006; Wang 1999). According to some authors, (Clark 1999; Schafer 1956) written records may prove that for several thousand years China has been on the forefront in using this natural resource in their territories (Erfurt 2012; LaMoreaux and Tanner 2001; Wang et al. 2018).

One of the oldest developed hot spring sites in China are the famous *Huaqing Hot Springs* near the city of Xi'an, which are recorded as dating back to the Western Zhou Dynasty (1050–771 BC). Huaqing is known as a favoured destination of several Chinese emperors, their families and entourage, who are said to have spent long periods of time at Huaqing, even carrying out official government business from a palace built close to the hot springs. Following political uprisings and rebellions, Huaqing Palace lay abandoned for nearly a thousand years (Chen and Han 2019; Erfurt-Cooper and Cooper 2009, 2010; Erfurt 2012; Schafer 1956; Wang 1999; Wang et al. 2018), until several decades ago, in 1982, the remnants of five of the royal Huaqing hot spring pools were rediscovered, excavated and preserved (Fig. 5.6). Today the hot springs of Huaqing are protected as a 'Major Historical and Cultural Site at National Level' and have the status of a 'National Key Scenic Area' (Chen and Han 2019; China Holiday 2019; Erfurt 2012).

An early edition entitled 'Waterways Classic' (Shui Jing 水經), attributed to either Sang Qin (桑欽) from the Three Kingdom Period (220–280 AD) or Guo Pu (郭璞), a scholar from the Jin Dynasty (265–420 AD), is said to document not only China's rivers, but also recorded ancient hot springs and

Fig. 5.6 One of the rediscovered Huaqing hot spring pools. *Source* Alex Kwok (2006) (Public Domain). Image modified by Author

their locations. While the original book has been lost, the later compiled 'Commentary on the Water Classic', also translated as 'Annotation on Water Scripture' (Shui-ching Chu or Shui-jingzhu 水經注) by Li Daoyuan, a geographer from the Wei Dynasty (386–534 AD), added considerably to the initial work about China's hydrogeology by recording 41 hot spring locations, together with a definition of different types of hot springs (ChinaKnowledge.de 2019; Erfurt-Cooper and Cooper 2009; Erfurt 2012; Hao et al. 2011; LaMoreaux and Tanner 2001; Schafer 1956). In more recent history, in 1956, the geologist Zhang Hongzhao recorded 927 hot springs in his summary of the 'Distribution of Hot Springs in China' (Hao et al. 2011; Wang et al. 2018). Several decades later, in 1982 at the First National Geothermal Conference, over 2600 hot springs were reported in China (Hao et al. 2011).

5.2.3 Taiwan (ROC)

While natural hot springs are widely distributed throughout mainland China, islands such as Hainan and Taiwan also have numerous hot spring areas, which have long attracted local and foreign visitors. Taiwan's hot springs were first mentioned in 1697 AD in a document, the *Beihai Jiyou* (裨海紀遊). The actual development of hot spring areas such as *Beitou* in the north of Taiwan did not start until 200 years later around 1893 (Erfurt 2012; Schafer 1956; Fig. 5.7), when the springs were accidently discovered by a businessman from Germany and the first hot spring bathhouse in the district was built. A few years later, in 1895, Taiwan came under Japanese rule and was occupied until 1945. During this time, many hot springs in Taiwan were established by the Japanese as facilities for their use, with the unique *onsen* culture common in Japan greatly influencing the local hot spring development (Erfurt 2012). However, according to the Taipei Hot Springs Association, the potential of Taiwan's hot springs as a natural resource was not fully realised until after the Second World War was over.

Surrounded by a park, Beitou's original bathhouse today is a popular museum telling the story of the history of hot springs in Taiwan (Fig. 5.8; Appendix 5.3; Figs. 9.6 and 9.7). Apart from the geoheritage of the country's hot springs, Taiwan also has a history in mining sulphur, a geothermal by-product, with the earliest recorded sulphur extraction taking place circa 800 years ago during the *Yuan* Dynasty (Wang 1999).

Fig. 5.7 Hot spring pool in the old bathhouse in Beitou (Taiwan) at the turn of the previous century, now a museum. Photo of a poster in the museum

5.2.4 Korea

Early civilisations on the Korean peninsula also developed indigenous bathing traditions. Documented in 173 articles contained in old history books and annals is the traditional use of natural hot springs for health and recreation by the people of ancient Korea. According to these records the Korean hot spring heritage goes back to the early part of the Three Kingdom Period (57 BC–668 AD). Historical records further corroborate that until the Japanese occupation of Korea in 1910, hot springs were primarily reserved for the king, his family and the nobility, although 'common' people are said to have been able to use some hot springs for health reasons. This changed completely after the arrival of the Japanese, who developed Korea's hot springs into typical onsen style bathing facilities with access for the general population (Erfurt 2012; Yum 1999).

The history of the *Suanbo Hot Springs* for example suggests that they existed under King Hyongjong (Koryo Empire) in the year 1018 AD, when they attracted not just scholars, politicians and the aristocracy, but also members of the public to 'take the waters' (Erfurt 2012). Because access to hot springs is important, King Sonjong (reign 1083–1094 AD) ordered an exploration to search for hot springs closer to the capital. In return for the discovery of suitable hot springs, individual citizens received social and monetary benefits, as well as exemptions from labour and other duties. The actual management of hot spring facilities and their maintenance was controlled by the governors of individual territories, who oversaw the legislative side and who documented the use of the springs (Erfurt 2012; Yum 1999). According to Korean history, another ancient hot spring site, *Onyang* (Chungcheong Province), has been in use for over 600 years and was frequently visited by King Sejong the Great (born 1397 AD) and the royal family for health and recreation during the early Joseon period (1392–1910 AD) (Garcia 2007; Hann 1996). And while Korean hot springs have been used for centuries, they were only more recently developed into advanced hot spring resorts and spas to attract international visitors (Erfurt 2012).

5.2.5 Japan

Japan's traditional hot spring or onsen bathing possibly found its way from China through Korea, although there is no definite evidence to confirm this, as historical records only go back to the time after the Chinese and Koreans

Fig. 5.8 Beitou's original bathhouse is now a hot spring museum where visitors can learn about several centuries of hot spring use in Taiwan (see also Appendix 3, Images 3, 4; Figs. 9.6 and 9.7)

started trading with Japan (Clark 1999). Supported by an abundance of volcanic hot springs, Japan gradually developed a unique bathing culture, largely based on recognising and appreciating the curative powers of the mineral-rich hot springs (Erfurt 2012).

While many hot springs were used at least for several centuries, there are signs indicating that this custom may go back much further. For example, bathing in natural hot springs might have been practiced at *Suwa-ko* (Lake Suwa, Nagano prefecture) as early as 6000 years ago, which appears to be established by excavated ruins of ancient settlements (Sekioka and Yoshi 2000). Based on the archaeological evidence, it therefore may be safely assumed that early humans in Japan settled around hot springs, seeking their warmth for bathing and cooking and to protect against the cold. The actual development of hot springs dates back at least to the Heian period (794–1185 AD), but as mentioned above could be much older, considering that the chronicled 'History of the Kingdom of Wei' refers to the practice of ritual bathing towards the end of the Yayoi period circa 297 AD (Clark 1999). Historical reports further reveal that Japanese feudal lords and noble warriors (daimyō, shoguns, samurai) enjoyed bathing in natural hot springs to recover after battle and heal their injuries (Hotta and Ishiguro 1986; Talmadge 2006). It was also not uncommon that rich and powerful people had hot water from popular springs delivered to their homes to be able to enjoy the mineral-rich waters on demand (Erfurt 2012).

Later, during the *Edo* Period (1603–1867 AD), the custom of hot spring bathing became more prevalent all over Japan, with access to natural hot springs reaching all levels of the population (Clark 1999; Erfurt 2012), including hydrothermal areas with their fascinating geysers and mud pools (Appendix 9.1). Also, there was no segregation between the genders and mixed bathing remained common until the year 1853, when a ship from America arrived and the captain and his crew discovered men and women together in the hot springs. This appeared to have shocked the Americans, who indignantly started to push for the segregation of genders, although it is not clear how and why they made it their business to interfere in local customs. However, over the following decades the separation of genders was gradually established and by 1870 mixed bathing was finally banned by law (Hotta and Ishiguro 1986). Ever since, apart from remote and rural hot springs or private baths, public onsen and their related *sento* and *furo* have largely remained separated, usually with individual entrances (Fig. 5.9).

Fig. 5.9 One of the countless traditional community onsen in Japan, which are typically separated by gender and often free of charge

To this day traditional onsen bathing is maintained by most of the Japanese population to maintain good health and as an informal way to relax as well as an opportunity to socialise with family, friends or colleagues (Erfurt 2012).

5.2.6 India

The ancient Hindu system of medical science *Ayurveda* describes water as the main element to cure many illnesses. In the third century BC the Greek ambassador *Megasthenes* visited India and in his travel journals he described hot mineral springs as *rarefied water* to emphasise the remarkable powers of the water (Chandrasekharam 1999; Erfurt 2012). To find reliable data associated with the historical use of Indian hot springs is rather challenging, although in many areas natural hot mineral springs are thought to be gifts from the gods and 'overflow' with divine powers. This was also confirmed in reports from early pilgrims, who described temples that were located close to hot springs (Chandrasekharam 1999), which parallels the frequent proximity of religious buildings and natural hot springs in other countries such as Greece, Germany, Japan and the UK, where hot springs were dedicated to local deities. With more than three hundred natural hot springs in India, the majority of the currently available literature seems to focus more on technical subjects such as generating geothermal energy, than on the historical use of hot springs for health and recreation (Chandrasekharam 1999; Chaterji and Guha 1968; Erfurt 2012).

5.3 European Origins of Hot Spring Use

5.3.1 The Greek Hot Spring Heritage

The practice of thermal bathing in natural hot springs for healing purposes in Greece was recorded by several ancient historians including Aristotle, Herodotus, Hippocrates, Plato, Plutarch and Strabo (Tables 5.1 and 5.2). Grounded in

Table 5.2 A timeline of early hot spring use in Greece shows the connections with Roman and Ottoman empires

Chronology of the Greek hot spring culture	
7000–3100 BC	Neolithic Greece—presumed ancient water worship and possible hot spring use in areas with hydrothermal resources
1600–1100 BC	Minoan Greece—Bronze Age—Greeks are settling in some parts of Turkey (1200 BC)
1100—46 BC	The era of Ancient Greece
500 BC	The Greek historian *Herodotus* describes hydrothermal manifestations Thermal baths are built near hot springs
Between 427 and 347 BC	The benefits of hot spring bathing are mentioned by *Plato*
Between 490 and 377 BC	*Hippocrates of Kos* is treating medical conditions with balneotherapy at his Aesclepian Centre on the Greek island Kos
Fourth century BC Between 396 and 371 BC	Hot springs and their healing properties encountered on his travels are mentioned by *Xenophon*
Fourth century BC	The ancient city of *Thermae* (now Loutraki) is reported as one of the first spa towns with a hot spring health resort
Between 384 and 322 BC	*Aristotle* refers to natural hot springs in *Aedipsos* (Edipsos, Edipsus)
Between 350 and 290 BC	*Megastenes* writes about hot springs in India and mentions their medicinal value
328 and 168 BC	Roman conquest and colonisation of parts of Greece—Romans are discovering and using hot springs in Greece
146 BC and 324 AD	Roman-Greece era following the Roman conquest
Between 46 and 119 AD	*Ploutarchos* (Plutarch, Greek biographer) describes the hot springs of Aedipsos
First century AD	Aedipsos is becoming a hot spring destination for health and recreation
324–1453 AD	Byzantine-Greece until the fall of Constantinople (Turkey)—hot spring use remains a part of life in hydrothermal areas
1453–1821	Ottoman-Greece—various hot spring cultures have been developed by now
1821–current	Modern Greece—natural hot springs continue to be used for health and recreation

this knowledge, modern scientists assume that hot springs in Greece were used for health and recreation long before the first written records and may have started in many locations in prehistory and long before the later partial occupation by the Romans.

Because of their reputation as health giving, hot springs in the ancient Greek mythology were revered as gifts from the gods of Olympus, and temples were built around the springs, which were dedicated to worship and healing (Chap. 6). In fact, archaeological evidence of human settlements close to hot springs in Greece confirms the early use of this natural resource, although obviously not the mythological connection. Because of their healing properties, natural hot springs were often dedicated to the god of medicine Asclepios (Fig. 5.1), the goddess of health Hygieia, and Heracles, the son of Zeus. In keeping with the Greek mythology, the springs were under the protection of nymphs (water spirits of divine origin), who were supposed to preserve the healing properties of the springs and safeguard them from human contamination (Fytikas et al. 1999).

During the fifth century BC, Hippocrates treated patients from the Mediterranean region at his famous Asclepian Centre on the Greek island of *Kos* with balneotherapy (Cataldi et al. 1999; Erfurt-Cooper and Cooper 2009; Erfurt 2012; Monti 2003). Besides practicing medicine, Hippocrates wrote an encyclopaedia covering not only all aspects of medical developments at the time, but he also included a classification of natural hot springs and their known characteristics (Erfurt 2012; Fytikas, Leonidopoulou and Cataldi 1999; Katsambas and Antoniou 1996). Based on the writings of Ploutarchos (Plutarch), hot spring-based health resorts and spas were already visited during the first two centuries AD by people with health conditions and others who pursued recreational goals. One of the destinations for health and recreation at the time was *Aedipsos*, which offered several facilities for health and wellness treatments, and as a result Aedipsos became a popular holiday retreat (Erfurt 2012; Katsambas and Antoniou 1996).

Another ancient Greek city, famous for its natural hot springs, was *Thermae* (Θερμαί, Greek for hot springs), which is considered as one of the first historic spa towns offering health and recreation facilities to travellers. The name Thermae further relates to the goddess Artemis Thermia, who was responsible for the protection of natural hot springs. The town Thermae was first mentioned by Xenophon, a Greek historian, in his account of the Corinthian War

(395–387 BC) (Erfurt-Cooper and Cooper 2009; Erfurt 2012). During excavations it was discovered that later the Romans had built magnificent thermal bathing complexes for their personal health and recreation. Today, Thermae is known as *Loutraki*, a centre for hot spring tourism with health resorts and spas attracting visitors to the healing waters (Erfurt-Cooper and Cooper 2009; Erfurt 2012). Greece has countless other historical hot spring sites, where healing waters that were considered sacred in ancient Greece, are still flowing and have been used for health and recreation for many centuries. To this day, spa towns are flourishing as a result of their access to natural hot springs and continue to be a key attraction for health and wellness tourists.

5.3.2 The Turkish Hot Spring Heritage

The region of Anatolia has a long tradition of hot spring use for health and recreation and, similar to other countries, most likely dates back to a prehistoric era between 7000 and 2000 BC. Early civilisations who inhabited Anatolia include the Hattians (2500–2000 BC), the old Hittite Kingdom (1750–1450 BC) as well as the Hittite Empire (1450–1200 BC) (Table 5.3). The consensus is that Hittite societies already used hot springs for health and recreation, which can be related to the fact that Anatolia has more than 700 natural hot spring areas (Erfurt-Cooper and Cooper 2009; Erfurt 2012; Özgüler and Kasap 1999).

Perhaps the best-known hot spring destination in Turkey with a rich natural and cultural geoheritage is the ancient city of Hierapolis, high above the town of Pamukkale. During the second century BC (circa 190 BC) Hierapolis was established by Eumenes II, King of Pergamon, and by the end of the second century BC the first thermal spa for health and recreation was established (UNESCO 2019). People were attracted to the health-giving mineral-rich hot springs in the hope to heal their diseases, while at the same time they were seeking the assistance of the on-site priests to advance their spiritual wellbeing. An important place within the city of Hierapolis was the *Plutonium* (Fig. 5.10), a site interpreted by Pliny (the Elder) and Strabo as a gate to the underworld (Pluto's Gate to hell) and which was used to worship the god Pluto (Özgüler and Kasap 1999; Pfanz et al. 2018). Below the Plutonium, the water of a natural hot spring is flowing in channels underground and to this day the Plutonium is known for its toxic gas (CO_2) emissions. In ancient times the lethal effect of the deadly fumes was resourcefully, although rather cunningly used by priests on large and small animals to demonstrate their own invincibility, when supposedly in contact with the poisonous vapours. Using various ruses and deceptive actions to avoid inhalation of the gas, they proclaimed their superiority as miracles, which served to elevate their status in the community (Erfurt 2012). According to modern science, the mystery of the Plutonium is based on seismic activity, which enables degassing from the Earth's mantle. To this day, deadly concentrations of CO_2 can be measured at the point of emission from the subterranean cave that contains the hot spring (Pfanz et al. 2018).

After being war-ravaged and destroyed several times by earthquakes the city of Hierapolis was finally abandoned in 1354 AD. The remaining ruins include large Roman bath facilities from the second century AD (Fig. 5.11), which are part of an expansive open-air museum directly above the 200 m high white travertine terraces Pamukkale is famous for.

Amidst the ancient ruins of Hierapolis another popular historical centrepiece is the so-called 'sacred pool', also referred to as 'Cleopatra's Pool'. This thermal pool is fed by hot springs and is littered with submerged fragments of ancient marble columns, some of them partly exposed (Fig. 5.12). Below the travertine cliffs the township of Pamukkale serves as a popular destination for health and wellness tourism specialised in balneotherapy using local hot springs (Erfurt 2012; Özgüler and Kasap 1999).

In Turkey's northwest, *Bursa* was once a capital city of great importance in the heart of the former Ottoman Empire. Numerous natural hot springs in this area were used for centuries and documented at least since the Eastern Roman Empire (Byzantine Empire 330 AD–1453 AD) (Imbach 1993). Records of the earliest hot spring tourists include the Byzantine Empress Theodora (500 AD–548 AD), who visited Bursa to 'take the waters'. For many centuries thereafter Bursa's health-giving hot spring waters have been used for their therapeutic value (Anatolia.com 2019). During the Ottoman Empire under Murat I. (1326–1389 AD) a large complex of domed baths was built to provide facilities for health and recreation. Today the historic architecture of Bursa, together with the natural hot springs, remain some of Turkey's most popular tourist attractions.

Another historic hot spring spa town is *Balçova* (Province of Izmir), which is located on the coast 368 km southwest from Bursa. According to ancient legends as well as being referred to in Homer's and Strabo's writings, the local hot springs were used as far back as 1200 BC by the legendary king *Agamemnon* and his soldiers, as they were following the advice of a prophet, who told the king about the waters' healing powers. This seems plausible when compared with the Roman and Japanese military's use of hot springs to heal their warriors' injuries, as well as providing some form of recreation for battle-scarred troops. Over time, bathhouses and pools were built in Balçova and the *Agamemnon Thermal Springs* have been used for health and recreational purposes ever since. Today the thermal facilities of Balçova

Table 5.3 The timeline of the earliest users of hot springs in Anatolia shows the connection to the Roman bathhouse culture

Chronology of the Turkish hot spring culture	
7000–2000 BC	Prehistoric period—the people of Anatolia most likely have used geothermal resources including hot springs—no written records
2500–2000 BC	The Hattian people are inhabiting central Anatolia
1750–1450 BC	Old Hittite Kingdom
1680–1193 BC	Hittite Empire in the region of Anatolia
1200 BC	Greeks settle in Turkey
Second century BC	Kings of Pergamon founded Hierapolis and used the local hot springs
129 BC	Hierapolis is administered by Roman proconsuls
300 BC to 395 AD	Roman era in Anatolia
Second century BC	Roman bathhouses are established at Hierapolis (Fig. 5.11) using local hot springs
Fourth century AD	Hierapolis is now an important religious centre—the Apostle Phillip died there
359 AD	Hierapolis comes under the rule of the Eastern Roman Empire
1299–1922 AD	Ottoman Empire
1354 AD	Hierapolis is abandoned after a major earthquake
1359–1389 AD	The Ottomans are building large thermal baths at Bursa
Early nineteenth century	Kangal hot springs first discovered
1988	Hierapolis/Pamukkale listed as UNESCO World Heritage Site

Fig. 5.10 The Plutonium at Hierapolis. Today the entrance to the plutonium is bricked up for safety reasons with only a small opening through which the rushing of the water below can be heard

Fig. 5.11 According to historical reports there were two Roman thermae in Hierapolis. The remains of the second Roman bathhouse, the Basilika Baths, are located closer to the travertine terraces and some of the hot springs. The majority of the buildings were constructed from locally quarried bedded travertine. The text inserts are copies of the original signboards at the Basilica Baths

provide high quality services to a flourishing health tourism industry (Visit İzmir 2019).

In more recent times, the *Kangal hot springs* (Central Anatolia, Sivas Province) were first observed in the early nineteenth century by a local shepherd, who had injured his foot but noticed that the wound healed quickly after bathing in the warm spring waters. As a result of the 'miraculous' healing, people started to become interested in the hot spring water and established pools and built other basic facilities (Erfurt-Cooper and Cooper 2009; Erfurt 2012). The springs of Kangal are said to be unique for another reason as well; small fish are inhabiting the hot springs and are said to bring remarkable relief for people suffering from the skin disorder psoriasis. Research carried out over time by dermatologists on patients with psoriasis appears to confirm the success of this unique hot spring therapy (Erfurt-Cooper and Cooper 2009; Erfurt 2012; Kangal Fish Balikli Kaplica Treatment Center 2019; Kangal Balikli Kaplicasi 2019; Özçelik et al. 2015; see Chap. 8, Example 8.5; Figs. 8.12 and 10.3). Many of Turkey's hot springs that were used since time immemorial are still destinations for therapeutic treatments today. Thermal bathing is an important Turkish tradition with several public baths [hammams] in most modern cities in Turkey, either using natural hot water or water that is artificially heated (Erfurt 2012; Özgüler and Kasap 1999).

5.3.3 The Roman Hot Spring Heritage

> *Baths, wine and sex corrupt our bodies, but baths, wine and sex make life worth living (Roman inscription from the first or second century BC, Corpus Inscriptionum Latinarum, 6.31241)*[1]

The Romans obtained their knowledge about hydrothermal manifestations from the Etruscans, who are believed to have used hot springs for thermal bathing dating back to the eighth century BC (Cataldi and Burgassi 1999). The Romans also learned about the benefits of this natural resource from the Greeks during the Macedonian Wars and from other nations they invaded during their conquering days, which verifies that they did not invent thermal bathing as is commonly understood (Erfurt-Cooper and Cooper 2009, 2010;

[1]The *Corpus Inscriptionum Latinarum* (CIL) is a compilation of ancient Latin inscriptions from the Roman Empire and occupied territories. From personal to public, the inscriptions highlight many different aspects of Roman life throughout the historical duration of their Empire.

Fig. 5.12 The 'Sacred Pool' at Hierapolis is also known as 'Cleopatra's Pool', who is said to have visited this site. The pool contains a submerged section of a Roman street with columns that toppled over during earthquakes

Table 5.4 Brief overview of the Roman hot spring history

Rise and decline of the Roman hot spring culture	
753 BC	Rome is founded by Romulus and Remus—unconfirmed legend
Fifth century BC	Romans are gradually becoming aware of hot spring use in Etruscan territories
437 BC–237 BC	Roman wars with Etruria—adoption of hot spring bathing from the Etruscans
Second century BC	Systematic development of a thermal bathing culture throughout the Roman Republic
First century BC 27 BC	The Roman Empire supersedes the Roman Republic—colonisation of many other countries—hot spring use becomes more widespread
First century BC–third century AD	Peak hot spring use—the Romans are building thermal bathhouses, usually together with temples dedicated to deities such as Minerva
Fourth century AD	Hot spring use is starting to decline in western Europe during the second half of the fourth century
c. 476 or 480 AD	Fall of the Western Roman Empire
Fifth–sixth century AD	Hot spring use is rapidly declining followed by inactivity until the end of the first millennium
Nineteenth century–current	Redevelopment of Roman style bathhouse architecture throughout Europe in countries/towns with a history of Roman occupation

Erfurt 2012; Melillo 1995). However, being aware of the advantages of establishing military settlements near hot springs, the Romans became very focused on the surrounding environment for signs of hydrothermal activity, and from the second century BC the use of hot springs progressively took hold in the Roman Empire (Table 5.4). In fact, the Roman's preference for natural hot springs generated a demand by the wealthy elite for authentic hot spring water to be delivered to their homes, comparable to the Japanese nobility and most likely other countries' upper classes as well (Erfurt-Cooper and Cooper 2009; Erfurt 2012; Talmadge 2006).

Thermal bathing and balneotherapy for medical purposes were increasingly developed by the Romans with hot spring use spreading across the Roman Empire (Cataldi and Burgassi 1999). Military settlements were often determined by the availability of natural hot spring water; for example, Aachen (*Aquae Granni*) and Wiesbaden (*Aquae Mattiacorum*, Example 5.1) in Germany or Bath (*Aquae Sulis*) in England (Erfurt-Cooper and Cooper 2009; Erfurt 2012).

Example 5.2 Wiesbaden, Germany

When the Romans arrived in the area of today's spa town Wiesbaden, they soon discovered the local (26) natural hot springs and utilised the healing benefits of the mineral-rich waters for health and recreation. Before the Romans arrived, these hot springs had already been used as a place for healing and leisure by Celtic tribes. Wiesbaden's Roman settlements are mentioned for the first time in 121 AD as *Aquae Mattiacorum*, the Latin name for 'Waters of the Mattiaci', a Germanic or Celtic tribe, who lived near the hot springs prior to the Romans' arrival.

Source Erfurt (2012), Selke (1936).

Due to their fondness of thermal bathing and relaxing in their sprawling bathhouse establishments, the Romans also artificially heated the water when they had no access to natural hot springs (e.g. Cilurnum (Chester), UK; Augusta Treverorum (Trier), Germany). From a long list of historic Roman settlements that developed into spa towns (Table 5.5) it becomes obvious that the Romans definitely preferred to set up their headquarters around natural hot springs. Initially all thermal bathing facilities were reserved for members of the Roman army in occupied countries, where they provided a place for rest and recreation as well as medical treatments to assist in the healing of injuries sustained during battle (Erfurt-Cooper and Cooper 2009; Erfurt 2012). During this time the Romans designed their thermal bathing facilities specifically for their requirements, although later they made their facilities available to the civilian population as well. At the end of the Roman occupation of dozens of countries all over Europe, North Africa and the Middle East (Fig. 5.13) the retreating legions of the Roman Empire left behind a sophisticated thermal spa culture and an impressive bathhouse architecture (Fig. 7.18; Appendices 5.4 and 6.2). Classic examples of abandoned Roman thermae include spa towns from Spain to Iraq and from Egypt to England.

A significant number of Roman thermal baths from the past remain in use today throughout Europe; often after undergoing extensive renovations over recent times to preserve the historical significance and the unique architectural legacy of this ancient bathhouse culture. The special quality of these facilities is still a main feature in spa towns in countries including Bulgaria (Kyustendil), the Czech Republic (Karlovy Vary and Teplice), France (Aix les Bains and Rennes les Bains), Germany (Aachen and Baden Baden), Hungary (Budapest), Italy (Bagno di Romagna), Portugal (Chaves), Spain (Archena) and of course in the United Kingdom (Bath) (Erfurt-Cooper and Cooper 2009; Erfurt 2012; Table 5.5).

The rise of the culture of thermal bathing in the Western Roman Empire was followed several centuries later by the inescapable decline due to the fall of the Roman Empire. The rather sophisticated traditions introduced by the Romans were soon forgotten as the leadership changed, the clergy insisted on different moral viewpoints, and hygiene was no longer considered a priority. In the Eastern Roman Empire, the use of hot springs remained unchanged over a considerably longer period of time, although eventually other successor states such as the Turks took possession of what the Romans had left behind (Erfurt 2012). Remnants of thermal bathhouse architecture (Fig. 5.14) throughout the former territories of the ancient Roman Empire, and in many cases their restoration and current use as a tourist attraction are reminders of the powerful Roman influence dating back more than 2000 years ago.

In **Switzerland**, the spa town of *Baden* has been a well-known destination for health and recreation going back to the Roman occupation of the country. To distinguish the Swiss Baden from other spa towns, the Romans named it *Aquae Helveticae* (Erfurt 2012). In neighbouring *Austria*, the discovery of the hot springs of Baden (*Aquae*) is understood to go back as far as 58 BC, when the Romans, aware of the healing properties of natural springs, built an impressive thermal spa centre. Baden was primarily set up as a hot spring bathing facility for the Romans, who built altars near the springs and dedicated them to various divinities. The sulphur springs were credited with healing powers and visitors showed their appreciation for successful treatments by worshipping individual gods and goddesses, as well as leaving donations or votive offerings (Erfurt 2012; Lüscher 1946). As a result, the extent of the Roman bathhouse culture and architecture constitutes a significant contribution to the geoheritage of natural hot springs.

5.3.4 Bath, England—Epitome of a Roman Spa Town

The Romans had no natural explanation for hot water pouring out off the ground; to them the Spring was the work of the gods. A sacred place.
(Poster in the Hot Spring Museum in Bath, UK)

The city of Bath in the south of England owes its foundation to the local hot springs, which seem to have sustained

Table 5.5 Historic Roman spa towns in Europe and North Africa, which are still popular hot spring destinations for health and recreation

Country	Current names	Latin names
Austria	Baden (near Vienna)	*Aquae Cetiae, Aquae Pannonicae*
Algeria	Hammam Guergour Hammam R'Irha Hammam Essalihine Hammam Meskhoutine	*Ad Sava Municipium* *Aquae Calidae* *Aquae Flavianae* *Aquae Thiblitanae*
Belgium	Chaudfontaine	*De Calida Fontana*
Bosnia and Herzegovina	Ilidža	*Aquae Sulphurae*
Bulgaria	Kyustendil Thermopolis or Burgas	*Pautalia* *Aquae Calidae*
Croatia	Varazdinske Toplice	*Aquae Iasae, Aquae Viva*
England	Bath Buxton	*Aquae Sulis* *Aquae Arnemetiae*
France	Aix en Provence Aix les Bains Bagnères-de-Luchon Chaudes-Aigues Dax La Chaud Néris-les-Bains Sceaux-du-Gâtinais Vichy	*Aquae Sextiae Salluviorum* *Aquae Gratianae, Aquae Allobrogum* *Ilixon* *Calentes Aquae, Calentes Baioe* *Aquae Tarbellicae* *Calidum* *Aquae Neri* *Aquae Segetae* *Aquis Calidis, Aquae Solis*
Germany	Aachen (Aix-la-Chapelle) Baden Baden Badenweiler Wiesbaden	*Aquae Grannii, Aquis Granum, Aquis Villa* *Aurelia Aquensis, Aquae Aurelia* *Aquae Villae* *Aquae Matticae*
Greece	Loutra, Edipsos Salonika	*Aidepsos* *Thessalonica*
Hungary	Budapest	*Aquincum*
Italy	Abano Terme[a] Acqui Terme Bagno di Tivoli Campiglia Marittima Cutilia Santa Cesarea Terme Saturnia Termini Imerese	*Aquae Patavinae, Aponi Fons* *Aquae Statiellae* *Aquae Albule* *Aquae Populoniae* *Aquae Cutiliae* *Litus Leuternum* *Aurinia* *Thermai Himerai, Thermae Himerenses*
Morocco	Moulay Yacoub	*Aquae Dacicae*
Portugal	Chaves	*Aquae Flaviae*
Romania	Băile Felix Băile Herculane	*Termae Varadiensis* *Aqua Herculis, Ad Mediam*
Spain	Alange (Spa) Caldes de Malavella Caldes de Montbui Villa Fasila	*Augusta Emerita* *Aquae Voconis* *Aquis Calidae* *Aquarius Vicus*
Switzerland	Baden	*Aquae Helveticae*
Tunisia	Henchir el Hammam Korbous	*Aquae Septimianae* *Aquae Calidae Carpitanae*

Some locations had more than one Latin name
Source Compiled by Author
[a]The name *Abano* is linked to Aponus, the god associated with thermal springs and healing virtues, as well as to the Greek *à ponos* (αφαιρώντας τον πόνο), which means confronting and removing pain

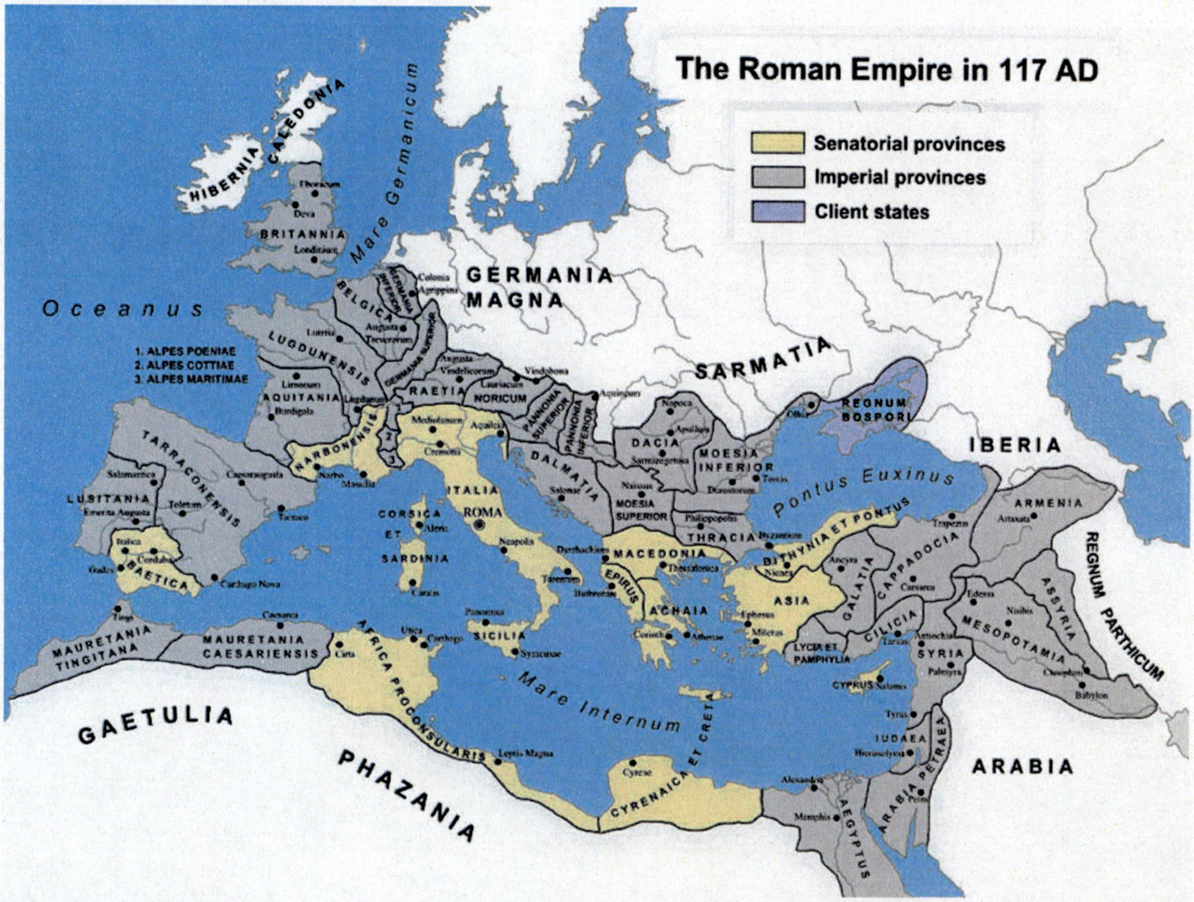

Fig. 5.13 The extent of the Roman Empire during the reign of Emperor Trajan in 117 AD. *Source* Lee (2014) (Public Domain). Map modified by Author

their large flow rate (more than 1.4 million litres per day) for thousands of years. Bath has a long tradition of hot spring use, and since ancient times has been an important centre of religion, healing and pilgrimage (Bowman 1998; Haley et al. 2005). This was based on the belief held by native Britons that water was a gift from the gods, which included lakes, rivers and streams, as well as natural hot and cold springs (Ottaway and Cyprien 1987). The original site of the hot springs of Bath was already considered a sacred site by the Celts long before the Roman invasion (Erfurt 2012; White 2000, 2003, 2006; Table 5.6). However, Bath's history is inextricably linked to the Romans and their occupation, based on existing records stating that during their time in the region the local hot springs were developed to provide facilities for rest and recreation for battle-weary Roman legions. This included the establishment of a sophisticated bathing complex (Appendix 5.4) as well as temples by the year 70 AD (Erfurt-Cooper and Cooper 2010; Erfurt 2012; White 2000, 2003, 2006).

Based on local legends, the original founding of Bath goes back much further to circa 860 BC, when Prince Bladud, who suffered from a skin condition thought to have been leprosy, was banned from his father's court. The prince retreated into the swampy marshes where today the city of Bath is located. To pass the time, he took to herding swine and watched them wallow in the warm swamp waters. The legend claims that Bladud followed the animals into the water repeatedly and noticed that his skin condition improved to the point that he was completely cured, whereupon he returned to the court and later became king and founded the city of Bath (Bowman 1998; Bullard 2004; Erfurt-Cooper and Cooper 2010; Green 2004; Havins 1976; White 2000, 2003, 2006). As this remains a legend, it is not clear beyond any doubt, when and by whom the hot springs of Bath were originally used (Lambert 2019).

Between the narrative of King Bladud and the later Roman occupation of the Bath area, the Celts, people of the Iron Age whose traces are found all over Europe, are said to

Fig. 5.14 This large marble basin was used for hot baths in the caldarium at the Stabian Baths in Pompeii. The floor is covered with mosaic tiles with two marble steps leading up to the bath

have known about the hot springs and dedicated the main spring to their goddess *Sulis*. From the second century BC the Celts' presence dominated the region until Britain was conquered by the Romans early in the first century AD. Soon after their arrival, the Romans discovered the potential of the natural hot springs and started building increasingly sophisticated thermal bathhouses (Fig. 5.15; Appendix 5.4) based on their customary *thermae* architecture (Erfurt 2012; White, 2000, 2003, 2006). Between 60 and 70 AD the Romans progressively developed the settlements around the hot springs into a municipality and called it *Aquae Sulis*, although they dedicated the springs to their own goddess *Minerva*.

The gradually decreasing influence of the Roman empire caused a decline of the thermal bathing culture, and by the time the Roman occupation of England ended in the fifth century AD (they left around 410 AD), the thermal facilities became neglected and fell into disrepair. Nevertheless, based on the reliable chronology of the Romans, their hot spring culture remains closely connected to the geoheritage of Bath. Following the Romans, towards the end of the sixth century AD, the Anglo-Saxons took over but failed to maintain the highly developed Roman culture of thermal bathing. This is mainly attributed to the suspicion that the Saxons were not interested in hygienic activities, possibly due to their dislike and distrust of anything related to the Romans (Erfurt 2012; Havins 1976). They named the area *Hat Batha* (Hot Baths), which led to the current name Bath. Several centuries of decline and neglect followed the Anglo-Saxon rule, until the Normans conquered Britain in 1066 AD, and a newly appointed Bishop of Bath ordered three large new baths to be built. They are still known as the King's Bath, the Hot Bath and the Cross Bath (Fig. 5.16). During this time, Bath gained a reputation for its healing waters throughout Britain.

Under the reign of Henry IV (1399–1413 AD) the city of Bath, including the thermal bathing facilities, again fell into

Table 5.6 The history of Bath's hot spring facilities shows the phases of their development and their decline alternating over time

Bath, the history of a Roman Spa Town	
860 BC	According to a legend Bath was founded by King Bladud after he was cured of leprosy bathing in the hot springs
Second century BC	The Celts, considered people of the Iron Age, dedicated the hot springs to their goddess Sulis
Early first century	Romans becoming aware of the hot springs and start using them for the military
First century (43 AD)	Romans conquering England and start building thermal baths and call their new military settlement *Aquae Sulis* (Fig. 5.15)
c. 60 to 70 AD	The Romans further develop the hot springs into a bathing complex, build temples dedicated to their goddess Minerva and a spa town starts to develop
Fifth century AD (410 AD)	End of the Roman occupation of Britain, Roman legions leave the thermal facilities of Bath deserted after retreating and the facilities subsequently fall into disrepair
Sixth century AD (577 AD)	Anglo-Saxons conquer the area and name it *Hat Batha* (Hot Baths)
—	**Several centuries of decline**
1066 AD	Conquered by the Normans—under the appointed Bishop of Bath three large new baths are built—known to this day as the King's Bath, Cross Bath and Hot Bath, and gaining a reputation for healing throughout England
1399 to 1413 AD	During the reign of Henry IV (1399–1413) the city of Bath falls again into decline, including the thermal bath facilities
1558–1603 AD	Short revival of Bath as a spa town
1574 AD	Following a visit by Queen Elizabeth I the popularity of the medicinal waters increased. By then the city of Bath maintained five large baths supplied with hot water sourced from the local springs
1661 AD	Spring water from Bath is bottled and sold
1706 AD	The first Pump Room is built but is replaced later in 1795
1738 AD	The Royal Mineral Water Hospital is founded in Bath (Fig. 8.14)
Eighteenth century	Redevelopment of Bath's thermal spas after decline (Appendix 5.5)
Nineteenth century	Bath is surpassed by other European spa towns in its importance as a health resort, and its heyday is over
1889 AD	The King's and Queen's Baths are restored and rebuilt, and the old Roman bath, which was largely preserved, is discovered during construction (Appendix 5.4)
1978	The thermal baths are closed down due to public health concerns over the water quality of the hot springs
1987	City of Bath becomes inscribed on the UNESCO World Heritage list
1997	Launch of the Bath Spa Project to restore Bath once again as a leading spa town
2006	The new Thermae Bath Spa opens
2011	The original Roman Baths undergo a £5.5 million redevelopment to improve accessibility and aid its preservation for the next hundred years

Source Various

decline until the Elizabethan era (1558–1603 AD), when Bath experienced a short revival as a spa town in conjunction with improvements of the thermal pools. Additional renewals of the urban infrastructure began to attract members of the aristocracy who spent the once so fashionable 'Saison' in Bath. More ups and downs ensued until the eighteenth century, when the hot spring baths were redeveloped and once again attracted visitors to the city of Bath for health and recreation (Appendix 5.5). At the beginning of the nineteenth century Bath's standing as a health resort was finally surpassed by other European spa towns and its prime time was over until 1889 AD, when the King's and Queen's Baths were restored and rebuilt. In the year 1978 the public hot spring pools were eventually closed and bathing was prohibited over health concerns about the water quality due to a suspected contamination of the hot springs.

Nearly twenty years later, in 1997, the 'Bath Spa Project' was launched to restore Bath once again as a leading spa town with the new *Thermae Bath Spa* (Fig. 5.17) finally opening for public use in 2006 after the hot spring facilities

Fig. 5.15 An impression of the first Roman hot spring bath under construction at the location of today's city of Bath in the south of England. *Source* Drawing by Ivan Lapper (with permission)

remained closed for 28 years. To achieve this mammoth task, new wells were drilled beneath the King's Spring and the Cross Spring to access unpolluted spring water for the redevelopment (Erfurt 2012; Thermae Bath Spa 2019; White 2000, 2003, 2006). The new Thermae Bath Spa is only a short distance from the original Roman Baths, which have used hot spring water from the same underground reservoir for nearly two thousand years. The 'old' Roman Baths (Fig. 5.18) are now a museum, maintained by the Bath & North East Somerset Council (2019) and their Heritage Services, and are a popular attraction for visitors.

The Roman history of Bath is significant and has influenced and shaped this ancient spa town for close to 2000 years. When Bath became a World Heritage site in 1987 this recognition of the significant geoheritage set the direction for the future of the city and surrounding areas as an important and unique tourist attraction. Walking through the centre of the old city the sense of the past is intense and the impressions can be overwhelming to a perceptive observer. It takes time and more than one visit to get a glimpse of the fascinating history of Bath. Many other spa towns with Roman heritage in Europe, north Africa and the Middle East have equally interesting stories to tell—more than enough material for another book.

5.4 The Hot Spring Heritage of Western Europe

Europe has an exceedingly rich hot spring heritage with far too many traditional hot spring spa towns to include in this chapter. However, some locations and areas have been

Fig. 5.16 The Cross Bath in the early 1800s (left). The restored building (right) is today part of the Thermae Bath Spa with a hot spring pool inside (Fig. 6.10; Fig. 8.15b). At this site the Celts already worshipped their goddess Sulis and today the Cross Spring is officially recognised as a sacred site

selected to illustrate aspects of their geodiversity and their historical background. To highlight key points and different stages of development in the larger European region, examples of hot spring destinations from different countries are used to present a wide-ranging overview of the last 2000 years.

5.4.1 Italy

Italy is a country with a long history of hot spring use dating back to the Etruscans (Monti 2003). Later, under the Romans, Italy developed many hot springs as destinations for health and recreation, with four examples chosen to briefly illustrate the geoheritage of hot springs in this country. The historical spa town *Montecatini Terme* (Tuscany, Province of Pistoia) with its thermal baths has been known since the fourteenth century (Fig. 5.19). The thermal waters were recommended by Ugolino de Montecatini (1348–1425 AD) for medicinal purposes in 1417 AD (Altman 2000; Montecatini 2020), and from 1773 AD the hot springs of Montecatini were famous for their therapeutic value, although the spa town never achieved international recognition until the late nineteenth century (Erfurt 2012).

The area of *Salsomaggiore* (Emilia-Romagna, Province of Parma) was already known in ancient times to Celtic tribes, followed by the Romans, who used the saline springs to extract salt for domestic purposes. The actual spa town of Salsomaggiore Terme came into being in 1839, when initial experiments with the warm spring waters for thermal therapies were carried out. The historical spa architecture, which is dating back several centuries, adds a focal point for people who are visiting for health and wellness reasons (Emilia-Romagna Turismo 2019; Terme di Tabiano 2019).

In the north-east of Italy, circa fifty kilometres west of Venice, the spa town *Abano Terme* (Province of Padua) is one of Italy's most popular hot spring destinations. The name Abano seems to be primarily linked to the god *Aponus*, who was associated with thermal springs in the Paleovenetian age (Città di Abano Terme 2019b). However, Abano may also be connected to the Greek term *à ponos* (αφαιρώντας τον πόνο), which refers to the removal of pain. Attested by archaeological finds, Abano's origins are believed to go back to the eighth/ninth centuries BC, when ancient tribes living in the area made use of natural hot springs to perform rituals and to recover from illness. As in various other countries, Abano's hot springs were considered as significant for religious and cultural purposes, with later archaeological discoveries including a temple dedicated to Aponus, together with supplies of pottery for internal consumption of the health-giving waters. Written chronicles about the initial development of Abano into a centre for health and recreation based on the local hot springs date back to the twelfth century AD (Erfurt 2012), although some sources claim that earlier records containing information about Abano were kept by the parish church of San Lorenzo from the tenth century (Città di Abano Terme 2019b). With 78 hotels offering high quality spa treatments and 120 thermal pools surrounded by parks and gardens, Abano Terme today is considered as one of the most important thermal resorts in Europe (Città di Abano Terme 2019a).

The nearby spa town *Montegrotto Terme* and its long tradition of hot spring use is another historical example, with the knowledge about the curative properties of the

Fig. 5.17 The architecture of the new Bath Spa Thermae is in striking contrast to the traditional buildings reflected in the glass panels of the modern hot spring baths

mineral-rich hot springs being placed as early as the Iron Age (onset circa 1380 BC). The Romans certainly knew about the natural hot springs, as is evidenced by the remains of an original Roman bathing complex, the *Terme Romanae*. This might explain the origin of the name Montegrotto, which is derived from the Latin *Mons Aegrotorum* and translates into a 'mountain of sick people'. The spa town has been popular for many centuries and Montegrotto Terme

Fig. 5.18 The original Pump Room in Bath c. 1820 (left) and today. While changes were made over time, parts of the building have remained exactly the same

Fig. 5.19 An old etching by C. Zocchi of Montecatini Terme in Italy's Tuscany (n.d.). *Source* Montecatini (2020)

today is described as a 'haven for health and well-being' with therapeutic waters and thermal mud treatments (European Historic Thermal Towns Association 2019).

Other examples of traditional hot spring use can be found on the volcanic islands Ischia and Sicily, which have developed their own individual hot spring geoheritage. While some of Ischia's hot springs date back to Roman times, during the sixteenth century AD other springs were utilised by a physician named Giulio Iasolino from Calabria, who developed Ischia into one of the richest thermal areas (Erfurt 2012; Rando 2003). In Lipari, on the island of Sicily, stone-lined pools were discovered, which may have been used circa 1430 BC for thermal bathing (Cataldi et al. 1999). And in the region between Taormina and Messina

Fig. 5.20 The original Geysir in Iceland in a painting by Mayer, c. 1839

sulphur-rich volcanic hot springs are still used for health and recreation (Erfurt 2012).

5.4.2 Iceland

Iceland is a country rich in geothermal resources and natural hot springs (Fig. 5.20), although their wider use seems to go back only to the beginning of the twentieth century. However, reports exist that around the time of Iceland's official conversion to Christianity in the year 1000 AD, people were opposed to being baptised by immersion in cold water and instead chose natural hot springs for the baptismal ceremony (Erfurt 2012; Frazier 2000; Fridleifsson 1995; Hjálmarsson 1993). Later, bathing in volcanic hot springs was frequently mentioned in the Icelandic Sagas, documenting life in the twelfth and thirteenth centuries (Fridleifsson 1995; Kristmannsdóttir 2010). But without reliable records about their use in earlier times, it can only be assumed that natural hot springs near settlements would have been used for washing, cooking and bathing to provide some form of comfort and to counteract the climate (Erfurt-Cooper and Cooper 2009; Erfurt 2012).

Some Icelanders however, like the powerful chieftain *Snorri Sturlusson* (1179–1241 AD), who was also an educated scholar, politician and historian, have used hot springs for health and recreation in the past, although such activities did not become integral to the Icelandic culture until the twentieth century (Fridleifsson 1995), when thermal pools were built and access to natural hot springs became easier. Many swimming pools constructed in the early part of the century were located close to natural springs, which reduced the effort of accessing sufficient amounts of hot water from a permanent source (Haraldsson and Lloret 2014). In more recent times, in 1976, the famous Blue Lagoon near Keflavik was unintentionally created (Gudmundsdóttir et al. 2010) after effluent geothermal water, discharged from the Svartsengi power station, flooded the adjacent lava fields. In the following years, while the lagoon spread its milky-blue water across the barren landscape, people suffering from psoriasis discovered the beneficial effects of bathing in the mixture of warm sea- and groundwater that originates from a depth of 2000 m (Fig. 1.5; Chap. 4; Figs. 4.11, 4.12 and 10.13). Over the decades since its formation, the Blue Lagoon has grown into a major Icelandic tourist attraction with resort and spa facilities and a special clinic for psoriasis patients (Haraldsson and Lloret Cordero 2014).

5.4.3 Germany

Germany's recorded history of hot spring health resorts and spas spans nearly two thousand years. In 77 AD Pliny the Elder (Table 5.1) wrote about hot spring areas he had visited while travelling through Western Europe including the areas of Germany's spa towns *Aachen*, *Wiesbaden* and *Baden Baden* (Freedman and Waugh 1996). Long before the Romans arrived in the first century AD and started to establish thermal baths for their armed forces, Aachen (Aix-la-Chapelle) was settled by Celts, although this was not recorded in writing (Erfurt 2012). Centuries later the rise of Christianity caused doubt of the benefits of thermal bathing to the point, that people abandoned the idea of health and hygiene in favour of improved morality. However, Charlemagne, Emperor and King of the Franks (aka Karl, the Great, 742–814), kept the bathing culture alive during the early Middle Ages, perhaps because he suffered from rheumatism and appreciated the relief hot springs provided. After the fall of the Western Roman Empire and a long hiatus in the pursuit of health through water, at the end of the Middle Ages hot spring bathing once again became popular in Germany. In 1370 AD sixteen bathhouses were in operation in Wiesbaden, and by 1800 AD, this number had risen to twenty-three bathhouses. As thermal spa towns like Wiesbaden and Aachen (Fig. 5.21) continued to grow, other visitor attractions including extensive recreational facilities such as casinos, parks and golf courses were added (Erfurt 2012; Appendix 5.6; Chap. 8; Fig. 8.4).

Even though the main purpose of hot spring bathing was for health and hygiene, the use of water in general was frowned upon and considered as an immoral and unsafe practice (Maier and Fiedler 2004). Nevertheless, during the fifteenth century, natural thermal bathing facilities in Germany were still frequented and supported by clerics and members of the aristocracy despite a potential risk to their wellbeing (Rumpf and Sollner 2006). Annual visits to hot spring destinations again increased in popularity among Germany's more affluent class, when a few centuries later (eighteenth/nineteenth century) the hygienic aspects of thermal bathing were re-evaluated and recommended as an important factor for general health and wellbeing. And so, the culture of thermal bathing was re-introduced, and hot spring destinations were again places to visit for health and recreation (Asensio 2002). Since this time physicians have recommended and prescribed balneotherapy for its health benefits, which was created as a new science discipline and taught at most European medical schools (Erfurt 2012; Leavy and Bergel 2003).

5.4.4 Spain

In Spain natural hot springs were known for their healing properties since ancient times with their historical use in many Spanish hot spring provinces dating back to the second century BC. Natural hot springs were preferred destinations for pilgrims from various cultures, including Romans, Moors and Ottomans, who developed religious centres and thermal health and recreation facilities (Erfurt 2012; Ledo 1996). Always attracted to natural hydrothermal resources, the Romans also established municipalities such as *Aquae Calidae*, which is now known as the hot spring spa town Caldes de Montbui. Another example is *Archena*, where the origins of hot spring use most likely occurred as early as during the fifth and sixth centuries BC; at a time, when ancient Iberian tribes are thought to have settled in the area. When the Romans extended their reach further into Spain, it did not take them long to recognise the potential of the local hot springs, which resulted in the development of thermal bathhouses for their own use. Much later, towards the end of the eighteenth century in 1785 AD, King Carlos III of Spain ordered the redevelopment of Archena. This led to the growing significance of the town as a hot spring spa destination, and to this day Archena is continuously expanding owing to the popularity of its hot spring-based health resorts and spas (Erfurt 2012).

5.4.5 Portugal

Like many other European countries, some of Portugal's hot spring geoheritage is linked to the typical bathhouse culture established by the Romans in most of their occupied territories. The spa town Caldas da Rainha however has a different history. In the year 1484 AD Queen Leonore is said to have discovered natural steaming ponds while travelling. The Queen recognised the therapeutic value of the natural hot springs for treating certain health conditions and within one year she financed and established a hospital under her personal management. Even though the therapeutic benefits of hot spring therapy were known since the time of Queen Leonore's reign, it was only later, under King João V during the eighteenth century, that the therapeutic benefits of natural hot springs were officially recognised (Erfurt 2012; Sousa Dias 2017). Today, Caldas da Rainha still is a widely popular hot spring destination for health and recreation (Michelin 2002). Being part of Portugal the volcanic islands of the Azores have a long history of attracting visitors to a

Fig. 5.21 The spa centre (Kurhaus) of Burtscheid/Aachen with the colonnades of the Victoria Fountain and the spa gardens in 1911. *Source* Postcard, unknown author (1911) (Public domain)

wealth of natural features all over the nine islands. Especially the hydrothermal manifestations of the Caldeira das Furnas in São Miguel have not lost their appeal over the centuries (Appendix 5.3, Images 1, 2).

5.4.6 Belgium

The Belgium city of Spa has a special place in the more recent spa history, based on the discovery of a natural spring with curative properties in the fourteenth century. Later, during the seventeenth century, the city of Spa slowly developed by building on its reputation of natural springs with healing benefits and became accessible to a larger market. However, a thorough review of the literature discussing the history of thermalism and the geology of Belgium indicates that apart from *Chaudfontaine* no natural hot springs are located in the country. The famous ferrugenious springs of the city of Spa discharge only cold water, which is artificially heated to 33 °C for use in the local thermal bathing establishments (Thermes de Spa 2020). Therefore, the geoheritage of Belgium's natural springs primarily refers to cold springs, although many parallels with the history of other European spa towns exist.

5.4.7 France

In neighbouring France many historical Roman hot spring spa towns still exist and include well-known places such as *Aix les Bains*, *Rennes les Bains*, *Aix en Provence*, *Chaudes-Aigues* and *Vichy* (Table 5.5). The same natural hot springs the Romans were drawn to, still flow over 2000 years later and are used mainly for health and recreational purposes. An important point is that in France the use of natural hot springs is supported by a positive attitude of medical professionals and the health sector, as well as by the country's government and the tourism industry (Chaps. 7 and 8). Therefore, the medical spa tradition based on natural thermal resources (Fig. 5.22) still plays an important role in the twenty-first century (Erfurt 2012; Erfurt-Cooper and Cooper 2009; Mackaman 2007; Weisz 2001).

5.5 Central Europe, Eastern Europe and Russia

Hot spring destinations in central and eastern Europe often have historic links that confirm the influence of Greek and Roman cultures. At those times, when pilgrimage was the

Fig. 5.22 The thermal establishment at La Bourboule circa 1920. While La Bourboule does not go back to Roman times, an abundance of volcanic hot springs has led to the successful development of the resort town. *Source* Office National du Tourisme (1921)

equivalent of modern-day travel, the healing powers of natural hot springs were sought for health and recreation. More recently, after the Second World War, the health spa industry based on hot springs had come to a halt in many western European countries. By contrast, hot spring spas in central and eastern Europe continued to develop, as they were sponsored by the state and supported by trade unions (Bachvarov 2004; Erfurt 2012).

5.5.1 Hungary

Budapest

The area around Budapest presumably was inhabited by Celtic and Dacian tribes since the Neolithic era, a period, that while not consistently defined, is generally placed between 9000 and 300 BC. Based on archaeological evidence the town of Budapest was originally established by Celtic tribes, who named their settlement *Ak Ink*, which implies an abundance of water and most likely referred to the local hot springs. Around the middle of the first century, between 41 and 54 AD, the Romans, under Emperor Vespasian, arrived and set up an initial military base. Gradually the Romans expanded their Hungarian stronghold, renamed the city *Aquincum* and established thermal bathing complexes for health and recreation (Cohut and Árpási 1999; Encyclopaedia Britannica 2019; Erfurt-Cooper and Cooper 2009; Erfurt 2012). Opinions vary whether Aquincum was based on the original Ak-Ink or whether the name is derived from *Aquae Quinque*, a Latin reference to five water sources —perhaps a combination of both.

During the fifth century AD the Huns, a nomadic people, arrived from the central Asian Steppes to conquer parts of Europe and at the same time defeated the Romans. Much later, during the sixteenth and seventeenth centuries, large parts of Hungary, including Budapest, were taken over by Ottoman tribes arriving from Turkey, who built hammams and thermal bathing centres using the local hot springs to supply their facilities (Ákoshegyi and Árpási 2005; Nagy 2016). Like the Romans before, the Turkish invaders established amenities that were important to their individual way of life, with some of their domed bathhouses like the *Király* baths still in use today (Fig. 5.23).

Lake Hévíz

Located in the Western Transdanubian region of Hungary, *Lake Hévíz* is one of only a few thermal lakes worldwide and was already known to the Romans around the second century AD while they occupied the area (LaMoreaux and Tanner 2001). Using the thermal waters for health and recreation, the Romans left behind evidence of their presence including ancient coins and what appears to be the

Fig. 5.23 The construction of the Turkish bath Király Fürdö, located in the centre of Budapest, started in the year 1565 AD and has been in use ever since

remains of an altar. Following the Romans, Germanic and Slavic tribes apparently inhabited the area around Lake Hévíz, which is based on archaeological evidence in the form of excavated objects that indicate their presence during the Migration Period (375–568 AD) (Budapest.com 2019). After the Mongol invasion of Hungary (1241–1242 AD), first written records of Lake Hévíz appear, dating back to 1328 AD. Later, the area was destroyed in an advance by the Ottoman Empire, which lasted from 1541 to 1699 AD (Böcskei and Szita 2008; Budapest.com 2019). The first documented scientific analysis of the hot and cold springs feeding into Lake Hévíz was carried out by Ferenc Szláby in the year 1769 AD, and by the end of the eighteenth century Lake Hévíz was put on the map as a destination for health and recreation (Hungarian Digital National Archive 2019). In its present form 'taking the waters' at Lake Hévíz has a history of over 200 years. During this time the town of Hévíz was developed with the curative effects of the lake supporting a popular and lucrative health tourism industry (Budapest.com 2019; UNESCO 2020). As part of the *Balaton Uplands Cultural Landscape* the area of Lake Hévíz together with its cultural heritage, is tentatively listed as a future UNESCO World Heritage site (UNESCO 2020).

Hungary's current hot spring health spa destinations were largely established during the eighteenth century, when scientific research of the mineral content of the thermal waters was first carried out to determine the healing potential of Hungary's hot springs (Nagy 2016). At the turn of the eighteenth and nineteenth centuries spa towns such as Hévíz, *Harkány* and *Balatonfüred* developed into leading centres for balneotherapy (Ákoshegyi and Árpási 2005), while spa resorts gained in popularity through visits from royals and their entourage, who hailed the natural hot spring waters as beneficial for their health and established summer residences close to the springs. Despite its turbulent historical and political past, Hungary is widely known today for the therapeutical benefits of the country's natural hot springs and mud products, which are used at most health centres and contribute to the flourishing of the country's health tourism industry (Altman 2000; Erfurt 2012).

5.5.2 Romania

Located to the east of Hungary and bordering on the Black Sea and Ukraine, Serbia and Bulgaria, the country of Romania also has a Roman legacy going back to the time of their invasion of Dacia in 106 AD. As in many other countries, the Romans began to use the local hot springs to develop their customary thermal bathhouses. Long before the Romans arrived though, natural springs throughout Romania were used by the Dacians, which is attested by traces of settlements near local hot springs (Cohut and Árpási 1999; Erfurt 2012). Later, the natural hot springs at *Băile Herculane* (Aqua Herculis) were developed by the Roman military to accommodate health and recreational needs during their occupancy. Studies of the Romanian spa resort heritage discovered inscriptions and votives, which were dedicated to various Greco-Roman deities and nymphs that were worshiped at Băile Herculane, *Băile de la Geoagiu* (Germisara) and *Băile Aquae Călan*. The uncovered votives revealed references to Artemis, Asklēpiós, Hercules, Hygeia, Hypnos and Odrysta (Boda et al. 2017), confirming the significant religious influence suggested by various researchers based on their investigations (Piso 2018). When the Romans left Romania, the region was invaded by barbaric tribes, which resulted in the deterioration of the Roman thermal architecture (Cohut and Árpási 1999). Gradually, spa towns such as Băile Herculane recovered and were frequented by royals and their followers, which generated public interest and turned Băile Herculane into an attractive destination that has lasted to this day as one of the popular Romanian centres for health and recreation (Cohut and Árpási 1999; Erfurt 2012).

5.5.3 Bulgaria

The hot spring heritage of Bulgaria resembles that of many other European countries and is unquestionably linked to the Roman occupation of the region, which was conquered by legions of the Roman Empire in 46 BC. However, based on remnants of Thracian settlements found near Bulgaria's natural hot springs, the areas of ancient spa towns such as *Pautalia* (Kyustendil) and *Serdika* (Sofia) were in all likelihood used by Thracian tribes and go back to Neolithic times (4000 BC) (Tsankov and Kamarashev 1996). During their occupation of Bulgaria the Romans developed several hot spring spa centres, which to this day follow the traditional use of hot springs for health and recreation (Erfurt 2012). Other spa towns such as *Pavel Banya* first started the use of hot spring water for thermal bathing in 1923 and established a considerable tradition in the treatment of many different medical conditions. In more recent years, many Bulgarian spa resorts have rapidly developed as a result of renovating existing hot spring centres and through building new accommodation and entertainment facilities to cater for a growing demand (Hristov et al. 2010). In the village of *Ovoshtnik* a recreational centre, the Waterland Aquapark, was built in 2006 next to the local hot springs, which supply all the pools with warm mineral water (Guide-Bulgaria.com 2019). Although Bulgaria is not a large country, it boasts several hundred natural hot springs; each of them with their individual physical features and chemical characteristics (Erfurt 2012; Tsankov and Kamarashev 1996). Today Bulgaria's spa towns are active treatment centres for health and recreation based on an unwavering recognition of the curative value of natural hot springs. Out of about 230 hot spring health resorts and spas a quarter are promoted as resorts providing balneotherapy (Vassileva 1996), which once again confirms the long-lasting interest of humans in the use of hot springs for health purposes (Erfurt 2012).

5.5.4 Poland

Poland's thermal bathing facilities have an extensive history with the first written records of the use of hot springs dating back to the thirteenth century. At this time, local tribes at several hot spring locations started to use basic forms of balneotherapy with places such as *Cieplice Śląskie-Zdrój*[2] and *Lądek Zdrój* developing into spa towns over time (Erfurt 2012; Dryglas and Hadzik 2016; Kępińska 2002, 2004, 2010; Lund and Boyd 2015; Omulecki et al. 1996; Pawlikowska-Piechotka and Piechotka 2014). The oldest spa resort in Poland is Ladek Zdrój, with first accounts of thermal water use going back to the year 1242 AD. But it was not until later, at the end of the fifteenth century, that the first bathhouse was constructed. Historical records from the year 1281 AD indicate that in Cieplice hot spring water was used to treat people with medical conditions (Kępińska 2004). Like in other parts of Europe hot spring-based spas and health centres in Poland were abandoned around the fifteenth century, based on the assumption that thermal bathing was contributing to the spreading of various diseases (Dryglas and Hadzik 2016) and was responsible for the deterioration of moral values.

Once upon a time, health tourism and a visit of Poland's hot spring spas was a luxury reserved for a selected few, who had the time and financial means to afford such indulgence. Only during the 1920s visits to Polish health resorts became available for the general public (Pruska 2014). And while the number of Poland's contemporary health resorts and spas without access to natural hot springs declined in the 1990s due to a decrease in visitor numbers by fifty percent, most of the traditional thermal health centres continued to

[2]Zdrój = Spa in Polish.

Fig. 5.24 The Mill Colonnade in Karlovy Vary contains five drinking fountains at temperatures between 52 and 65 °C. A total of 124 columns are supporting the large structure, which is over 130 m long. Left image source: Ivanhoe (2011); (right) Bobak Ha'Eri (2011) (Public Domain)

attract visitors and even increased their popularity in recent times (Bachvarov 2004; Dryglas and Hadzik 2016; Erfurt 2012). It is interesting though, that the Romans never invaded Poland and therefore did not have a hand in the development of the local hot springs and their cultural heritage.

5.5.5 Czech Republic

The Czech Republic's leading hot spring spa town almost certainly is *Karlovy Vary* (Karlsbad, near Prague; Fig. 5.24), which was founded in 1370 AD by Karl IV, Holy Roman Emperor and King of Bohemia and therefore named after him (Erfurt 2012; Lund 2000). Even though the first settlements in the area where Karlovy Vary is located are dated from around the middle of the fourteenth century, archaeologists believe that people have lived in the vicinity of the hot springs prior to this time (Burachovič 2010, 2019; Erfurt 2012; Karlovy Vary 2019).

In 1401 AD, under King Wenceslas IV, the spa town of Karlovy Vary was granted the privilege of peace and a ban on carrying weapons in the town was put into effect. In the year 1707 AD, during the reign of Emperor Joseph I, Karlovy Vary was confirmed as a free royal town. Despite being destroyed several times by fires and floods, as well as being ransacked by plundering hordes, Karlovy Vary eventually began the commercial use of the saline thermal waters in 1769 AD. During the next centuries the town gradually developed into a thermal health centre, which became a favourite with visitors seeking health and recreation and who embraced the traditional atmosphere (Burachovič 2010, 2019; Erfurt 2012; Karlovy Vary 2019). Other well-known historical spas in the Czech Republic include *Jansk Lazne*[3] and *Teplice,* with additional health resorts and recreational destinations such as *Termály Velké Losiny*, Aqualand Moravia in *Pasohlavky* and *Benesov nad Ploucnici* as well as thermal spas in *Brno* (Lund and Boyd 2015).

5.5.5.1 Slovakia and Slovenia

The country of Slovakia reportedly has over 1100 natural hot springs with their healing qualities documented since 1147 AD. Many of the original health centres practicing balneotherapy are still in use today and are catering for the growing demand for health and wellness facilities. *Piestany*, one of the leading health resorts in Slovakia, has a history of hot spring use that goes back to Celtic and Germanic tribes as well as to ancient Slavic civilisations, who inhabited the region of Slovakia long before the Romans arrived (Erfurt 2012; Lund 2000).

Further to the east, in Slovenia, the use of natural healing resources for health, wellness and recreation also has a centuries' old tradition. Archaeological evidence indicates that the history of many health resorts and spas dates back to Roman times about 2000 years ago (Erfurt 2012; Slovenian Spas 2015). Today, 18 thermal health resorts and spas as well as several recreation centres in Slovenia are still using traditional treatment methods in combination with the healing effects of the country's 87 natural thermal springs (Lund and Boyd 2015; Slovenia.si 2019).

[3]Lazne = Spa in Czech.

5.5.6 Russia

In Russia the situation was different, with health spas only starting to develop on a broader basis after the Russian revolution in 1919, although from this time onwards they played an increasingly important role in the national health system (Altman 2000; Erfurt 2012; Simkins 1986). As the Soviet era came to an end in 1991, circa 2500 health spas were state-run with a capacity for up to 500,000 patients (Oddy 1999). However, after the USSR was dissolved, many health spas went into decline, even though in numerous areas of the former Soviet Union natural hot springs have been used for bathing, cooking and washing for centuries (Erfurt 2012). According to a country update from 2010 about the utilisation of geothermal resources worldwide, about 150 thermal health resorts and spas remain in Russia (Svalova 2010).

Crimea

Going back in history, during the first century AD, the Crimea in the south of Russia was under the influence of the Roman Empire, and the existence of thermal springs on the Crimean Peninsula were mentioned by Pliny the Elder in his book *Natural History* (Svalova 2000).

Georgia

In Georgia, a former part of the Soviet Union, natural hot springs played an important part in the founding and the naming of the capital Tbilisi (*tbil-i* means warm in Georgian) by king Vakhtang Gorgasal in the fifth century AD (Lerner 2001). Since the 1940s the therapeutic qualities of the natural hot springs have been used by the *Tbilisi Balneological Health Resort,* one of the major centres for health and recreation (Melikadze et al. 2010). After nearly 2000 years Tbilisi is still famous for its hot springs, although political unrest and instability in some areas of Georgia so far have discouraged an increase of international visitor numbers (Erfurt-Cooper and Cooper 2009).

5.6 Hot Spring Use in the Middle East

Jordan's hot springs, located in the Jordan Valley, along the Wadi Araba and the Dead Sea, were known to the inhabitants of the region for thousands of years, who used them for various purposes, including cultural and religious traditions (Schäffer and Sass 2014). Between the first century BC and the sixth century AD the Jordan Valley hot springs were mentioned repeatedly in the classical works of Byzantine and Roman scholars of geography and history (e.g. Epiphanius, Eusebius and Antoninus of Placentia, Josephus Flavius, Pliny the Elder, Ptolemaeus and Strabo). These reputable chroniclers reported in detail on the curative powers of the Jordan Valley hot springs and their reputation as a successful treatment method for different diseases (Jaffé et al. 1999). The natural hot springs of *Callirhoe* and *Zarqa Ma'in* for example, are believed to have been used in ancient times by locals and pilgrims, who valued the thermal springs for health and recreation (BiblePlaces.com 2019; Oz 2007). *Hamat Gader* in the Yarmouk River valley (now under Israeli control) also has a long history as a hot spring resort with remnants of Roman architecture (Eck 2016). Ancient documents indicate that the healing powers of hot springs in the Dead Sea area further south were also known well before the Romans arrived and established their customary bathing facilities (Oz 2007).

5.7 The Hot Spring Heritage of Oceania

5.7.1 New Zealand

New Zealand and Australia are two countries with only more recent historical records available in writing. But even though both countries' hot spring use prior to the arrival of European settlers was never documented, it is evidenced in oral transmissions of legends, folklore and dreamtime stories. We can therefore safely assume that the traditional inhabitants of both countries have known about the healing hot spring waters and have used them for domestic, ceremonial and recreational purposes.

Since the beginning of their settlement circa 1100 AD, Māori tribes used selected springs for sacred ceremonies, which was an integral part of their culture (Chap. 6). In similar ways as people in other countries the Māori considered natural resources to be a treasure received from the gods. They believed that certain hot pools had spiritual guardians, and reserved them for important rituals (Erfurt 2012; Jackson 2007; Severne 1999; Swarbrick 2006), while other less sacred springs were used for cooking, bathing, heating, washing and for therapeutic bathing to treat various diseases. For example, centuries prior to the arrival of European settlers, the Māori people visited places known to them for their healing waters such as *Waiwera* north of Auckland and *Wai Ora* near Rotorua (Erfurt 2012). With the arrival of European settlers, public bathhouses were established, and in 1880 the Government of New Zealand started to build hot spring resorts and spas in *Rotorua* (Fig. 5.25), *Te Aroha* and *Hanmer Springs* (Fig. 6.2) for health and recreation. As a negative (but not unusual) effect of New Zealand's colonisation, the Māori were expected to use separate pools, which frequently generated conflict caused by the new settlers' disregard for the spiritual significance natural hot springs had for the Māori (Erfurt 2012; Swarbrick 2006). Further disagreements arose over the disrespect for the sacred rules of *tapu*, a system used by the Māori to protect natural resources from unsustainable exploitation and unacceptable misuse (Severne 1999).

Fig. 5.25 The old Sanatorium in Rotorua was the official centre of the Thermal Springs District on New Zealand's North Island, where hundreds of hot springs are located with very different chemical characteristics and at a wide range of temperatures up to 100 °C. The Sanatorium was a Government institution under qualified medical supervision and the curative properties of the baths were used to treat various medical conditions. *Source* Glimpses of New Zealand (1896)

In Rotorua, a flourishing tourism industry gradually developed around hot springs such as the *Polynesian Spa* and the *Rotorua Bath House* (Fig. 6.2). Guided by the success of European hot spring resorts, New Zealand's government continued to build health resorts and spas throughout the country to generate revenue, although after the 1920s the government's interest started to fade. Factors such as high maintenance costs, effects of the First World War and the Great Depression, together with New Zealand's remote location, were reasons why the hot spring tourism sector never reached the level of the European spa scene in those early years (Jackson 2007). Nevertheless, the Queen Elisabeth Hospital (QE Health) in Rotorua has served as a rehabilitation centre using local hot springs since 1942, when it was established as a 'Services Convalescent Hospital' to provide treatment to soldiers returning from the Second World War. The QE Health today is considered as one of the largest thermal facilities in the South Pacific offering therapeutic spa treatments dedicated to health and wellness (QE Health 2019).

Maruia Hot Springs on the South Island of New Zealand has been recently redeveloped, but its origins go back to the early 1900s when the first bathing huts were built (Appendix 5.3, Images 7, 8). Having changed owners and joined forces with the Peninsula Hot Springs in Australia a few years ago, the Maruia Hot Springs retreat (Fig. 6.3) has undergone significant improvements and is now offering the ultimate outdoor experience based on the wellness philosophy that made the Peninsula Hot Springs the multi-award-winner of the 'Best Mineral Springs Spa in the World' four years in succession.

5.7.2 Australia

The Australian continent is known as extremely dry on the surface but has vast subterranean reservoirs of geothermal water stored in deep artesian basins, which underlie much of the central and eastern inland of the country (Fig. 5.26; Chap. 4). While the discovery of the *Great Artesian Basin* (GAB) dates to around 1880, to this day only a few thermal resources are used for hot spring bathing (Ah Chee 2014; Habermehl and Pestov 2002). Even less are used to generate geothermal energy, although there are many instances where artesian springs emerge at the surface at high temperatures and large discharge rates (Fig. 2.20). Even though, this could

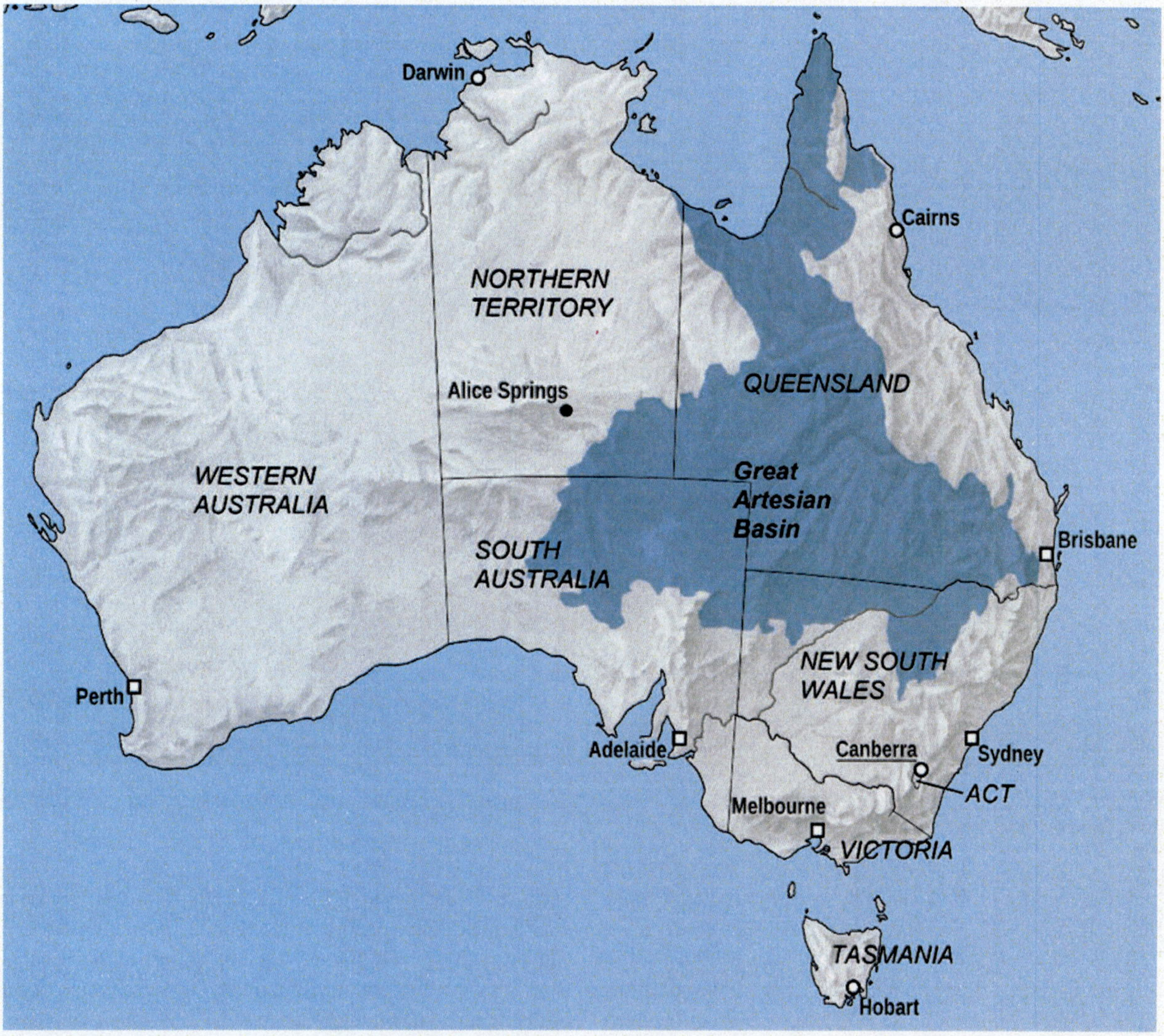

Fig. 5.26 The approximate size and location of the Great Artesian Basin in Australia, the origin of countless hot springs. *Source* Tentotwo (2013) (Map modified by Author)

change with the new Australian Geothermal Association (AGA) formed in 2016 with the intent of promoting the development of geothermal energy in Australia (Beardsmore et al. 2020; Chap. 11). Nonetheless, the Indigenous people of Australia knew about natural hot springs many thousand years ago and refer to them in their dreamtime stories as healing the sick and giving strength to the healthy (Ah Chee 2014; Anon 1910; Griggs 2013; Chap. 6). On the recreational side, at present numerous public swimming pools throughout the country, including the island of Tasmania and a few locations in Western Australia, are using locally sourced geothermal water (Beardsmore and Hill 2010).

One of the few historical examples of developed natural hot springs in Australia is the town of Moree in northern **New South Wales** (Erfurt 2012). In 1895, drilling into the artesian basin was carried out to access groundwater for irrigation purposes (Fig. 5.27a) but the geothermal resources were found to be so plentiful that a large pool was built and the residents of Moree were able to enjoy an Australian version of 'taking the waters'. By 1898 the first Moree Baths complex was completed with further extensions added in 1913 to provide more comfort for visitors of the open-air pools. After undergoing recent modernisations, the *Moree Hot Mineral Baths* are now known as the *Moree Artesian Aquatic Centre* (MAAC) and are a popular hot spring destination (Fig. 5.27b). The artesian hot springs are especially valued by people of European descent, who are aware of the health benefits of the mineral-rich waters and travel long

Fig. 5.27 The original fountainhead (**a**) of the artesian hot water bore that supplies the aquatic centre in Moree, New South Wales (Australia) and (**b**) one of the thermal pools in Moree. *Source* Tim Harrison 2018

distances to Moree for health and recreation (Erfurt 2012; Moree History 1995; Tourism Moree, 2018).

The documented geoheritage of **Queensland's** hot springs is rather limited. By 1910 only three locations, *Helidon Spa*, *Innot Hot Springs* and *Muckadilla*, were known as places where people with certain health conditions could 'take the waters'. By 1950, locations with artesian hot springs that were recognised for their therapeutic benefits had risen to six, including *Barcaldine*, *Dalby* and *Aramac* and were visited for health and recreation. The infrastructure however remained relatively low-key and was generally limited to bathing pools, small bathhouses and some buildings for visitor accommodation (Griggs 2013).

The Innot Hot Springs in Queensland's tropical north were originally discovered by J. Atherton in the early 1870s but disregarded until they were rediscovered in 1885 when it became known that the local Indigenous people considered them as healing waters (Cairns Post 1940). Emerging at over 70 °C the hot spring water is cooled after entering Nettle Creek and while seeping up through the sandy creek bed (Fig. 5.28). The Innot Hot Springs were used for recreational purposes since their discovery until the early 1900s and because of the presumed health benefits, the spring water was bottled and exported to Europe (Lottermoser and Cleverley 2007). To this day the hot springs attract visitors from all over the country, although the surrounding local infrastructure does not appear to show any major improvements since a last (research) visit several years ago.

In the south of Queensland, the explorer A. Cunningham discovered and mapped the Helidon Hot Springs, although the local indigenous tribes were familiar with the hot springs and regarded them as sacred in their dreamtime legends (Anon 1910; Erfurt 2012; Griggs 2013; Pearn and Little 1998). In 1879 the Helidon Spa was established and promoted as 'Australia's Wonderful Mineral Water' on a par with the best springs in Europe and, like the spring water from Innot, Helidon water was exported overseas. By 1966 the Helidon Spa claimed to have the largest manmade swimming pool in Australia. Unfortunately, in the 1980s concerns about bacterial contamination arose and the water was required to be treated with chemicals to prevent risks to human health. This was deemed as a contradiction to the attractiveness of natural springs and the spa declined over time (Pearn and Little 1998).

Fig. 5.28 Nettle Creek in northern Queensland is heated by water from the Innot Hot Springs (temperature ~75 °C), which enter the creek further upstream as well as seeping up through the sand banks

In the southern States of Australia, hot and cold natural springs most certainly would have been known to the local Indigenous people long before European settlers arrived but again, without written records this remains only a logical assumption (Erfurt 2012; Menadue 1972). In Victoria, early developments of natural spring spas never led to a successful health and wellness industry, most likely because the natural springs were not warm enough to attract visitors and had to be heated. However, in more recent years the Peninsula Hot Springs Centre, located on the Mornington Peninsula south of Melbourne, was developed by *Charles Davidson* after travelling through Japan and becoming inspired by the pleasant experience of onsen bathing. Together with his brother Richard he was determined to establish something similar in Australia (Swart 2006; Richards 2008) and started the process of exploration and drilling, ultimately being rewarded after eight long years with an abundant supply of natural hot water (Fig. 5.29). The development of the Peninsula Hot Springs started with the first stage opening in June 2005, followed by continued investment and the addition of new facilities with further plans for the future (Appendix 5.7). This hot spring spa is the only one in Victoria with natural hot water and has become a popular destination for hot spring enthusiasts from all over the world (Davidson et al. 2011).

Today the Peninsula Hot Springs complex reflects a multi-cultural bathing experience based on research into hot springs and spas in many countries (Davidson 2006).

5.8 Historical Hot Spring Use in North America

In the United States, Canada and in the Latin American countries, local native tribes have long used natural hot springs for health and recreation as well as for ceremonial purposes (Erfurt 2012; Table 6.1). This widespread belief in the healing benefits of the 'medicinal' water was later

Fig. 5.29 Charles and Richard Davidson are overjoyed after successfully tapping into a thermal reservoir that has the capacity to supply 4.3 million litres per day at temperatures of around 50°Celsius. After years of planning, exploration and hard work the Davidson brothers' dream of finding natural hot water on the Mornington Peninsula has come true. From these moments to the first stage of the Peninsula Hot Spring Resort meant more years of determination and long hours to fulfil this dream (see Appendix 7)

adopted by early settlers, who made yearly visits to hot springs, comparable to the annual sojourns to thermal spa resorts in Europe (Bullard 2004).

5.8.1 Canada

Canada's documented hot spring history is relatively recent. The first planned development of hot spring destinations in Canada started around 1883 after European visitors arrived in the year 1859. In 1885 the area around *Cave and Basin Hot Springs* (Banff, Canada) was reserved for the development of a national park that included a European style hot spring resort (Kurillowicz 1995). In 1886 the original Grand View Villa with a hot spring bathhouse was built at the Banff Upper Hot Springs. Although initially privately owned, the buildings were later brought under the management of the Canadian Government (Erfurt 2012; Lund 2003).

Long before the first Europeans arrived in Canada, the *Radium Hot Springs* in British Columbia were already used by the local Kootenai Indian tribes (Zieroth 1978). While the Radium Hot Springs were initially developed under private ownership, they were later incorporated into the Kootenay National Park under the administration of Parks Canada (Lund 2003; Parks Canada 2019). Other natural hot springs were also used by local Indian tribes well before settlers from Europe arrived; for example, the *Miette Hot Springs* in the Jasper National Park in Alberta (Erfurt 2012). In British Columbia the area around the hot springs, which now supply the *Fairmont Hot Springs Resort,* was first inhabited by the Nez Perce, the Shoshones and the Flathead, who established a small community close to the springs, known as *Medicine Waters* in the past (Lienau 1993). Canada's recorded history only dates back to the time when the country was colonised by British and French explorers, although unwritten accounts about hot spring use have been passed on from the first indigenous nations and are similar to other American countries with natural hot springs. Canada's hot springs include undeveloped 'wild springs' in remote areas (Fig. 5.30) as well as luxury hot springs resorts (Woodsworth and Woodsworth 2014).

5.8.2 North America

The native peoples of North America considered natural hot springs as sacred and believed in the healing powers of the springs with many hot springs in the United States recorded as used by local tribes for rituals and ceremonies (Erfurt 2012; Frazier 2000). By the year 1888 the number of recorded hot springs in the United States alone had reached 8843, with 634 hot springs used as spas for health and recreation, which were subsidised by the government (Bischoff 2019; Erfurt 2012; Waring 1965; Fig. 5.31). However, during the middle of the following century the administrative support decreased, and most of the hot spring-based health resorts and spas were forced out of business and had to close their doors (Bischoff 2019; Erfurt 2012).

Faywood Hot Springs and *Ojo Caliente* (New Mexico) are two examples of early North American hot spring spas, with written records documenting the first thermal health facility as being established at Ojo Caliente in the year 1880. Long before that time, in 1525 AD, Spanish conquerors, who were searching for gold and the fabled 'fountain of youth' in the area, discovered the natural thermal springs and named them Ojo Caliente (Witcher 2002a), which means

Fig. 5.30 Hot springs in Canada are advertised as natural destinations to attract visitors. *Source* Screen capture of promotional Internet images

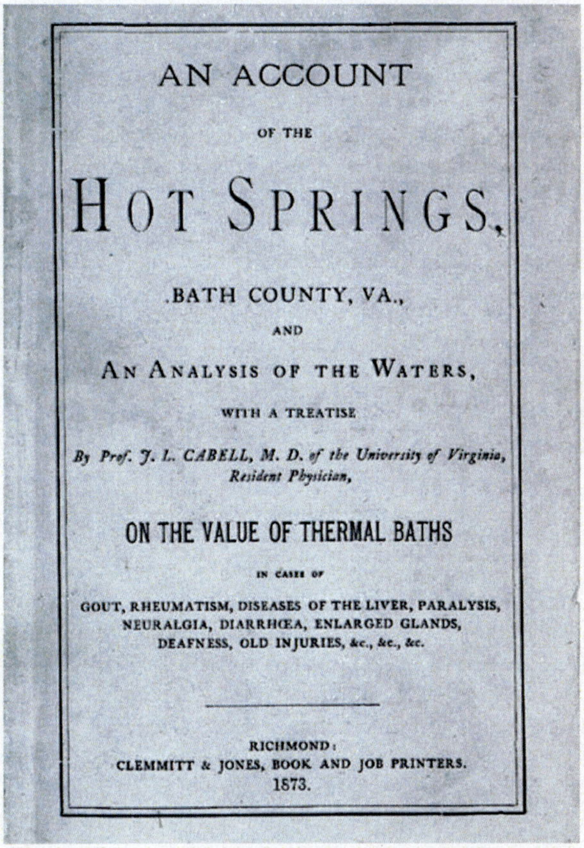
AN ACCOUNT

OF THE

HOT SPRINGS,

BATH COUNTY, VA.,

AND

AN ANALYSIS OF THE WATERS,

WITH A TREATISE

By Prof. J. L. CABELL, M. D. of the University of Virginia, Resident Physician,

ON THE VALUE OF THERMAL BATHS

IN CASES OF

GOUT, RHEUMATISM, DISEASES OF THE LIVER, PARALYSIS, NEURALGIA, DIARRHŒA, ENLARGED GLANDS, DEAFNESS, OLD INJURIES, &c., &c., &c.

RICHMOND:
CLEMMITT & JONES, BOOK AND JOB PRINTERS.
1873.

Fig. 5.31 Literature from 1873 confirms that hot springs were once held in high regard and valued for their curative properties in the north American state of Virginia. *Source* Cabell (1873)

'hot eye'. However, according to local history, the hot springs were already used in ancient times by native tribes, who built pueblos with terraced gardens overlooking the springs (Erfurt 2012; Witcher 2002a). Nowadays Ojo Caliente is a well-known hot spring health resort and spa.

Located in the desert of New Mexico, the hot springs at Faywood first attracted native American peoples such as the Mimbres Culture and the Apaches. Later Spanish explorers came, followed by buffalo hunters and gold miners. Without written accounts covering the time before the Spanish explorers and other European pioneers arrived during the first half of the sixteenth century there are no reliable records available. Interestingly, the website of the Faywood Hot Springs Resort includes a reference that the Celts possibly may have been the first western visitors to the hot springs, which is based on Celtic petroglyphs found in the area (Faywood Hot Springs 2019). In more recent times the hot springs at Faywood were commercially developed as a hot spring resort, which offers a wide range of activities for natural health purposes and for recreation (Erfurt 2012; Witcher 2002b).

The natural hot springs of the *Hot Springs National Park* (HSNP) in the State of Arkansas have been valued for their health benefits and their recreational value ever since they were discovered. As their importance was recognised in the year 1832, the first Federal Reservation was established to protect the natural resources for the public and future generations. The hot springs subsequently attracted settlers and businesses to the area, which lead to the development of the city of *Hot Springs* (Erfurt 2012; Yeatts 2016). With at least

43 natural hot springs in the area (numbers vary according to the source), various accounts of hot spring use from before 1877 report that some springs were fenced off with walls and covered by masonry arches to protect them from contamination (Erfurt 2012; Scully 1966). Today the city of Hot Springs is still a world-famous spa town with a rich geoheritage and is promoted as 'America's First Resort', a destination for health and recreation (Visit Hot Springs 2019).

5.9 Hot Springs in Central America

Researching the history of natural hot spring use in Latin American countries and the Antilles is a complex task because not much literature is available in English, with some sources incomplete, inaccurate or distorted and have become vague over time (Erfurt 2012; Hernández-Galán et al. 1999). The following section includes two samples of the hot spring geoheritage in this part of the world.

Mexico

In Mexico, 576 hot springs are recorded with many of them identified in the past and recognised for their therapeutic benefits (Brooke 2019). Some of these natural hot springs were used by the early Aztecs for health and recreation (Lund and Freeston 2001), with leaders such as Montezuma I and II, together with their warriors, using natural thermal spas to unwind after royal responsibilities and battles (Erfurt 2012). One of the hot springs, *Agua Hedionda*, which means 'smelly water', was later developed by the Spaniards into a fashionable spa (Suárez Arriaga et al. 1999; Brooke 2019; Lund 2002; Salgado-Pareja 1988). Other hot springs were developed towards the end of the 1700s in *San Bartolo Agua Caliente*, although it is universally believed that the use of the hot springs goes back much further (Hodgson 2004), as it is generally accepted that the ruling class of the Aztecs were aware of the healing properties of natural hot springs and used them for the treatment of various diseases (Brooke 2019).

Guatemala

The small Central American country of Guatemala south of Mexico was once inhabited by the Maya civilisation. Evidence from anthropological research and archaeological discoveries at *Lake Amatitlan* indicates that the Maya may have used natural hot springs as early as 500 BC (Erfurt 2012; Altman 2000). In more recent times the hot springs of Guatemala are used for a variety of purposes including health, wellness and recreation, as well as industrial and commercial usage such as heating of greenhouses and for fruit drying processes (Mariniet al. 1998; Merida 1999).

5.10 South America's Hot Spring Geoheritage

On the South American continent, natural hot springs were historically used for healing and for water worship. South America has an abundance of these springs, which derive from volcanic activity in large parts of the continent with a geoheritage that goes back to ancient times, long before explorers and conquerors from Europe arrived. One of the pre-Incan cultures, the *Caxamarca*, built a whole city near their local hot springs. Later, when the *Incas* gained regional influence, this city would become known as *Baños del Inca*. The hot spring baths developed into an estate that was the preferred residence of Inca chiefs, who ruled before the arrival of the Spanish conquerors. It was at Baños del Inca where, according to historical sources, the Inca ruler *Atahualpa* (1502–1533 AD) and his court first became aware of the Spanish invasion (1531–1532 AD), with some sources claiming that he was relaxing in the hot springs when he received the news (Erfurt 2012).

Until the invasion by Europeans, the largest south American territory consisted of the empire of the Incas, who ruled over the Argentinian uplands, Bolivia, the north of Chile, the south of Colombia, Ecuador and Peru (Calderón 1999). Numerous Inca palaces and temples were built near natural hot springs and were well-appointed with bathing facilities supplied with hot and cold water via a system of pipes and conduits (Haraldsson and Lloret Cordero 2014). However, written records again only date back to the arrival of the Spaniards, who came to explore and exploit South America's natural resources. While searching for gold and other valuable resources, they came across many natural hot springs, which they used for health and recreational purposes (Erfurt-Cooper and Cooper 2009). Today Cajamarca, together with the geoheritage preserved in the *Hot Spring Village of the Incas*, is one of the most significant tourist destinations in Peru (Calderón 1999; Termasworld 2016).

5.11 Hot Spring History in a Nutshell

The extent of the geoheritage of natural hot springs clearly demonstrates that their usage has worldwide origins, reaching back to ancient societies and most likely even further to prehistoric civilisations. Early writings support the view that the most popular places for healing and recreation were those with access to natural hot springs. By taking into consideration the wealth of evidence left behind by Greek, Roman, Japanese and Chinese scholars, a significant amount of the Asian and European hot spring history can be reliably traced back several thousand years. In fact, in many countries worldwide and throughout history natural hot springs have been regarded as sacred spaces and sites for worship of divinities, places of healing and of course as an opportunity for recreation.

The use of natural hot springs for health and recreation is embedded in the history of all countries, where such springs were considered as a blessing and a health-giving contribution to the communities who settled in their vicinity. Different traditions and bathing cultures (Chap. 6) as well as treatment practices such as balneotherapy (Chap. 8) were developed over time and have been passed on through generations. Archaeological evidence informs us that the establishment of healing centres near sources of natural hot water took place in the earliest settlements in history. Owing to the success rate in treating a wide range of medical conditions and the apparent lack of side effects, hot springs have maintained their reputation as medicinal waters with curative powers throughout history to this day. Therefore, historical hot spring spa towns still exist in many European and Asian countries, where they have either developed new facilities or preserved the remnant architecture of ancient thermal healing facilities in museums and interpretive visitor centres.

Apart from documenting the magnitude of hot spring use worldwide, the main aim of this chapter was to establish some reference points related to the origins of hot spring customs and traditions, as well as to develop a realistic timeframe of the stages of hot spring development. A further objective was to draw attention to hydrothermal environments in different countries and the parallels of their history. Despite the distance between continents and the lack of communication in the past, many similarities exist in the way how people perceived the natural phenomenon of hot springs emerging from the ground and how they utilised them for their personal benefit. The next Chap. 6 explores the cultural geoheritage as it presents itself to visitors of hot spring destinations, followed by Chap. 7, which takes a closer look at the natural geoheritage of hot springs.

Hydrothermal tourism over a century ago in São Miguel, Azores

5.12 Appendices

Appendix 5.1 Chronology of the Use of Hot Springs and Their Development Worldwide

The following timeline includes reference points for the plausible use of hot springs during parts of the Holocene (circa 11,700 years to current), as well as historically documented developments from a variety of countries. Included at the beginning are also believable and for many people acceptable timeframes that, while undocumented in writing, have been orally transmitted through many generations. Archaeological findings in hydrothermal areas all over the world have frequently helped to determine early users of natural hot springs. However, without written proof the prehistoric use of natural hot springs cannot be established beyond any reasonable doubt, although it can be assumed that ancient civilisations were naturally drawn to warm water sources.

Dreamtime ~60,000 BC	Australia	Dreaming stories from ancient times have transmitted the traditional knowledge about the close connection of the Aboriginal peoples with natural water sources across Australia
Neolithic age quaternary	Worldwide	Start of Neolithic times estimated at 12,000 years ago—no written records available—based on the rationale that humans were always drawn to natural hot springs, these were most likely used when and where people had the opportunity
~14,000–300 BC Jōmon Period	Japan	Possible use of hot springs for the first onsen during this period—the Dōgo Onsen is suggested by unconfirmed sources to possibly date back to as early as 14,500 BC although this remains speculation
~4000 BC	China	Hot springs are most conceivably used by early Chinese peoples based on their occurrence throughout the country
~3000–1700 BC	Indus Valley, Pakistan	The Indus Valley civilizations (e.g. Mohenjo-Daro and Harappa) are likely to have used natural hot springs that were flowing at the time. The Indus Valley includes areas abundant with hot springs, which are still utilised today, although seismic events may have relocated these sources of water over time[a]
~Third millennium BC	Italy	Cults and rituals related to hydrothermal manifestations are starting to appear
Third to second millennium BC	Greece	Hydrothermal by-products are extracted and traded in Greece and other Mediterranean areas[b]
1680–1193 BC	Anatolia, Turkey	The Hittite Empires are said to have used hot springs for recreation and therapeutic treatments
1500 BC	Greece	Ancient Greeks are visiting hot springs for hygienic purposes and start building public bathing facilities where medical conditions are also treated
1430 BC	Lipari—Sicily, Italy	Stone-lined ponds fed by hot springs were most likely used for therapeutic thermal bathing
1050–771 BC	Huaqing, China	Huaqing was the favourite hot spring destination of emperors of various dynasties and appears to have been reliably documented[c]
1000–500 BC	Meso-America, Central America	The Mayan empire dates back to about 1000 BC and included Mexico in the Yucatán Peninsula, Guatemala and parts of Belize and Honduras. The abundance of hot springs in these countries leads to the assumption that the Maya used them for various purposes as they were a highly advanced society
~1000 BC	China	Hot springs are used extensively for health and recreation
~1000 BC	Western Europe	West European tradition based on worshipping of sacred and healing springs
Seventh century BC	Greece	Homer refers to thermal springs in his writings
Eighth century BC	Italy	Pompeii was founded around this time and thermal water was used to heat buildings and baths within the city
Eighth century BC	Loutraki, Greece	The thermal waters of *Thermae* (Loutraki) were 'revitalised' around this time for athletic games held at Corinth

(continued)

863 BC	Bath, England	The legend of the founding of Bath tells that Bladud, father of King Lear, cured his skin condition thought to have been leprosy by immersion in the warm springs he discovered while tending to a herd of swine
Eighth to sixth century BC	Etruria, Central Italy	In northern Italy the Etruscans have used thermal bathing and geothermal by-products (750–500 BC). At this time the use of hot springs in areas under Etruscan authority is spreading systematically but with only limited development in other Mediterranean regions[d]
~700 BC	Shikoku, Japan	Dōgo Onsen (Ehime Prefecture on the island of Shikoku) is one of the oldest and best known Onsen in Japan and may have been used even earlier during the Jōmon and Yayoi periods
~420 BC	Greece	Hippocrates of Kos was the first author who systematically classified the waters, including hot springs, and their distinctive properties
Fifth century BC	Greece	Hippocrates of Kos treated patients from all over the Mediterranean by using balneotherapy in his famous Asclepian Centre
Fifth century BC	Spain	Archaeological evidence indicates that the thermal springs of Archena may have been used by early Iberian settlers
Fifth to fourth century BC	Etruria, Etruscan Territories, Italy	Hot spring bathing and the use of hydrothermal by-products is widespread in Etruscan territories—methodical development of hot spring use with some initial interest by the Romans in hot spring locations in Italy for strategic purposes[e]
Fifth to fourth century BC	Roman Republic, Italy	The Roman Republic is engaged in battles with Etruria—Roman soldiers visited hot springs for thermal bathing on military expeditions in Etruscan held territories
Fifth to fourth century BC	Greece	The benefits of hot spring bathing are mentioned by several classical Greek authors: Hippocrates, Plato and Aristotle (Tables 5.1 and 5.2)
Third century BC	İzmir, Turkey	The hot springs of Agamemnon (Balcova Termal) are mentioned in writings by Philostratos, a Greek teacher of philosophy
Third century BC	India	Megastenes, an ambassador from Greece, mentions the medicinal value of hot springs in India
Third to second century BC	Greece	Greece and other Mediterranean countries trade in geothermal by-products, e.g. kaolin, sulphur, pozzolana, etc
Third to second century BC	Etruria, Italy	After the annihilation of the Etruscans by the Romans the use of hot springs temporarily declines in the Etruscan territories—at the same time the Roman elite starts visiting hot springs
Second century BC 133 BC	Hierapolis (Pamukkale), Turkey	Hierapolis is established by Eumenes, King of Pergamon, who develops the local hot springs—in 133 BC the city is bequeathed to the Romans who build large baths near the hot springs
Second century BC	Roman Republic, Italy	Systematic development of thermal bathing culture throughout the Roman Republic – balneotherapy is becoming more prevalent in Roman society
206 BC to 220 AD	China	During the Han Dynasty salt was extracted from hot springs in the Sichuan Province[f]
~120 BC	Aix les Bains, France	The Romans construct comfortable thermal baths and call them *Aquae Grantianae* after taking over the region
84 BC	Aedipsos, Greece	The Roman general Sulla undergoes thermal treatment after which he supports the construction of a 'hydrotherapeutic thermal establishment', the Thermae of Sulla
49 BC	Lisbon, Portugal	The Cassian Spas are built using natural hot mineral springs under Quinto and Lucio Cassio, who represent Julius Caesar during that time[g]
First century BC	Greece	Strabo describes the main hydrothermal manifestations in the Mediterranean area
First century BC	Jordan	Jordan Valley Springs are described in the classical literature by Roman and Byzantine historians (between first century BC and sixth century AD) as having healing properties, especially for leprosy

(continued)

~First century BC	Thermopolis, Wyoming, USA	Big Horn Hot Springs, known as Smoking Waters (Bah-gue-wana) by the Native American Shoshone tribe, are used for their curative value as healing waters. It is assumed that other tribes also made use of hot springs where available and that their use goes back much further than is documented
~First century BC	Palmyra, Syria	The Efca sulphur spring is used for health purposes since 'Biblical Times', but is thought to have either ceased to flow due to over-extraction[h] or has been destroyed in recent wars in the Middle East
First century BC 27 BC	Rome, Italy	The Roman Empire supersedes the Roman Republic—thermal bathing is popular with and without hot springs as the Romans have developed ways of heating their large bathhouses more efficiently—but natural hot springs are the preferred option
First century BC	Pompeii, Italy	The Forum Thermal Baths are established under the Roman dictator Lucius Cornelius Sulla (138–78 BC)
First century BC	Croatia	The first Roman thermae is built in Varaždinske Toplice and named *Aquae Iasae*
First century BC to third century AD	Roman Empire	Development of balneotherapy throughout the Empire—rapid spread of natural hot spring use for health and recreation
First century AD	Italy	Pliny the Elder (23–79 AD) refers to hydrothermal manifestations in his writings
43 AD	Bath, England	The initial development of the hot springs of Bath is thought to coincide with the Roman invasion of Britain in the year 43 AD
50 AD	Austria	Baden near Vienna is founded by the Romans and named *Aquae*
70 AD	Bath, England	The Romans build a more sophisticated bathing complex at Bath including temples and name it *Aquae Sulis*
81 AD	Vizela, Portugal	Roman baths are built under the reign of Titus Flavius
Second century AD	Yarmouk River Valley, Middle East	The springs at Hammat Gader are first exploited by the Romans who convert the site into thermal baths
Second century AD	Tiberias, Israel	The Romans develop the hot springs and build thermal baths
Fourth to fifth century AD	Europe	General decline of hot spring bathing—but still practiced locally at reduced levels
Fifth century AD	Caucasus, Georgia	The city of Tbilisi is founded by King Vakhtang on the site of a natural hot spring with healing powers
~476 or 480 AD	Roman Empire (West)	The fall of the western Roman Empire is initiating a rapid decline of hot spring use as a cause of immorality
552 AD	Japan	Introduction of Buddhism and purification through immersion using thermal springs
~700 AD	Ishikawa Prefecture, Japan	The Yamanaka Onsen of Kaga is founded near a temple—the historical significance is recorded on a scroll depicting the origin of Yamanaka Onsen[i]
720 AD	Matsuyama, Japan	The Dōgo Onsen is mentioned in the Japanese chronicles *Nihon Shoki* (Nihongi)
765 AD	Aachen, Germany	First written record of Aachen, previously named by the Romans *Aquis Villa*
870 to 930 AD	Iceland	The time of Iceland's settlement—according to the Sagas hot springs are used for washing and bathing
900 AD	Lagkadas, Greece	A thermal spa is built and offers medical services as well as recreation to its visitors
~800 to 1200 AD	Kumamoto Prefecture, Japan	The Yamaga Onsen on Kyushu Island is said to have a history of 1200 years going back to the Heian Period (794–1192)

(continued)

1137 or 1175 AD	Cieplice, Poland	Cieplice Śląskie is mentioned in historical records and regarded as Poland's first health resort. However, the legend holds that the hot spring was accidentally discovered by Prince Boleslaus the Tall [decades later] in the year 1175 AD while chasing a wounded deer, which regained its strength from the warm water
1178 AD	Budapest, Hungary	The establishment of a settlement called Felhéviz (Upper Hot Spring) is mentioned in records—descriptions also refer to a hospital being built on the site of today's Hotel Gellért
Twelfth century AD	Budapest, Hungary	Knights of the order of St. John are curing the sick in the area of today's Lukács Bath
1281 AD	Cieplice, Poland	Historical record of the use of hot springs by the Knights of St. John of Jerusalem to cure diseases
1294 AD	Iceland	First written account in the chronicle Oddaverjaannjáll of the original Geysir connected to a series of earthquakes during which large new hot springs appeared and others vanished[j]
~Fourteenth century AD	Belgium	A health resort is founded in the city of Spa
1300s AD	New Zealand	Māori are using hot springs for healing and recreation since they colonised New Zealand between 1320 and 1350
1349 to 1350 AD	Czech Republic	According to legends the hot springs of Karlovy Vary (Karlsbad) are discovered during a deer hunt with settlements emerging close to the thermal springs
1359 to 1389 AD	Bursa, Turkey	The Ottomans build a large complex of domed baths during the reign of Sultan Murat I
Fourteenth to sixteenth century AD	Europe	Early development of spas on sites of ancient medicinal hot springs
1417 AD	Italy	Ugolino de Montecatini (1348–1425), the founder of balneotherapy in Italy, recommends the use of the mineral waters of Montecatini
1458 AD	Asan, South Korea	The Oonyang Spa is in use since the kings of the Joseon Dynasty visited the hot spring area for health and recreation
1485 AD	Portugal	Caldas da Rainha hospital is founded by Queen Leonore, who discovers natural hot springs with healing properties
Fifteenth century AD	Germany	Annual visits to hot spring therapy centres are becoming *en vogue* among the wealthier population
Fifteenth century AD	Budapest, Hungary	Early records about the miraculous hot springs at the site of today's Hotel Gellért (which was built much later)
1500s AD	North America	First recorded history of European style hot spring spas in North America
1502 to 1533 AD	Peru, South America	The hot springs of Cajamarca are used by the Inca ruler Atahualpa and his court as indicated in oral historical accounts
1522 AD	Karlovy Vary (Karlsbad), Czech Republic	In the sixteenth century the water of the Mill Spring is used mainly for bathing—and around 1522 the drinking cure of Karlovy Vary is introduced
1541 AD	Arkansas, North America	The Spanish explorer Hernando deSoto is claimed to be the first European visitor of Arkansas Hot Springs, which then were named the Valley of the Vapours
1545 AD	Brazil, South America	The hot springs of Caldas Novas are first mentioned in Spanish publications
1550 AD	Peru, South America	First European (Spanish) written records of hot springs and their use in Peru
1558 to 1603 AD	Bath, England	During the Elizabethan era Bath is revived as a spa with improvements to the thermal baths and begins to attract members of the aristocracy, who spend the fashionable 'Saison' in Bath

(continued)

1565 to 1570 AD	Budapest, Hungary	Construction of the Király Thermal Bath is started by Arslan, the Pasha of Buda
1566 AD	Budapest, Hungary	The Rudas Thermal Bath is built under Turkish occupation
Seventeenth century AD	Hainan, China (PRC)	Guantang hot springs are already known during the early part of the Qing Dynasty (1644–1911)
1697 AD	Taiwan, China (ROC)	Hot springs are first mentioned in a manuscript (Beihai Jiyou), but not fully developed until 1893
Seventeenth century AD	Europe	Hot spring spas exist throughout much of the European continent
Seventeenth century AD	China	Physicians evaluate and classify several medicinal springs
1709 AD	Japan	First Japanese medical study of hot springs
1722 AD	Caldas Novas, Brazil	Bartolomeu Bueno da Silva encounters hot springs while searching for gold, and settlers start using the thermal water
1737 AD	Belgium	A text about 'Demonstrations on the usefulness of the Mineral Waters of Spa' is published
1741 AD	Avéne, France	The first thermal baths are built
1776 to 1784 AD	Aix les Bains, France	The King of Sardinia, Victor Amedée II, establishes thermal baths, named the 'Etablissement Royal des Bains'
1785 AD	Archena, Spain	King Carlos III orders the complete redevelopment of the spa town Archena, after many facilities were damaged by floods
1786 AD	Villatoya, Spain	Friar Benito Marco, a Franciscan monk from the convent of Mahora, provided one of the first testimonies proclaiming the medical significance of the waters of Villatoya together with a precise map of the springs in the region
1790 AD	New York State, North America	Saratoga Hot Springs are offering spa treatments and accommodation
1798 AD	Villatoya, Spain	The thermal spa Baños de la Concepción is mentioned in written documents
Early eighteenth century AD	Budapest, Hungary	First water analyses are carried out in Budapest on the orders of Maria Theresia (Archduchess of Austria, Queen of Hungary and Bohemia, Holy Roman Empress)
Nineteenth century AD	North America	Hot spring therapy is becoming popular
Nineteenth century AD	Furnas—São Miguel Island—Azores, Portugal	The spa resort of Furnas has become a popular tourist destination, although locals most likely used the volcanic hot springs since the Azores islands were first settled during the fifteenth and sixteenth centuries
Early 1800s AD	Austria	Baden near Vienna is redeveloping thermal bathing traditions and establishes new spa town policies
Early 1800s AD	South Africa	Europeans are visiting South Africa in search of a miracle cure[k]
1826 AD	Hawaii, North America	An early missionary and historian, William Ellis, describes the use of a hot spring-fed crater lake in the Kapoho area
1832 AD	Hot Springs—Arkansas, North America	The area of Hot Springs becomes the first national reserve in the United States
1834 AD	Chile, South America	Jahuel Hot Springs, which is considered one of Chile's oldest thermal centres, is visited by Charles Darwin in 1834
1851 AD	South Africa	The first research on hot springs is conducted in South Africa
1859 AD	Hanmer Springs, New Zealand (South Island)	The hot springs at Hanmer Springs are discovered by an early pioneer but were long before used by travelling Maori[l]
1873 AD	Namibia, Africa	The potential of the hot springs of Warmbad (Bela Bela) is discovered by white settlers, but the thermal waters were known earlier to local tribes who used it for its curative benefits

(continued)

1876 AD	Uganda, Africa	The African explorer HM Stanley visits Mtagata Hot Springs in Uganda—local inhabitants appear to have used thermal springs for their curative properties (see Fig. 5.2)
1878 AD	Rotorua, New Zealand	The Priest Spring is discovered by a catholic priest, who cures his arthritis in the hot spring
1879 AD	Helidon, Australia	Initial development of the Helidon Spa in the state of Queensland
1882 AD	Banff, Canada	The Banff Hot Springs are discovered and used mainly by European visitors
1884 AD	Budapest, Hungary	The Lukács Bath Spa Hotel is built
1891 AD	Bulgaria	The first 'Law for the preservation of mineral springs' is enacted in Bulgaria and springs are declared a national property by law
1891 AD	Issyk-Ata, Kyrgyzstan	The Issyk-Ata sanatorium is established as a thermal health facility
1892 AD	Portugal	The first official regulation concerning the Portuguese waters resources is published
1893 AD	Matsuyama, Japan	The current main Dōgo Onsen Honkan building, a public bathhouse, is constructed to join seven original buildings together
1895 AD	Moree—New South Wales, Australia	The first bore into the Great Artesian Basin in NSW is completed with the water gaining fame for its use in Moree's hot artesian baths, which are said to heal numerous ailments and are used for health and recreation
1895–1945	Beitou, Taiwan (ROC)	Taiwan's hot springs are developed under Japanese occupation—opening of first hot spring resort in Beitou
Early 1900s	Kangal, Turkey	The Kangal hot springs attract initial attention when a shepherd hurt his foot, only to see it healed by the warm spring water
1911	Karlovy Vary (Karlsbad), Czech Republic	Karlovy Vary is reaching the highest number of spa guests in its history with 70,935 people undergoing a cure in the year 1911
1918	Budapest, Hungary	The Gellért Baths are established with indoor and outdoor pools
1925	Hajdúszoboszló, Hungary	Natural thermal springs with healing powers are discovered at Hajdúszoboszló while drilling for oil
1929	Bulgaria	The Bulgarian Balneological Society is set up and unites specialists from different scientific fields
1931	Japan	Japan starts to carry out more scientific research programs into hot springs
Early 1930s	Budapest, Hungary	Budapest is awarded the title 'Spa City' as the capital with the most healing hot springs in the world
1937	Budapest, Hungary	The first International Balneological Congress takes place in Budapest with the seat of the international Balneological Association established at the Gellért Thermal Bath
1940	North America	Most American hot spring spas and resorts went into decline around this time
1940s	Tbilisi, Georgia	Establishment of the Tbilisi Balneological Health Resort
1945	Taiwan (ROC)	Hot spring culture went into decline after the Japanese occupation ended
1970	Aix les Bains, France	Restoration and enlargement of the thermal baths
1975	Tunisia, Africa	The Tunisian *Office du Thermalisme* is established, located in the capital Tunis
1976	Svartsengi (near Grindavík), Iceland	The Svartsengi geothermal power station is built, which creates a large lagoon of surplus geothermal water and shortly thereafter an employee notices the healing properties of the water
1978	Bath, England	The old thermal baths are closed down and bathing is prohibited due to public health concerns over the purity of the spring water
1979	Budapest, Hungary	A daytime hospital with complex thermal bathing facilities is established at the Lukács Thermal Bath

(continued)

1982	Huaqing (near Xi'an—Shaanxi Province) China	Ruins of the imperial hot spring pools in Huaqing are discovered and restored
1987	Blue Lagoon (near Grindavík), Iceland	The first public bathing facilities of the Blue Lagoon open next to the Svartsengi geothermal power station
1987	Bath, England	The city of Bath is inscribed in the UNESCO World Heritage list
1989	Karlovy Vary (Karlsbad), Czech Republic	A new era of developing balneotherapy centres starts in the Thermal Spring Valley
1994	Portugal	Forty (40) health spas throughout Portugal are using natural hot mineral springs. Of these springs, seventeen (17) are permitted by government regulation to treat dermatologic diseases
1994	Hainan, China (PRC)	The National Mineral Storage Resource Committee verifies and approves natural hot springs in Guantang as suitable for tourism and medical treatment
1995	Japan	Hospitals integrate spa medicine using volcanic hot springs
1997	Argentina, South America	After successful drilling for hot water the first thermal spa in the northeast of Argentina opens in 1997
1999	Taiwan	Hot spring culture is making a comeback due to large-scale promotion by the Taiwanese government
2004	Mývatn, Iceland	In the northeast of Iceland, the Mývatn Nature Baths (*Jarðböðin við Mývatn*) open
2005	Victoria, Australia	The first stage of the Peninsula Hot Springs Centre, a new development with natural hot springs, opens in the Mornington Peninsula south of Melbourne
2005	Blue Lagoon, Iceland	A new clinic for psoriasis patients opens, using natural geothermal seawater for treatment
2006	Bath, England	Opening of the new Thermae Spa Bath after years of redevelopment of the existing natural hot springs
2018	Victoria, Australia	Major expansion of the Peninsula Hot Springs in response to increasing visitor demand—further development projects are planned (Appendix 6.2)
2019	Beppu, Japan	New large hotel development—the ANA InterContinental Beppu Resort & Spa offers hot spring heritage combined with luxury accommodation
2020	Húsavík, Iceland	Opening of GeoSea, the geothermal seawater baths on the coast with views over the Arctic Ocean

Sources include [a]Erfurt-Cooper and Cooper (2009)
[b]Cataldi and Chiellini (1995),
[c]Schafer (1956)
[d,e]Cataldi et al. (1999)
[f]Wang (1999)
[g]Picoto (1996)
[h]Waltham (1996)
[i]Yamanaka Onsen (2019)
[j]Hroarsson and Jonsson (1992)
[k]Spicer and Nepgen (2005)
[l]Hanmer Springs Thermal Pools & Spa (2019)

Appendix 5.2 Hierapolis and the Travertine Terraces of Pamukkale

The original painting shows the ruins of Hierapolis, then known as Pambouk Kalesi (meaning cotton castle), which is now called Pamukkale. It was painted by Thomas Allom on the way from Laodicea circa 1838. Later reproductions were coloured in (compare image at the beginning of this chapter. *Source* Allom and Walsh (1838)

Appendix 5.3 Hot Spring History—Then and Now

1, 2—Caldeira das Furnas (São Miguel, Azores) c. early 1900s and the same location in 2014; 3, 4—Inside the bathhouse of Beitou (Taiwan) when it was still in use and the empty pool now; 5, 6—The hydrothermal phenomena of Beppu (Japan) then and now; 7, 8—Maruia Hot Springs (New Zealand) and the first bathing huts c. early 1900s. On the right a contemporary view of one of the new buildings

Appendix 5.4 Plan of the Roman Bath Complex in Bath, UK

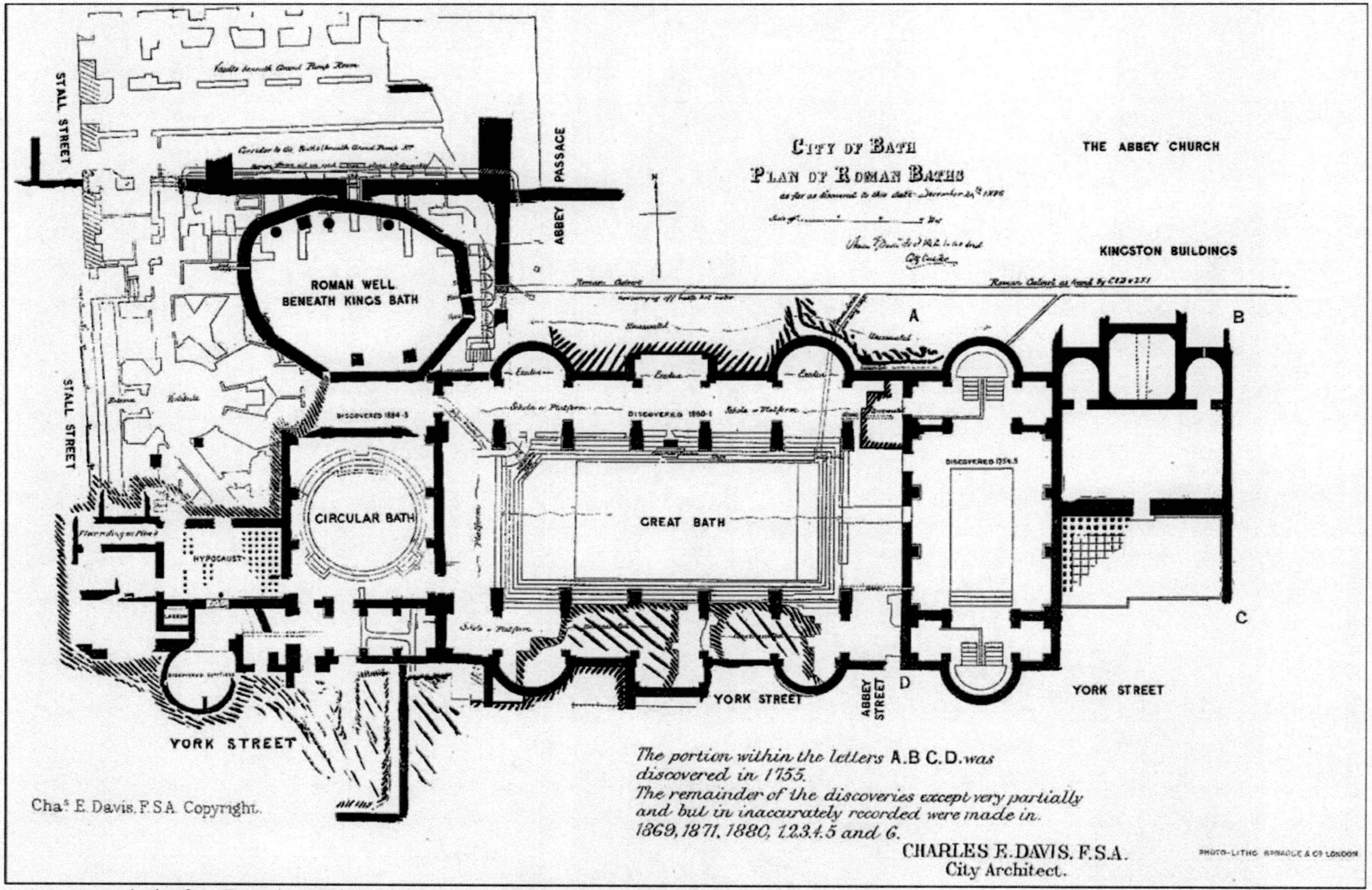

A plan from December 1886 shows the excavated sections of the Roman Bath Complex in Bath, UK

Appendix 5.5 The King's Bath in Bath, UK

1—A satirical painting titled the 'Comforts of Bath' by Rowlandson (1798); 2—The King's Bath in a painting by John Nixon from 1801—both pictures include King Bladud watching over the healing waters (see Chap. 6; Fig. 6.23)

Appendix 5.6 Bad Aachen (Aix-La-Chapelle, Aquisgrana) and Burtscheid/Aachen, Germany

1—The Emperors Bath (Kaiserbad) in Aachen with floating trays for Food and wine. *Source* Luyken (1682) (Public Domain); 2—*The Spa Park (Kurpark) in Burtscheid/Aachen. Source* Unknown author (1822) (Public Domain)

Appendix 5.7 Peninsula Hot Springs, Victoria, Australia

1—The original development site of the Peninsula Hot Springs started as a horse paddock in 1997 on the Mornington Peninsula in Victoria (Australia); 2—The aerial view shows the Peninsula Hot Springs in 2019 —the location is the same as the image above and has been transformed into a hot spring paradise with over sixty thermal bathing pools. In the foreground is a stage for cultural performances, which take place on a regular basis; 3, 4—the shows on the stage can be observed from the hot pools, which were built into the hillside to create a natural amphitheatre

A drawing by Joseph Latour from the nineteenth century depicts visitors at a ferruginous (chalybeate) spring in Luchon, France. Today Bagnères-de-Luchon is a fashionable health resort and famous for its thermal springs (up to 65 °C) and its Roman heritage

References

Ah Chee D (2014) Indigenous people's connection with Kwatye (water) in the Great Artesian Basin. Presented at the Australasian Spa Conference 2014

Ákoshegyi G, Árpási M (2005) Multipurpose utilisation of thermal water in hungary prospect and future. In: Proceedings World Geothermal Congress, Antalya, Turkey, 24–29 Apr 2005

Alam U, Sahota P, Jeffrey P (2007) Irrigation in the Indus basin: a history of unsustainability? Water Sci Technol Water Supply 7 (1):211–218. https://doi.org/10.2166/ws.2007.024

Allom T, Walsh R (1838) Constantinople and the scenery of the seven churches of Asia Minor. Drawings from Nature by Thomas Allom, Historical Account by Robert Walsh. Fisher, Son & Co., London

Altman N (2000) Healing springs—the ultimate guide to taking the waters. Healing Arts Press, Rochester

Amin A, Ahmed I, Salam N, Kim B-Y, Singh D, Zhi X-Y, Xiao M, Li W-J (2017) Diversity and distribution of thermophilic bacteria in hot springs of Pakistan. Microb Ecol 74(1):116–127. https://doi.org/10.1007/s00248-017-0930-1

Anatolia.com (2019) Bursa Hot Springs. Retrieved from https://anatolia.com/activities/spas/bursa/index.html

Anon (1910) Australia's wonderful mineral water Helidon Spa. Established 1879, Supreme Today. Unknown Publisher

Asensio P (2002) Spa & Wellness Hotels. teNeues Publishing Group, New York

Bachvarov M (2004) Spas in Central-Eastern Europe between decline and revitalisation. University of Lodz, Lodz

Bakht MS (2000) An overview of geothermal resources of Pakistan. Geological Survey of Pakistan. In: Proceedings World Geothermal Congress 2000 Kyushu—Tohoku, Japan, 28 May–June 10

Bath & North East Somerset Council (2019) The Roman Baths. Retrieved from https://www.romanbaths.co.uk/

Beardsmore GR, Hill AJ (2010) Australia—country update. In: Proceedings World Geothermal Congress 2010. Bali, Indonesia, 25–29 Apr 2010. Paper no. 0150

Beardsmore G, Davidson C, Ricard L, Pujol M, Larking A, Bendall B (2020) Current directions for geothermal energy development in Australia. In: Proceedings world geothermal congress 2020, Reykjavík, Iceland, 26 April–2 May, 2020

BiblePlaces.com (2019) Callirhoe. Retrieved from https://www.bibleplaces.com/callirhoe/

Bischoff MC (2019) Touring hot springs New Mexico: the state's best resorts and rustic soaks, 3rd edn. Falcon Guides, Guilford

Böcskei E, Szita BG (2008) Bath culture and subsidization policy in Hungary. In: Proceedings 1st International Conference on Tourism and Hospitality Management, pp 375–386, 13–15 June 2008, Athens, Greece

Boda I, Timoc C, Bunoiu V (2017) Population dynamics at the spas of Roman Dacia. Case study: the population of Băile Herculane. Studia Antiqua et Archaeologica 23(1):173–185

Bowman M (1998) Belief, legend and perceptions of the sacred in contemporary Bath. Folklore 109:25–31

Brooke B (2019) Taking the Waters in Mexico. Retrieved from https://bobbrooke.com/mexnaturalspas.htm

Budapest.com (2019) History. Retrieved from https://www.budapest.com/hungary/cities/heviz/history.en.html

Bullard L (2004) Healing waters—Missouri's historical mineral springs and spas. University of Missouri Press, Columbia

Burachovič S (2010) Chapters in the history of Karlovy Vary. Retrieved from https://lavernaromana.com/app-content/files/karlovy-vary/history-of-karlsbad-en.pdf

Burachovič S (2019) Journey to the hidden secrets of the most famous Czech spa Karlovy Vary. Translated from Czech. Retrieved from https://tinyurl.com/y52h62xp

Cabell JL (1873) An Account of the Hot Springs, Bath County, Va., and an Analysis of the Waters with a Treatise. University of Virginia. Richmond: Clemmitt & Jones, Book and Job Printers. Historical Collections at the Claude Moore Health Sciences Library. Retrieved from https://exhibits.hsl.virginia.edu/springs/hotcabell/

Cairns Post (1940) Innot Hot Springs. Cairns Post (Qld.: 1909–1954). Retrieved from https://trove.nla.gov.au/newspaper/article/42232374

Calderón GR (1999) Andean cultures and geothermal phenomena in historical chronicles. In: Cataldi R, Hodgson SF, Lund JW (eds) Stories from a heated Earth—our geothermal heritage, Chap. 34, pp. 554–569. Geothermal Resources Council, International Geothermal Association, Sacramento

Cataldi R, Burgassi PD (1999) Flowering and decline of thermal bathing and other uses of natural heat in the Mediterranean area, from the birth of Rome to the end of the first millennium. In: Cataldi R, Hodgson SF, Lund JW (eds) Stories from a heated Earth—our geothermal heritage, Chapter 10. Geothermal Resources Council, International Geothermal Association, Sacramento, pp 147–163

Cataldi R, Chiellini P (1995) Geothermal energy in the Mediterranean area before the Middle Ages. In: Proceedings of the World Geothermal Congress, Florence, Italy

Cataldi R, Hodgson SF, Lund JW (eds) (1999) Stories from a heated Earth—our geothermal heritage. Geothermal Resources Council. International Geothermal Association, Sacramento

Chandrasekharam D (1999) A Prehistoric view of the hot springs of India. In: Cataldi R, Hodgson SF, Lund JW (eds) Stories from a heated Earth—our geothermal heritage, Chapter 22. Geothermal Resources Council, International Geothermal Association, Sacramento, pp 356–365

Chaterji GC, Guha SK (1968) The Problem of origin of high temperature springs in India. In: Proceedings of symposium II—International Geological Congress, Genesis of Mineral and Thermal Waters, Report of the Twenty-Third Session Czechoslovakia 1968

Chen O, Han D (2019) Rediscovering the idea of cultural heritage and the relationship with nature: four schools of essential thought of the ancient Han Chinese. Heritage 2:1812–1834. https://doi.org/10.3390/heritage2030111

China Holiday (2019) Huaqing Pool, Xi'an. Retrieved from https://www.chinaholiday.com/scenic-of-xian/xian-huaqing-pool-xian/

ChinaKnowledge.de (2019) Shuijingzhu 水經注. Retrieved from https://www.chinaknowledge.de/Literature/Science/shuijingzhu.html

Città di Abano Terme (2019a). Abano, 'acqua Termale e il Fango. Retrieved from https://www.abanoterme.net/right/visitare-abano-terme/benessere-alle-terme.html

Città di Abano Terme (2019b) La storia—origine e sviluppo della città. Retrieved from https://www.abanoterme.net/right/visitare-abano-terme/storia.html

Clark S (1999) Japan, a view from the Bath. University of Hawaii Press, Honolulu

Cohut I, Árpási M (1999) Ancient uses of geothermal waters in the Precarpathian area of Romania and the Pannonian Basin of Hungary. In: Cataldi R, Hodgson SF, Lund JW (eds) Stories from a heated Earth—our geothermal heritage, Chapter 15. Geothermal Resources Council, International Geothermal Association, Sacramento, pp 238–249

Danino M (2018) Harappan Civilization. Lecture given. Retrieved from https://www.youtube.com/watch?v=_s5EDxuL0Vs

Dasgupta T, Chattopadhyay RN (2004) Ecological contradictions through ages: growth and decay of the Indus and Nile Valley

Civilizations. J Human Ecol 16(3), 197–201. https://doi.org/10.1080/09709274.2004.11905738

Davidson C (2006) Peninsula Hot Springs—A developer's story. GHC Bulletin, Sept 2006

Davidson C, Anderson T, Stanley DR (2011) Development of geothermal waters for recreational purposes—Mornington Peninsula Australia. Presented at Australian Geothermal Energy Conference 2011

De Cazanove O (2007). Etruria. In: Iles Johnston S (ed) Ancient religions. The Belknap Press of Harvard University Press, London, pp 220–224

Dryglas D, Hadzik A (2016) The development of the thermal tourism market in Poland (Rozwój rynku turystyki termalnej w Polsce). Geotourism/Geoturystyka 46–47:27–42. https://doi.org/10.7494/geotour.2016.46-47.27

Eck W (2016) Die römische Armee und der Ausbau der heißen Bäder von Hammat Gader. In: Schaps DM, Yiftach U, Dueck D (eds) When West Met East. The Encounter of Greece and Rome with the Jews, Egyptians, and Others. EUT Edizioni Università di Trieste, pp 117–130. https://hdl.handle.net/10077/14683

Einstein N (2005) Etruscan civilization map. Public domain, Creative commons—GNU-FDL. Retrieved from https://commons.wikimedia.org/wiki/File:Etruscan_civilization_map.png

Emilia-Romagna Turismo (2019) Salsomaggiore Terme. Official Tourist Information Site. Retrieved from https://tinyurl.com/t7k87sj

Encyclopaedia Britannica (2019) Aquincum—Ancient Settlement, Hungary. Retrieved from https://www.britannica.com/place/Aquincum

Erfurt-Cooper P, Cooper M (2009) Health and wellness tourism: spas and hot Springs. Channel View Publications, Bristol

Erfurt-Cooper P, Cooper M (2010) Volcano and geothermal tourism: sustainable geo-resources for leisure and recreation. Earthscan, London

Erfurt P (2012) An assessment of the role of natural hot and mineral springs in health, wellness and recreational tourism. Unpublished Doctoral Dissertation, School of Business, James Cook University, Cairns, Queensland, Australia

European Historic Thermal Towns Association (ehtta) (2019) Montegrotto Terme (Italy). Retrieved from https://ehtta.eu/portal/montegrotto-terme-italy/

Faywood Hot Springs (2019) History of the Faywood Hot Springs Resort. Retrieved from https://faywoodhotsprings.com/about-us

Frazier D (2000) Colorado's hot springs, 2nd edn. Pruett Publishing Company, Boulder

Freedman D, Waugh MA (1996) The spa and sexually transmitted diseases. Clin Dermatol 14(6):577–582. https://doi.org/10.1016/S0738-081X(96)00088-0

Fridleifsson IB (1995) Historical aspects of geothermal utilization in Iceland. In: Proceedings of the World Geothermal Congress 1995, Firenze, Italy, pp 389–394. Retrieved from https://www.geothermal-energy.org/pdf/IGAstandard/WGC/1995/1-fridleifsson.pdf

Fytikas M, Leonidopoulou GM, Cataldi R (1999) Geothermal energy in ancient Greece: from mythology to late antiquity (3rd Century A. D). In: Cataldi R, Hodgson SF, Lund JW (eds) Stories from a heated Earth—our geothermal heritage (Chap. 6). Geothermal Resources Council, International Geothermal Association, Sacramento, pp 69–101

Garcia CRA (2007) Escape to the hot springs. The Korea Times, Arts and Living. Retrieved from https://www.koreatimes.co.kr/www/news/art/2008/04/203_15863.html

Glimpses of New Zealand (1896) Beautiful Britain of the South. New Zealand Publishing Company Ltd., Christchurch

Green K (2004) Bath: valley of the sacred spring. Second Nature Press, Bath

Griggs P (2013) Taking the waters: mineral springs, artesian bores and health tourism in Queensland, 1870–1950. Queensland Rev 20 (2):157–173. Cambridge University Press. https://doi.org/10.1017/qre.2013.18

Gudmundsdóttir M, Brynjólfsdóttir Á, Albertsson A (2010) The history of the Blue Lagoon in Svartsengi. In: Proceedings of the World Geothermal Congress 2010, Bali, Indonesia. Retrieved from https://tinyurl.com/rpmh7b9

Guide-Bulgaria.com. (2019) Aqua park Waterland. Retrieved from https://visit.guide-bulgaria.com/a/100/aqua_park_waterland.htm

Habermehl R, Pestov I (2002) Geothermal resources of The Great Artesian Basin, Australia, Bureau of Rural Sciences. GHC Bulletin, June 2002. Geo-Heat Center, Oregon

Ha'Eri B (2011) Karlovy Vary Mill Colonnade. Retrieved from https://commons.wikimedia.org/wiki/File:2007-KarlovyVary-MillC-002.jpg

Haley AJ, Snaith T, Miller G (2005) The social impacts of tourism—a case study of Bath, UK. Ann Tourism Res 29(3):429–438. https://doi.org/10.1016/j.annals.2004.10.009

Hanmer Springs Thermal Pools & Spa (2019) Hanmer Springs History. Retrieved from https://hanmersprings.co.nz/about/our-story/our-history/

Hann S-K (1996) Mineral water and Spas in Korea. Clin Dermatol 14 (6):633–635. https://doi.org/10.1016/S0738-081X(96)00096-X

Hao W, Wang X, Xiang Y, Gu Li AM, Zhang X (2011) History of hot spring bath treatment in China. Europe PMC Repos 41(4):235–239 (PMID:22169492)

Haraldsson IG, Lloret Cordero A (2014) Geothermal baths, swimming pools and spas: examples from Ecuador and Iceland. Presented at "Short Course VI on Utilization of Low- and Medium-Enthalpy Geothermal Resources and Financial Aspects of Utilization", Santa Tecla, El Salvador, 23–29 Mar 2014

Harrison T (2018) Moree Artesian Baths. Image used with permission

Havins PJN (1976) The spas of England. Robert Hale & Company, London

Hernández-Galán JL, Lámbarri JG, Suárez Arriaga MC (1999) An overview of the historical aspects of geothermal influences in Mesoamerica. In: Cataldi R, Hodgson SF, Lund JW (eds) Stories from a heated Earth. Geothermal Resources Council, International Geothermal Association, Sacramento, pp 518–532

Hjálmarsson JR (1993) History of Iceland—from the settlement to the present day. Iceland Review/Edda Publishing, Reykjavík

Hodgson S (2004) A beautiful spa—thermal waters at San Bartolo Agua Caliente, Mexico. GHC Bulletin, June 2004. Geo-Heat Center, Oregon

Hotta A, Ishiguro Y (1986) A guide to Japanese hot springs. Kodansha International, Tokyo

Hristov V, Benderev A, Bojadgieva K (2010) Assessment of hydrogeological and hydrochemical conditions in Kazanlak Basin (Bulgaria). In: Proceedings World Geothermal Congress 2010. Bali, Indonesia, 25–29 Apr 2010

Hróarsson B, Jónsson SS (1992) Geysers and hot springs in Iceland. Málog Menning, Reykjavik

Hungarian Digital National Archive (2019) Bathing Culture—Hévíz. Retrieved from https://en.mandadb.hu/cikk/650474/Bathing_Culture

Imbach T (1993) Environmental hydrogeology of a Karst system with thermal and normal groundwaters: examples from the Bursa Region (Turkey). In: Hydrogeological processes in Karst Terranes (Proceedings of the Antalya Symposium and Field Seminar, October 1990). IAHS Publ. no. 207, 1993

Ivanhoe (2011) Karlovy Vary Colonnade. Retrieved from https://tinyurl.com/y85math2

Jackson S (2007) Hot springs of New Zealand. Reed Publishing, Auckland

Jaffé FC, Dvortjetski E, Levitte D, Massarwieh R, Swarieh A (1999) Geothermal energy utilisation in the Jordan Valley between Lake Kinneret and the Dead Sea: a view from antiquity. In: Cataldi R, Hodgson SF, Lund JW (eds) Stories from a heated Earth—our geothermal heritage, Chap. 4. Geothermal Resources Council, International Geothermal Association, Sacramento, pp 34–49

Kangal Balikli Kaplicasi (2019) Doctor fishes—treatment. Retrieved from https://kangaldoctorfishes.com/treatment/

Kangal Fish Balikli Kaplica Treatment Center (2019) A most unique place in the world, famous for its healing waters and effective psoriasis fish treatment pools. Retrieved from https://www.kangalfishbalikli.com/

Karlovy Vary (2019) The history of balneology in Karlsbad (Karlovy Vary). Retrieved from https://www.karlovy-vary.cz/en/about-karlovy-vary/balneology-and-springs

Katsambas A, Antoniou C (1996) Mineral water and spas in Greece. Clin Dermatol 14(6):615–618. https://doi.org/10.1016/S0738-081X(96)00093-4

Kenoyer JM (2019) Mohenjo-Daro, an ancient Indus Valley Metropolis. Retrieved from www.mohenjodaro.net/mohenjodaroessay.html

Kępińska B (2002) Thermal springs and spas in Poland. GHC Bulletin, Mar 2002

Kępińska B (2004) Lectures on geothermal resources and utilization in Poland and Europe. In: Geothermal training programme. Reports 2003, Number 2, pp 1–27. Retrieved from https://orkustofnun.is/gogn/unu-gtp-report/UNU-GTP-2003-02.pdf. ISBN 9979-68-148-9

Kępińska B (2010) Geothermal energy country update report from Poland, 2005–2009. In: Proceedings World Geothermal Congress 2010. Bali, Indonesia, 25–29 Apr 2010

Kristmannsdóttir H (2010) Prospects in balneology to boost the economy and employment of rural NE Iceland. In: Proceedings World Geothermal Congress 2010. Bali, Indonesia, pp 25–29

Kurillowicz K (1995) The Canadian Rockies. Irving Weisdorf & Co., Ltd., Canada

Kwok A (2006) Huaqing Lotus Pool. Public Domain. Retrieved from https://commons.wikimedia.org/wiki/File:Hua-Qing-Chi-Lian-Hua-Tang.jpg

Lambert T (2019) A brief history of Bath, England. Retrieved from https://www.localhistories.org/bath.html

LaMoreaux PE, Tanner JT (eds) (2001) Springs and bottled waters of the world—ancient history, source, occurrence, quality and use. Springer, Berlin

Leavy HR, Bergel RR (2003) The spa encyclopaedia: a guide to treatments & their benefits for health & healing. The Day Spa Association. Thomson Delmar Learning, New York

Ledo E (1996) Mineral water and spas in Spain. Clin Dermatol 14 (6):641–646. https://doi.org/10.1016/S0738-081X(96)00098-3

Lee T (2014) The provinces of Rome in 117. Public Domain. Retrieved from https://www.vox.com/world/2018/6/19/17469176/roman-empire-maps-history-explained

Lerner CB (2001) The "River of Paradaise" and the Legend about the City of Tbilisi: a literary source of the legend. Folklore 16. 10.7592/FEJF2001.16.tbilisi

Leydecker G (2011) Erdbebenkatalog für Deutschland mit Randgebieten für die Jahre 800 bis 2008 (Earthquake catalogue for Germany and adjacent areas for the years 800 to 2008). Geologisches Jahrbuch, Bundesanstalt für Geowissenschaften und Rohstoffe (BGR) Hannover

Lienau PJ (1993) Fairmont Hot Springs resort. GHC Bulletin, March 1993. Geo-Heat Center, Oregon. Retrieved from https://tinyurl.com/uufda2z

Livingston M, Beach M (2002) Steps to water: the ancient stepwells of India. Princeton Architectural Press, New York. ISBN 1-56898-324-7

Lottermoser BG, Cleverley JS (2007) Controls on the genesis of a high-fluoride thermal spring: Innot Hot Springs, north Queensland. Austral J Earth Sci 54(4):597–607. https://doi.org/10.1080/08120090701188988

Lund JW (2000) Balneological use of geothermal water in the USA. GHC Bulletin. Sept 2000. Geo–Heat Center, Oregon. Retrieved from https://tinyurl.com/ug592db

Lund JW (2002) Balneological use of geothermal waters. GHC Bulletin. Geo-Heat Center, Oregon. Retrieved from https://tinyurl.com/rr9dm3h

Lund JW (2003) Hot spring resorts in the Canadian Rockies. GHC Bulletin. Geo-Heat Center, Oregon. Retrieved from https://tinyurl.com/v6owrka

Lund JW (2005) Basic principles of balneology and examples in the United States. Proceedings World Geothermal Congress 2005. Antalya, Turkey, 24–29 April 2005.

Lund JW, Boyd TL (2015) Direct utilization of geothermal energy 2015 worldwide review. In: Proceedings World Geothermal Congress 2015. Melbourne, Australia, 19–25 Apr 2015

Lund JW, Freeston DH (2001) World-wide direct uses of geothermal energy 2000. Geothermics 30(1):29–68. https://doi.org/10.1016/S0375-6505(00)00044-4

Lüscher G (1946) Die Thermen von Baden" und die Mineral und Heilquellen der Schweiz. Buchdruckerei Neue Aargauer Zeitung, Aargau

Luyken J (1682) Aachen Kaiserbad 1682. Engraving. Retrieved from https://commons.wikimedia.org/wiki/File:Aachen_Kaiserbad_1682.jpg

Mackaman DP (2007) Leisure settings: Bourgois culture, medicine and the spa in modern France. University of Chicago Press, Chicago

Maier D, Fiedler HG (2004) Der große Thermenführer: Wellness und Badespaß in Österreich und den Benachbarten Regionen. NP Buchverlag, Wien

Marini L, Cioni R, Guidi M (1998) Water chemistry of San Marcos Area, Guatemala. Geothermics 27(3):331–360. https://doi.org/10.1016/S0375-6505(98)00006-6

Mayer MA (c.1839). Illustrations. In: Paul Gaimard M (ed) Voyage en Islande et au Groenland—Atlas Historique—Tome Deuxième. Arthus Bertrand, Paris

Mehnert MF (2008) Statue of Asclepius. Public Domain. Retrieved from https://commons.wikimedia.org/wiki/File:Asklepios_-_Epidauros.jpg

Melikadze G, Tsertsvadze L, Tsertsvadze N, Vardigoreli O, Barabadze T (2010) Country update from Georgia. In: Proceedings of World Geothermal Congress 2010. Bali, Indonesia, 25–29 Apr 2010. Paper no. 0167

Melillo L (1995) Thermalism in the ancient world. Medicina nei Secoli 7(3):461–483

Meltzer A, Sarker G, Beaudoin B, Seeber L, Armbruster J (2001) Seismic characterization of an active metamorphic massif, Nanga Parbat, Pakistan Himalaya. Geology 29(7):651–654. https://doi.org/10.1130/0091-7613(2001)029%3c0651:SCOAAM%3e2.0.CO;2

Menadue JE (1972) The mineral springs spa waters of Australia—a unique national resource and a challenge for development. Horticultural Press Pty. Ltd., Melbourne

Merida L (1999) Curing blocks and drying fruit in Guatemala. GHC Bulletin, Dec 1999. Geo-Heat Center, Oregon. Retrieved from https://tinyurl.com/ryvaqe9

Michelin (2002) Der Grüne Reiseführer: Portugal, Madeira, Azoren. Michelin Reise-Verlag, Karlsruhe, Germany

Montecatini (2020) Montecatina Terme, Tuscany, Italy. Etching by C. Zocchi after A. Terreni. Retrieved from https://tinyurl.com/wtzqpko

Monti S (2003) Thermalism between past and future. In: Proceedings of the conference 'the cultural turn in geography'. 18–20th of September 2003—Gorizia Campus. Part VI: Tourism, sustainable development and culture turn. Retrieved from www.openstarts.units.it/dspace/bitstream/10077/862/1/f1monti.pdf

Moree History (1995) History of the Moree mineral Baths. Centenary of Moree Bore Baths. Moree and District Historical Society. Yilaalu no 22–1995

Nagy H (2016) Review on the history of Hungarian Bath/spa culture. Sustainable development of tourism market: international practices and Russian experience. In: Book of proceedings of IV international scientific-practical conference, Stavropol, 11–16 Mar 2016, pp 67–72

Oddy J (1999) Sanatoriums: collective memory. Retrieved from www.independent.co.uk/life-style/sanatoriums-collective-memory-1123832.html

Office National du Tourisme (1921) The Spas of France. Ministère de Traveaux Publics, Paris

Omulecki A, Nowak A, Zalewska A (1996) Spa therapy in Poland. Clin Dermatol 14(6):679–683. https://doi.org/10.1016/S0738-081X(96)00104-6

Onsen Y (2019) The history of Yamanaka Onsen. Retrieved from https://www.yamanaka-spa.or.jp/global/eng/history.html

Ottaway P, Cyprien M (1987) A traveller's guide to Roman Britain. Routledge & Kegan Paul, London

Oz U (2007) Die Heilkraft des Toten Meeres von der Antike bis heute. Retrieved from https://archiv.ub.uni-heidelberg.de/volltextserver/8482/1/Oz.pdf

Özçelik S, Polat HH, Akyol M, Yalçin AN, Özçelik D, Marufihah M (2015) Kangal hot spring with fish and psoriasis treatment. J Dermatol 27(6):386–390 (2000). https://doi.org/10.1111/j.1346-8138.2000.tb02188.x. Originally published in 2000

Özgüler ME, Kasap A (1999) The Geothermal history of Anatolia, Turkey. In: Cataldi R, Hodgson SF, Lund JW (eds) Stories from a heated Earth—our geothermal heritage, Chap. 5. Geothermal Resources Council, International Geothermal Association, Sacramento, pp 51–67

Pakistan Water Gateway (2019) Water resources of Pakistan. Retrieved from https://waterinfo.net.pk/?q=node/19

Parks Canada (2019) Canadian Rockies hot springs—Banff, Radium, Miette. Retrieved from https://www.pc.gc.ca/en/voyage-travel/promotion/sources-springs

Pawlikowska-Piechotka A, Piechotka M (2014) Phenomenon of spa town as a health resort and leisure destination. Architecture civil engineering environment. The Silesian University of Technology, pp 23–32

Pearn J, Little V (1998) The taking of the waters: health springs and spa waters of high lithium content at Helidon, Queensland. Collected Papers of the Fifth Biennial Conference of the Australian Society of Medicine. Published in the series of Occasional Papers in Medical History Australia

Pfanz H, Yuce G, Gulbay AH, Gökgöz A (2018) Deadly CO_2 gases in the Plutonium of Hierapolis (Denizli, Turkey). Archaeol Anthropol Sci 11(4):1359–1371. https://doi.org/10.1007/s12520-018-0599-5

Picoto A (1996) Mineral water and spas in Portugal. Clin Dermatol 14 (6):637–639. https://doi.org/10.1016/S0738-081X(96)00097-1

Piso I (2018) Kleinasiatische Götter und Kolonisten in Dakien. Gephyra 15:37–70. Retrieved from https://dergipark.org.tr/tr/download/article-file/467488

Pruska I (2014) Development of health tourism and its impact on shaping healthy lifestyle behavior. In: Dias F, Oliveira S, Kosmaczewska J, Pereira Â (eds) New trends in tourism research—a Polish perspective. Part 4—Health and medical tourism, pp 178–190. GITUR—Tourism Research Group of Polytechnic Institute of Leiria, Portugal. ISBN 978-989-98965-2-9

Qayyum S (2014) Great bath view Mohenjo Daro. Public Domain. Retrieved from https://tinyurl.com/teax2g7

QE Health (2019) Wellness for every body - Our story. Retrieved from https://www.qehealth.co.nz/our-story

Rando E (2003) Ischia, Island of the Soul. Imagaenaria Edizioni Ischia, Ischia

Richards C (2008) Spa wars. The age newspaper—archives. Retrieved from https://www.theage.com.au/lifestyle/spa-wars-20080503-gdsbx8.html

Robbins Schug G, Blevins KE, Cox B, Gray K, Mushrif-Tripathy V (2013) Infection, disease, and biosocial processes at the end of the Indus Civilization. PLoS ONE 8(12):e84814. https://doi.org/10.1371/journal.pone.0084814

Robinson A (2016) The Indus: lost civilizations. Reaktion Books, London

Rowlandson T (1798) The comforts of Bath, 1st edn. London. Reprint Edition 1858, Bath, Somerset. Reprint Edition 1985, Ashton House Publications

Rumpf M, Sollner G (2006) Bädertempel & Kuroasen – Die 75 schönsten Termal- und Kurbäder mit Flair und Tradition in Mittel- und Osteuropa. Merian-Guide, Travel House Media, München

Salgado-Pareja JS (1988) Hydrothermal activity in Mexico—its utilization for heat generation and balneology. Geo-Heat Center Q Bull 1(2), Klamath Falls, Oregon

Schafer EH (1956) The development of bathing customs in ancient and medieval China and the history of the floriate clear palace. J Am Orient Soc 76(2):57–82. https://doi.org/10.2307/595074

Schäffer R, Sass I (2014) The thermal springs of Jordan. Environ Earth Sci 72(1):171–187. https://doi.org/10.1007/s12665-013-2944-4

Schweitzer PA (1892) A report on the mineral waters of Missouri, vol 13. Missouri Geological Survey and Water Resources, Jefferson City

Scully FJ (1966) Hot springs, Arkansas, and Hot Springs National Park. Pioneer Press, Little Rock

Sekioka M, Yoshi M (2000) Country update: report of geothermal direct uses in Japan. In: Proceedings of the World Geothermal Congress 2000. Kyushu-Tohoku, 28 May–10 June, pp 433–437

Selke AC (1936) Geographic aspects of the German tourist trade. Econ Geogr 12(2):205–216

Severne CM (1999) Traditional use of geothermal resources by New Zealand Maori. In: Cataldi R, Hodgson SF, Lund JW (eds) Stories from a heated earth—our geothermal heritage, Chap. 28. Geothermal Resources Council, International Geothermal Association, Sacramento, pp 435–449

Simkins KL (1986) Physical therapy and spa treatment. Med Anthropol Q 17(5):146–147. https://doi.org/10.1111/j.1937-6219.1986.tb01068.x
Slovenia.si (2019) Thermal Spas. Retrieved from https://www.slovenia.si/visit/slovenian-green/thermal-spas/
Slovenian Spas (2015) Healthy waters—Slovenian natural spas. Published by: The Slovenian Tourist Board (STB) and the Slovenian Spas Association (SSA)
Sousa Dias JP (2017) Mineral waters, spas and therapeutics in seventeenth and eighteenth centuries Portugal. Retrieved from https://research.unl.pt/ws/files/4190037/mineral_waters_and_spas_08.pdf
Spicer S, Nepgen J (2005) Holistic holidays in South Africa: health spas, hot springs, magical places and sacred spaces. Human & Rousseau, Cape Town
Staubwasser M, Sirocko F, Grootes PM, Segl M (2003) Climate change at the 4.2 ka BP termination of the Indus Valley Civilization and Holocene south Asian monsoon variability. Geophys Res Lett 30 (8):1425. https://doi.org/10.1029/2002GL016822
Suárez Arriaga MC, Cataldi R, Hodgson SF (1999) Cosmogony and uses of geothermal resources in Mesoamerica. In: Cataldi R, Hodgson SF, Lund JW (eds) Stories from a heated Earth—our geothermal heritage, Chap. 31. Geothermal Resources Council, International Geothermal Association, Sacramento, pp 498–516
Svalova V (2000) The history of geothermal resources use in Russia and the former USSR. In: Proceedings World Geothermal Congress 2000. Kyushu—Tohoku, Japan, May 28—June 10, 2000. Retrieved from https://www.geothermal-energy.org/pdf/IGAstandard/WGC/2000/R0647.PDF
Svalova V (2010) Mineral extraction from brines and geothermal resources complex use in Russia. In: Proceedings World Geothermal Congress, 2010. Bali, Indonesia, 25–29 Apr 2010. Paper no. 2822. Retrieved from https://www.geothermal-energy.org/pdf/IGAstandard/WGC/2010/2822.pdf
Swart G (2006) Massage messages—wet etiquette can raise a sweat. Sydney morning herald. Retrieved from https://www.smh.com.au/lifestyle/massage-messages-20060101-gdmpmn.html
Swarbrick N (2006) Thermal pools and spas. Te Ara—encyclopedia of New Zealand. Retrieved from https://teara.govt.nz/en/thermal-pools-and-spas
Talmadge E (2006) Getting wet—adventures in the Japanese Bath. Kodansha International, Tokyo
Tentotwo (2013) Great Artesian Basin. Retrieved from https://commons.wikimedia.org/wiki/File:Great_Artesian_Basin.png
Termasworld (2016) Excavations in Baths of the Inca. Retrieved from https://termasworld.com/content/view/146/45/.
Terme di Tabiano (2019) The history of Salsomaggiore—Salsomaggiore Terme. Retrieved from https://www.termeditabiano.it/en-US/the-history-of-salsomaggiore.aspx
Thermae Bath Spa (2019) Britain's original natural thermal spa. Retrieved from https://www.thermaebathspa.com/treatment-prices/spa-sessions-cross-bath/
Thermes de Spa (2020) The 3 mineral waters of the Thermes of Spa. Retrieved from https://www.thermesdespa.com/all-about-spa-water/
Tourism Moree (2018) Moree Plains—Artesian Water Country. Promotional literature
Tsankov NK, Kamarashev JA (1996) Spa therapy in Bulgaria. . Clin Dermatol 14(6):675–678. https://doi.org/10.1016/S0738-081X(96)00103-4
UNESCO (2019) Hierapolis-Pamukkale. World Heritage List. Retrieved from https://whc.unesco.org/en/list/485/
UNESCO (2020) Balaton Uplands Cultural Landscape—tentative lists. Retrieved from https://whc.unesco.org/en/tentativelists/6269/
Unknown author (1822) Ansicht des Burtscheider Kurgartens mit dem Trinkbrunnen, später als Victoriabrunnen umgebaut. Retrieved from https://commons.wikimedia.org/wiki/File:Burtscheid_Stich_Kurgarten_1822.jpg
Unknown author (1911) Burtscheid Kurhaus 1911. Old Postcard. Retrieved from https://commons.wikimedia.org/wiki/File:Burtscheid_Kurhaus_1911.jpg
Vassileva S (1996) Mineral water and spas in Bulgaria. Clin Dermatol 14(6):601–605. https://doi.org/10.1016/S0738-081X(96)00090-9
Verdel H, Pittler AP (2003) Kurbäder, Europa Erlesen. Wieser Verlag, Klagenfurt
Visit Hot Springs (2019) Spas and Baths—where unwinding is an art. The best ways to enjoy the spas and baths of Hot Springs. Retrieved from https://www.hotsprings.org/pages/spas-baths/
Visit İzmir (2016) Agamemnon thermal springs. Retrieved from https://www.visitizmir.org/en/district/balcova/where-to-visit/-162790
Waltham T (1996) Field meeting in Jordan and Syria, 14–25 April 1996. Proc Geol Assoc 108:71–77. https://doi.org/10.1016/S0016-7878(97)80007-7
Wang J (1999) Historical aspects of geothermal waters in China. In: Cataldi R, Hodgson SF, Lund JW (eds) Stories from a heated Earth—our geothermal heritage, Chap. 23. Geothermal Resources Council, International Geothermal Association, Sacramento, pp 367–377
Wang G-L, Zhang W, Ma F, Lin W-J, Liang J-Y, Zhu X (2018) Overview on hydrothermal and hot dry rock researches in China. China Geol 1(2):273−285. https://doi.org/10.31035/cg2018021
Waring GA (1965) Thermal springs of the United States and other countries of the world. In: United States Geological Survey Professional Paper, p 492
Webb C (2005) Brothers keep their dream afloat. Retrieved from www.theage.com.au/articles/2005/06/23/1119321850056.html
Weisz G (2001) Spas, mineral waters, and hydrological science in twentieth-century France. Isis 92(3):451–483. https://doi.org/10.1086/385278
Wellcome Library (2018) Mtagata hot springs, Central Africa. Wood engraving from 1878. No. 18398i. Copyright status: The Library of Congress is unaware of any copyright restrictions for this item. Retrieved from https://tinyurl.com/trzefh6
Wheeler M (1968) The Indus Civilization, 3rd edn. Book Club Associates, London
Witcher, J.C. (2002a) Ojo Caliente—America's oldest spa? In: GHC Bulletin, Dec 2002. Geo-Heat Center, Oregon. Retrieved from https://tinyurl.com/vnvcglg
Witcher JC (2002b) Faywood Hot Springs. In: GHC Bulletin, Dec 2002. Geo-Heat Center, Oregon. Retrieved from https://tinyurl.com/reqy9ub
White G (2000) Bath: a world heritage site—the Bath Spa Project. In: GHC Bulletin. Sept 2000. Geo-Heat Center, Oregon
White G (2003) Hot Bath: the story of the spa. Nutbourne Publishing, Bath
White G (2006) The hot springs of Bath: Geology, geochemistry, geophysics. Bath & North East Somerset Council
Woodsworth G, Woodsworth D (2014) A complete guide also includes some hot springs in Washington & Alaska. Gordon Soles Book Publishers Ltd., West Vancouver, Canada

Yeatts DS (2016) Characteristics of thermal springs and the shallow ground-water system at Hot Springs National Park, Arkansas. U.S. Geological Survey—Scientific Investigations Report 2006-5001. Prepared in cooperation with the National Park Service

Yum BW (1999) Historical review of hot spring waters in the Republic of Korea. In: Cataldi R, Hodgson SF, Lund JW (eds) Stories from a heated Earth—our geothermal Heritage, Chap. 24. Geothermal Resources Council, International Geothermal Association, Sacramento, pp 379–391

Zieroth D (1978) Nipika—a story of Radium Springs. Published by Supply and Services Canada, Quebec

Zitterland FWL (1836) Aachen's heiße Quellen—Ein Handbuch für Ärzte, so wie ein unentbehrlicher Rathgeber für Brunnengäste. Reproduction of a book published before 1923: Nabu Press

Hot springs in Japan are closely connected to the spiritual world

Hot Springs and Their Cultural Heritage

6

Contents

P. Erfurt, *The Geoheritage of Hot Springs*, Geoheritage, Geoparks and Geotourism,
https://doi.org/10.1007/978-3-030-60463-9_6

The geothermal system of the El Tatio (Chile) geyser field is part of a volcanic complex with more than 80 active geysers. *Source* Ministerio Bienes Nacionales, 2014 (Public Domain)

Taking the waters embodied a basic and important but fleeting relationship with the natural environment. People searched for the healing powers of nature and found them in mineral springs
(Valenza 2000:3)

6.1 Introduction—No Life Without Water

6.1.1 Cultural Aspects Related to Hot Springs

Since the dawn of human societies, water has been an inseparable symbol of life and the key to physical and emotional health as well as to spiritual wellbeing (King 1966; Ryrie 1999). Hot spring sites have played a consistent role in determining human settlements as they were not only important points of water supply, but they also provided a source of warm and mineral-rich water with healing benefits (Bullard 2004). Water in its pure form is a crucial element to all lifeforms on Earth because the essence of all beings is water (Emoto 2005), and therefore water and life are inseparable. Water is mentioned in the Bible on the very first page in Genesis 1:2, which indicates the significance of this natural element from the beginning of time. The Koran confirms that water is the source of all life on Earth and that by water everything lives (King 1966; Shiva 2002). Similar meanings are included in the Sumerian legends of creation, as well as in legends from the Nile and the Indus Valley civilisations. In Shinto and Buddhist texts and in the Hindu Sutras water is a prerequisite for all life and plays a main role in the creation of the Earth (Erfurt-Cooper 2009; Erfurt 2012).

Ancient cultures including the Chinese, Aztecs, Incas and African peoples believed that moving water had the ability to collect, store and transfer not only energy, but also physical information such as vibration. They also knew ways to store water to prevent its contamination and how to keep it fresh. The Chinese for example saved glacier water in jars made of jade, while the Aztecs and Incas kept water in jars made of

Fig. 6.1 One of the boiling springs at the Haukadalur geyser field, Litli Geyser (Iceland)

the volcanic stone obsidian (Ryrie 1999). Early human societies soon discovered that water is a flexible element that can be used in each of its distinct phases; as a liquid, as vapour (steam, gas) and in its solid state as ice. Presided over by local priests and tribal elders, natural hot and cold springs were held in high regard by the native peoples in many countries, who lived in close association with them. Depending on the region, the belief in gods and other spiritual beings was essential in explaining the diversity of hydrothermal manifestations, which include not only warm springs gently emerging from the ground, but visible and audible spectacles such as violently erupting geysers, boiling springs (Fig. 6.1) and lakes or bubbling mud pools.

The ferocity of high temperature hydrothermal steam eruptions was an opportunity for wily priests and other influencers to demonstrate their power and substantiate their status to maintain control within the community. As societies became more sophisticated, the veneration of water developed into more worldly activities such as health improvements and recreational interests (Valenza 2000), which were readily catered for at Roman *thermae,* Turkish *hammams* or Japanese *onsen*, eventually leading to a modern-day health, wellness and recreation culture spanning the globe (Fig. 6.2).

The cultural knowledge from early civilisations was passed on from one generation to the next and reflects the special cultural, mythological and religious connections different societies had with natural water sources. For example, the ancient Greeks recognised water as one of the four highest elements on Earth and the *Cherokee* Indians still consider water as the lifeblood of the Earth (Ryrie 1999). While the original onset of human interactions with hot springs remains speculative, many ancient traditions and customs are clearly linked to local legends and mythology associated with natural phenomena. Archaeological discoveries of temples, shrines and other signs of worship, which were built as components of ancient thermal bathing facilities, further support the remarkable cultural importance of natural hot springs (Erfurt-Cooper 2009).

Fig. 6.2 Once a popular hot spring spa, the former Rotorua bathhouse (New Zealand) today is home to a museum of art and history. This famous landmark was first opened in 1908 as a major investment in the hot spring tourism industry

This chapter looks at some of the cultural aspects related to natural hot springs and presents an overview of relevant customs and traditions, some of which date back to the earliest civilisations. One of the main objectives is the identification of settings, where hot springs are integrated in health and recreational activities to determine the cultural context of this resource for traditional developments. The analysis of the cultural background of hot spring sites includes examples from several countries, which were evaluated and compared based on their reputation and/or fame for the curative value of hot springs, the traditional ambience (e.g. architecture), and other noteworthy customs such as religious practices. Legends about the origins of hot springs and mythological aspects further provide an insight and an understanding of the importance of natural hot springs in a variety of socio-cultural settings. An assessment of the extensive use of natural hot springs on a global basis (Chap. 5) has substantiated different cultural traditions and customs, which are determined and shaped by the natural and physical environment at hot spring destinations.

6.2 Socio-Cultural Settings

For thousands of years in many societies hot spring sites were centres of cultural activities and communal gatherings. Commonly located in natural environments, hot springs provided, then and now, a place for socialising in a peaceful setting, a relaxing hot bath and a feeling of invigoration and rejuvenation (Erfurt-Cooper 2009; Erfurt 2012). Aside from the physical appeal of natural hot spring environments there are also cultural attractions, which offer unique experiences for visitors from different socio-cultural backgrounds. One of the attractions of hot springs surrounded by nature is the absence of a hierarchical society order, because sharing the experience of a hot bath provides an opportunity to be on equal terms with each other. Not surprisingly, there is a common trend in relation to natural hot spring environments, with the focus on either keeping these settings as natural as possible (Fig. 6.3) or designing hot spring features that replicate the natural landscape (Erfurt-Cooper 2010).

Fig. 6.3 Connecting with nature in the rock pools at Maruia Hot Springs in New Zealand. *Source* James White (2020) (with permission)

6.2.1 Social Settings—Meeting Places

In the northern parts of America, the indigenous peoples valued natural hot springs not just as sacred sites, but they reserved them as a place of peace, as neutral ground for meetings with other tribes and as an area for rest and recuperation after conflict (Birkby 2002; Bischoff 2001; Frazier 2000; Nelson 1999; Nickell 2005; Lund 1993, 2002, 2005). An examination of the American hot spring history implies that the faith in the healing powers of natural hot springs goes back several thousand years, and to this day native Indian peoples have maintained their connection with the sacred waters (Table 6.1), even though most hot springs are now accessible to visiting tourists. But during the off-season, members of the local communities go to these special social settings while there is privacy for their spiritual visits (Frazier 2000).

On the European continent, the Romans were known for the cultivation of thermal bathing to the point of decadence and over-indulgence. Their bathhouses are described as centres of activities that included eating and drinking, social meetings, the closing of business deals, and of course the spinning of political intrigues (Der Spiegel 1987; Erfurt 2012). This is not uncommon in other cultures either who use their thermal pools for meetings of various kinds. The Chinese and Inca rulers even received official visitors and enjoyed political discussions while relaxing in their hot springs (Erfurt-Cooper 2009; Schafer 1956). Turkish hammams, Japanese onsen and Icelandic hot tubs are commonly used for various types of communicative interaction including discussions, debating and the occasional gossip.

Centuries later, many of the original Roman spa towns had evolved into important socio-cultural settings with a unique character, which attracted the upper classes of Europe to these thermal spas. When it comes to prestigious and fashionable social settings the German spa town *Baden Baden* remains one of Europe's most recognised destinations for hot spring related health and recreation (Example 6.1). Located in Baden-Württemberg, Baden Baden is at the top of almost sixty (60) health and wellness resorts and spas in this state in the southwest of Germany (Sanner 2000).

Table 6.1 Native societies of the American continents (north, central, south) used hot springs for health and recreation as well as for cultural ceremonies and religious observance

Country, state	Natural hot spring	Indigenous peoples
Alaska	Kruzgamepa Hot Springs, Kilo Hot Springs, Chief Shakes Hot Springs	Eskimos, Tlingit Indians
Arkansas	Valley of the Vapours (*City of Hot Springs*)	Tunicas and Natchitoches Indians, later Cherokees and Quapaws
California	Agua Caliente	Cahuilla Indians
California	Calistoga in the Napa Valley	Wappa, Pomos and Mayacmas Indians
Canada	Miette (*Fairmont*) Hot Springs (*Medicine Waters*)	Nez Perce, Shoshones and Flathead Indians
Canada	Radium Hot Springs	Kootenai and Apache Indians
Colorado	Glenwood Hot Springs (*Big Medicine Spring—Yampah*)	Ute, Comanche and Arapaho Indians
Georgia	Warm Springs	Creek Indians and Iroquois Indians
Guatemala	Lake Amatitlan Thermal Springs	Maya
Mexico	Agua Hedionda (*smelly waters*)	Aztecs
Mexico	Tequesquiatlapan (*River with carbonated water*)	Nahua and Chichimeca Indians
Montana	Sleeping Child Hot Springs	Nez Perce Indians
New Mexico	Faywood Hot Springs	Mimbres Culture, Apaches
New York State	Saratoga Springs, High Rock Spring	Mohawks (Iroquois) and Tuscarora Indians
Oregon	Crater Lake, (*Mt. Mazama Caldera*)	Klamath Indians
Oregon	Warm Springs	Paiute Indians, Wasco Indians
Peru	Cajamarca (*Baños del Inca—Inca Baths*)	Inca
South Dakota	Minnekahta Springs	Sioux Indians
Utah	Pah Tempe Hot Springs	Paiute Indians and Navajo
West Virginia	White Sulphur Springs	Shawnee Indians
Wyoming	Yellowstone National Park, Smoking Waters (*Bahguewana*), Thermopolis	Arapaho and Shoshone Indians

Source Compiled by Author from various sources

Example 6.1 The Spa Town Baden-Baden, Germany

Over the last two thousand years the city of Baden Baden evolved from a magnificent thermal spa centre under the Romans into an international hot spring health resort and spa town (Fig. 6.4). While the European bathing culture underwent a change with a renewed appreciation of personal hygiene during the eighteenth century, in more recent times, traditional spa settings such as bathhouses and public pools, were regarded as a feature of the past. In the late 1980s Germany decided that *thermal bath-scapes*, *aquatic temples* and *water palaces* were trending in preference to old-fashioned health resorts. Advertised as *Bathing like the Romans* the spa town Baden-Baden praised the novel spa settings as an *Oasis of Happiness* and a *Bathing Paradise*.

Source Erfurt (2012) and Wettstein (2010)

Some European hot spring spa towns such as Aachen (Germany) and Budapest (Hungary) were visited by kings and emperors, who proclaimed the health benefits of the mineral-enriched thermal waters and proceeded to build summer residences near the hot springs, thereby endorsing a thermal bathing culture for several centuries (Erfurt-Cooper 2009, 2010; Erfurt 2012). In countries with a cooler climate one of the major appeals of public facilities is the pleasant temperature of thermal spring water. Even in summer most public swimming pools remain quite cold unless they are heated. Therefore, hot spring-fed pools and aquatic entertainment centres (Fig. 6.5) in northern European spa towns are a great attraction for recreational purposes (Erfurt-Cooper 2009).

In **France**, one of the key factors that encouraged the revival of traditional thermal spas was the preservation of a distinct socio-cultural environment of indulgence, which played an important psychological role among spa visitors (Boulangé 1982), and is mirrored in other traditional

Fig. 6.4 The screen capture taken from one of Baden Baden's websites depicts a historical scene of daily promenades in the, as it was then known, summer capital of Germany. *Source* Baden Baden (2020)

Fig. 6.5 Looking back, only two decades ago there was little information about hot spring destinations available on the worldwide web. Now every European hot spring spa is promoting health, wellness and recreation online

European spa towns. In **Iceland's** contemporary spa scene hot spring tubs are popular and most guesthouses and hotels make them available to their guests. Public pools fed by natural hot springs or heated with geothermal energy are common in towns all over the country. These are typical Icelandic settings where people socialise in a relaxed atmosphere. However, most tourists in transit from other countries prefer to visit the famous Blue Lagoon, usually on their way to or from the airport.

In **Turkey** and other Mediterranean countries, public hammams (with and without hot spring access), in following Islamic traditions were initially only available to men who used them for spiritual reasons as well as for recreation (Fig. 6.6). Later, women were also given the opportunity to visit local hammams for health and hygiene purposes (Erfurt-Cooper 2009).

Another interesting hot spring setting are the artesian hot springs in the **Australian** desert (Chaps. 4 and 5). Due to

Fig. 6.6 Similar to the Roman thermae a hammam has several areas at different temperatures but is a variant of the steam bath. Some hammams have access to hot spring water, although in general tap water is preferred for purification. Drawing of a Turkish hammam titled "The Bath". *Source* Walsh and Allom (1836)

their remoteness they can be difficult to reach and usually involve a long drive, often only possible with 4WD vehicles —weather permitting, or by plane. Besides, not all artesian springs are open to the public, as some of them are located on private properties including sheep and cattle stations and are used for domestic purposes only (Erfurt-Cooper 2009). Others, such as the Dalhousie Springs Complex (Fig. 6.7) are located in protected areas such as the Witjira National Park and are managed by the traditional owners, the Southern Arrente people and the Irrwanyere Aboriginal Corporation. A difficult task, as the springs were subjected to immense environmental degradation since the arrival of European settlers, who claimed the land for their cattle. Prior to this, for countless generations, the Arrente used the water of the springs for healing and spiritual purposes and as a food source. In fact, the Indigenous name *Irrwanyere* for Dalhousie Springs refers to healing springs. During the time landholders and cattle graziers claimed the use of the springs, they deprived the Arrente people not only of what they considered as their lifeblood, which they needed for their survival, but also of access to places of cultural significance (Ah Chee 2014). In more recent times there is a growing interest in visiting these unusual hot springs in the desert and a campground has been established for ecotourists and adventure seekers. Conservation efforts have also been made by local interest groups such as the Friends of Mound Springs (FOMS) to protect the fragile ecosystems of the springs from further deterioration by establishing fences and controlling invasive species (see Chap. 4).

The most unusual hot spring setting surely must be *Pendulum Cove* on Deception Island (**Antarctica**). Deception Island is an active volcano far down in the Southern Ocean in a somewhat unwelcoming and harsh environment. The submerged crater is surrounded by a black volcanic sand beach, which consists of weathered rock fragments and minerals. However, the unique attraction for cruise ship passengers is the hot water from an underlying hydrothermal system that is seeping out of the ground (after some digging) and creates warm pools in the black sand (Fig. 6.8). Visitors are encouraged to explore the shores and take a soak in the hot spring water, followed by a plunge in the icy seawater, which presents quite a challenge for those who brave the elements. Depending on the weather, this is deemed the highlight of an Antarctic cruise with exceptional photo

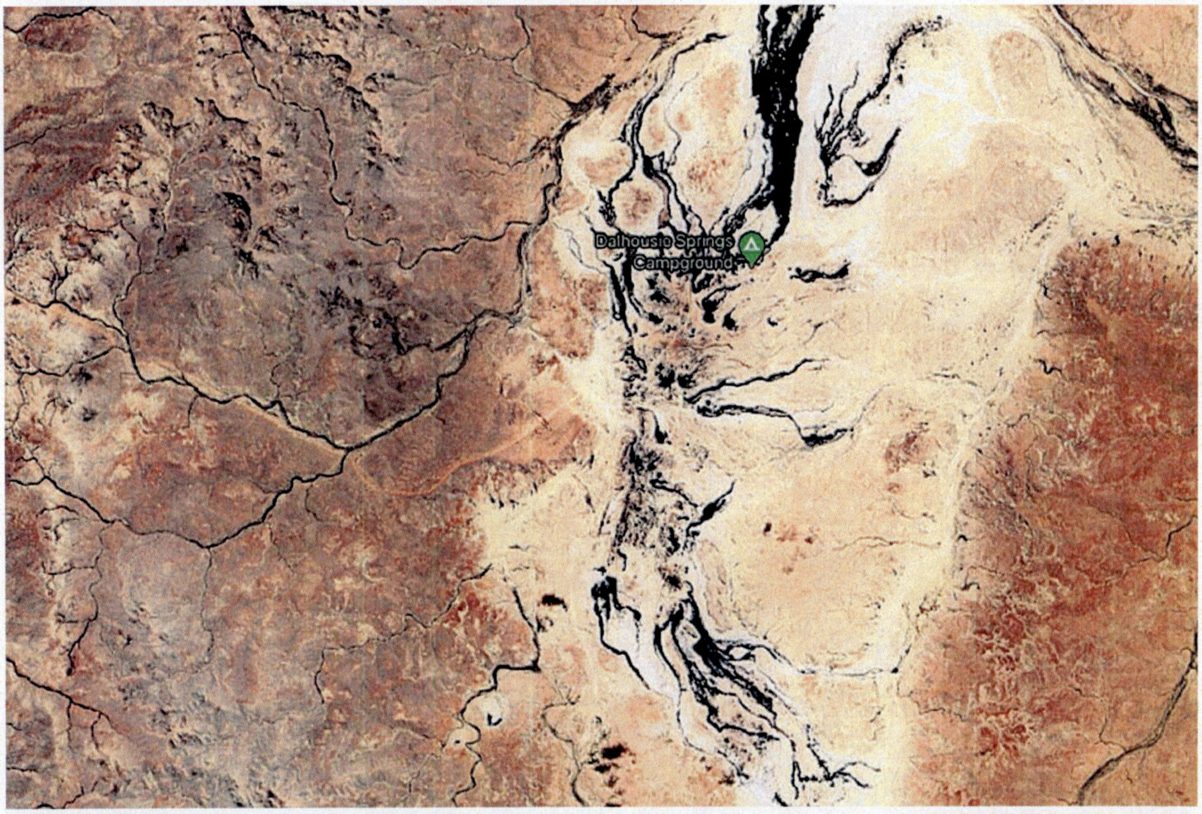

Fig. 6.7 The Dalhousie Springs Complex in the middle of the Australian desert. *Source* Google Map data Google Image © 2020 DigitalGlobe (see also Appendix 4.2)

opportunities. The remarkable environment also is a chance to study the remains of an abandoned whaling station, that briefly operated from 1906 to 1907. All that is left today of a once promising industry are ruins populated by penguins.

6.2.2 Historical Settings—Hot Spring Infrastructure

Expectations of thermal spa towns often include, apart from the actual hot springs, a significant architectural heritage together with a traditional atmosphere. While not every hot spring tourist expects a specific ambience surrounding a hot spring location, many visitors, especially in Europe, are looking for different attractions than backpackers or adventure tourists at wilderness hot springs. Table 6.2 specifies some general visitor expectations at hot spring destinations.

The built environment surrounding hot springs is usually composed of a wide range of structural features ranging from ancient temples and shrines to Roman style bathhouse complexes and modern waterparks. In between, region specific infrastructure combines various styles of architecture and other physical features, which can include hot spring museums or monuments and memorials (Fig. 6.9) associated with the heritage of hot springs.

Apart from purpose-built facilities, the physical surrounds also include the sublime features of a traditional atmosphere, a perception of quality, a sense of place, or a feeling of safety. This type of awareness is of great importance for instance at Japanese onsen, where maintaining the ambience of a natural environment is closely guarded and preserved, especially around external hot spring pools (Erfurt-Cooper 2009; Fig. 11.12). Notable examples of other hot spring settings that have maintained their ambience and their architectural heritage (Table 6.3) include renowned hot spring towns like Aachen (Germany), Budapest (Hungary), Huaqing (China), Rotorua (New Zealand), Kusatsu (Japan), Saratoga (United States), Banff (Canada) or Bath (England).

Because of the recognition of their historical significance, hot spring sites such as the old Roman spa town of Bath have retained much of their traditional atmosphere. The old hot spring architecture has been preserved in synergy with the new development of the Bath Spa Thermae facilities (Figs. 5.16, 5.17 and 6.10), where original walls and foundations add to the distinct ancient Roman hot spring heritage and create an extraordinary atmosphere. In addition, Bath

Fig. 6.8 Not a typical hot spring setting—cruise ship tourists at the hot spring beach on Deception Island. The contrast between the Antarctic temperatures and the warm water seeping from the sand is an unforgettable experience. *Source* Ivanov (2020) (Public Domain)

Table 6.2 Visitors of developed hot spring destinations expect more than just water and health treatments. Quality of service and a satisfactory infrastructure are equally important

Visitor expectations at hot spring settings	
Genuine hot springs	Authentic spring water containing beneficial mineral elements Thermal water especially in cold climates Free of added chemicals
Natural environment	Unique and idyllic landscapes Peaceful surroundings
Curative properties of hot spring water—medically respected	Health improvements—rehabilitation after illness Prevention instead of cure—boosting the immune system Healing treatments, e.g. balneotherapy and qualified medical staff Holistic approach to cater for body, mind and soul
Time-out away from home	Relaxation and stress relief—safe environment Rejuvenation, invigoration, purification
Supporting infrastructure	Fundamental public services—modern comfort, luxury Up-to-date treatment facilities Parks and gardens, traditional hot spring architecture Transport networks—ease of access
Service quality	Cleanliness, hygiene—welcoming atmosphere and friendly, well trained personnel
Recreational attractions	Sightseeing tours Hot spring history, e.g. World Heritage areas Entertainment opportunities (Table 6.4)
Hydrothermal attractions	Visits to hydrothermal sites with extreme hot springs such as geysers, mud pools, boiling ponds

Source Author

Fig. 6.9 The historic 'Rebecca Fountain' is located next to the Bath Abbey and close to the historic Roman hot spring Museum in Bath, UK. The statue dates to 1861 AD and was erected to promote water over drinking alcohol with two inscriptions: 'Take the Water of Life Freely' and 'Water is Best'

has been listed as a cultural World Heritage site in 1987 for its *Outstanding Universal Value* based on the Roman remains, the hot springs and the quality of architecture (UNESCO 2019).

The hot spring architecture of Europe consists to a large degree of traditional clinics for *taking the cure* or *taking the waters*, treatment centres for rehabilitation, hot spring health resorts and spas for medical treatment and recreation, as well as aquatic entertainment centres with interesting architectural features, aesthetic decor and state-of-the-art illumination. These establishments are in stark contrast to Zen Buddhist gardens in Japan, where hot spring visitors find a serene environment for meditation and reflective thoughts as they relax in the thermal waters. However, the original design of European spa settings always had a strong focus on natural environmental features such as expansive parks and gardens, which were usually designed by appointed architects together with the spa buildings (Bacon 1997; Erfurt 2012; Appendix 5.6).

Over the course of many centuries original Roman bath ruins frequently became buried under layers of soil or covered by other buildings, only to be discovered accidentally during later construction work. Wherever possible, historical settings such as former imperial bath complexes are preserved and exhibited in museums of ancient bathing culture, although some places are better preserved and protected than others (Wettstein 2010). Some historic hot spring spas that were left to decline, have been restored and/or redeveloped into modern day health resorts to maintain an aura of days

Table 6.3 Thermal treatment centres and bathhouses were established to take advantage of natural hot spring locations. Not all Roman or Turkish facilities had access to natural hot springs and used heated spring water and steam

Historical and contemporary hot spring settings	
Ancient Greece—hot spring sites with healing waters	Temples combined with thermal treatment centres—focus on worship and healing
Roman empire—purpose-built thermae	Luxurious bathhouse settings—focus on health, recreation and leisure
Ancient American hot spring settings	Considered as sacred places—focus on peace, healing, ceremonies and relaxation
Turkish hammams	Public and private bathhouse settings—focus on hygiene and relaxation (Fig. 6.6)
European spa towns	Thermal treatment centres—focus on health improvements, rehabilitation and opportunities to socialise
French cités thermale	Traditional hot spring spa towns—focus on health, rehabilitation and social recreation
Icelandic hot tubs and thermal pools	Public and private facilities—focus on relaxation, socialising and maintaining good health
Japanese onsen	Public and private hot spring facilities—focus on good health, relaxation, socialising and recreation
Chinese hot spring resorts	Luxury resort facilities—focus on good health, wellbeing, relaxation and recreation
New Zealand hot springs and bathhouses	Visitor attractions with focus on relaxation and wellbeing
Australian artesian pools and spas	Australian hot springs supply thermal pools in rural communities for health and recreational purposes
National Parks, world heritage areas and geoparks with hydrothermal manifestations	Major tourism drawcards worldwide—focus on visual attractions of hydrothermal areas

Source Author

Fig. 6.10 A view inside the restored Cross Bath, which is officially recognised as a sacred site where the Celts worshipped their goddess Sulis. The Cross Bath is part of the redeveloped Bath Spa Thermae but remains a place of contrast, which evokes the atmosphere of an ancient historic setting, surrounded by traditional architecture

Table 6.4 Most spa towns publish a calendar with cultural and social events that take place throughout the year. As hot spring destinations are relatively independent from seasonal changes, they attract visitors all year round

Typical events in European Hot Spring Spa Towns	
Cultural events	Concerts, Theatre and Cabaret, Musicals, Music Festivals, Matinées, Recitals, special Movie Showings
Educational events	Museum Visits, Exhibitions, Public Lectures, Presentations, Seminars, Literary Meetings, Yoga classes, Cooking classes
Outdoor events	Organised City Discovery Tours, Walking Tours Guided Bike Tours, Geocaching, Treasure Hunts
Other activities	Festivals, Fashion Shows, Casino Visits, Sports Events, Exploring Culinary Delights of the Region
Retail events	Arts and Crafts Markets, Christmas Markets (variations according to season)

Source Compiled by Author

Fig. 6.11 The pictogram of a steaming hot spring has become a cultural symbol that is instantly recognised worldwide

gone by, which is sometimes referred to as a contemporary *spa renaissance* (Erfurt 2012; Lee 2004). All over Europe health spa towns have retained a requisite historical ambience, which is an essential component of the experience of 'taking the waters' for health seeking visitors. Although, when the Romans left Turkey for example, many of their remaining thermal baths were destroyed during the following centuries through wars, earthquakes or negligence. Still, much of the Roman thermal heritage was gradually restored and redeveloped using typical Turkish architecture (Özgüler and Kasap 1999).

6.3 The Cultural Significance of Hot Spring Destinations

The use of natural hot springs by many different civilisations was and still is extensive, with cultural traditions at hot spring sites frequently linked to health treatments such as balneotherapy as well as social and recreational activities. The custom of cultural entertainment in spa towns is another important element and usually catered for in the form of an events calendar, which is presented to visitors on their arrival. Depending on the size of the spa town such a list can be quite inclusive of activities for all ages, abilities and interests (Table 6.4). Therefore, depending on individual hot spring destinations, social activities are another significant cultural element.

Over the course of many centuries natural hot springs have become cultural and natural icons worldwide (Fig. 6.11), even in remote locations like Antarctica, and have gained in international significance as tourist destinations. Besides the use of natural hot springs for health, wellness and recreation, most hydrothermal environments hold a cultural appeal, although with regional variations.

6.4 Customs and Traditions

Individual traditions related to hot springs are normally determined by the cultural background of different settings and environments. People in various countries developed their thermal bathing traditions in a range of ways that suited their personal needs (Example 6.2), although culturally influenced customs and traditions regarding the use of hot springs frequently show basic similarities between different countries (Erfurt-Cooper 2009, 2010; Erfurt 2012).

Example 6.2 Roman Bathing Tradition

A typical visit to a Roman bath would last several hours as bathhouses were places of social gathering, bathing and exercise. Apart from the actual bathing facilities, Roman thermae usually comprised of a gymnasium for training, reading rooms or a library, lecture halls, parks and gardens, restaurants and bars, a marketplace, even museums or theatres that featured recitals and acrobatic performances. Visitors of the baths proceeded from one area to the next at leisure, while enjoying the comfort of the well-appointed amenities and the company of their friends.

Source Cartwright (2013) and Erfurt-Cooper (2009).

In **Japan** the tradition of onsen bathing revolves around the abundance of volcanic hot springs in all parts of the country and amounts to an important part of daily life. During the *Edo* period (1603–1867 AD) the use of hot springs extended all over Japan, reaching every level of the population. This tradition of unwinding after work by relaxing in a hot bath as a form of stress relief at the same time provides an opportunity to socialise with likeminded people and has remained a standard practice to this day (Erfurt 2012). While the onsen tradition is not entirely dissimilar from other forms of hot spring use elsewhere, the wealth of volcanic springs in Japan makes thermal bathing for health and recreation an option for everybody (Erfurt-Cooper 2009). Even so, there is a tradition of rejecting onsen patrons if they have tattoos on their body (Fig. 6.12), which is seen as a reference to members of societies connected with organised crime synonymous with the *Yakuza*. But because of the growing popularity of tattoos in Japan, there is a trend towards tattoo-friendly onsen that cater for regular people who are fond of their body artwork.

Another widespread ancient custom is throwing coins into springs, wells and other bodies of water (Figs. 6.13, 6.14), which can still be observed in many parts of the world, e.g. the *Fontana di Trevi* in Rome, hot springs in Japan or New Zealand, water-filled fissures in Iceland or ancient Roman baths in Bath, UK (Appendix 6.1). Even though most people may not realise this and merely acknowledge it as a harmless form of superstition, this nevertheless points to the enduring practice of water worship. This widespread custom of water worship occurs around all types of water and motivates people to part with some small coins in the belief that it will bring good luck and good health. While people may have not explored this custom in more detail, it is closely related to the common wishing wells described in the folklore of various countries.

Fig. 6.12 Multilingual signs like this are common in Japan's public and private bathhouses and leave no doubt that visitors with even the smallest tattoo are unwelcome and will be asked to leave. *Source* Onsen Tipster (2015) (Public Domain)

Throwing a coin into a wishing well, together with a spoken wish or a prayer, is based on the expectation that a wish or a blessing will be granted by the divine guardian spirits in control of the water source.

During archaeological excavations in Jerusalem, coins have also been found in one of the pools of Siloam, which were mentioned in the Gospel of John as a gathering place for pilgrims. This is clearly a habit that has roots all over the world where people use or have used water sources as a connection to the spiritual world to achieve a benefit for themselves. On the downside, the amount of 'coppers and silvers' in some fountains and springs can cause serious damage and must be removed on a regular basis. A common practice is to donate the coins to charities, as the overall amount can add up to a significant sum. This simple ritual of water worship is not restricted to a small number of springs or wells—in fact, there are many thousands of them that are asked for good luck at the price of a few coins.

6.5 Religious Aspects of Hot Spring Use

Natural water sources such as hot and cold springs were usually associated with spiritual beings and divinities of the earth. This is why most countries have religious, spiritual and mythological links to natural springs, which had a strong influence on the development style and purpose of hot spring facilities. In ancient **Greece**, temples in the vicinity of hot springs were usually dedicated to Asclepios (Asclepius, Aesklepios, Asklēpiós), the god of medicine (Fig. 5.1). Here, medicine was practiced and the treatment as well as

Fig. 6.13 A collection of coins has accumulated where the original hot springs of Bath emerge. The surface of most coins has been completely smoothed by the constant flow of the hot water

purification rites were carried out by priests in charge of the purpose-built temples (Erfurt 2012; Katsambas and Antoniou 1996; Melillo 1995; Monti 2003).

The **Croatian** spa town *Varaždinske Toplice* (Aquae Iasae in Roman times) displays several reminders of a religious basis for its continuing existence, including a basilica and a large marble block with reliefs of Roman mythical scenes, including three nymphs as symbols for the power of water. Temples at the hot spring site were dedicated to Jupiter, Juno and Minerva and formed a *capitoleum* (Gamulin 2001). The influence of Roman and Greek deities was also exposed in **Tunisia** (North Africa) during construction work at the hot spring pools of *Hammam Lif,* where inscriptions were found that show dedications to Asclepios as well as to Saturn. At the *Hammam Jedidi* (also in Tunisia) archaeologists discovered a number of marble figurines of Asclepios and his daughter Hygieia (goddess of health) in a cistern, thought to be votive offerings by recovering patients who acknowledged the protection of the foreign Greek gods (Erfurt-Cooper 2009).

Under the guidance of local priests, druids, shamans, or other influential community elders, including healers, witchdoctors and oracles, native peoples in many parts of the world developed a close association with natural hot and cold springs. For example, the sacredness of water is reflected in the beliefs of the native Americans, who considered natural springs and flowing waters as entities full of life and as places on the Earth's surface where humans could connect with the spiritual and the divine (Arvigo and Epstein 2003; Erfurt 2012). Reaching out to the spiritual world in the form of religious ceremonies was practiced by the *P'urhépecha*, an indigenous community who still lives in Michoacán, **Mexico**. The P'urhépecha used the hot springs at *Los Azufres* (Lagoon of Sulphur) to carry out rituals and political ceremonies. At *Las Huaringas* in **Peru** (South America) are seven sacred hot spring-fed lakes, which have

Fig. 6.14 One of many hot springs in Japan that motivate people to part with some small change in return for a blessing in the form of good health or a wish come true

historically been used for rituals as well as for healing ceremonies dating back to the time before the first Spanish explorers arrived. To this day the sacred thermal lakes are considered to be powerful and magical healing waters, which are visited on a regular basis by faith-healers and shamans from other parts of Peru (Erfurt-Cooper 2009).

During their occupation of European territories, the Romans exploited hot springs located in the area of today's city of **Aachen** (Germany), were they established a military settlement and named it *Aquae Granni* after Grannus, the German god for water and healing. In keeping with their customary bathhouse constructions, the Romans typically included temple buildings near their thermal facilities all over the Roman Empire. One of the best-known examples of this approach by the Romans is perhaps Bath in **England**, where they developed *Aquae Sulis,* a centre of religion, healing and pilgrimage (Bowman 1998; Haley et al. 2005; UNESCO 2019). However, the hot springs of this site were already considered as sacred by the Celts, who built a shrine dedicated to the goddess *Sulis* before the arrival of the Romans. This supports the belief of the native Britons that water such as rivers, streams and all springs were a gift from the gods (Ottaway and Cyprien 1987; White 2000). Another reminder of the Roman bathhouse tradition is the spa town *Baden* near Vienna in **Austria**, which was called *Aquae* by the Romans. Set up primarily for the Roman armed forces, the facilities here too included altars near the hot springs, which were dedicated to various divinities. The same was the case in **Spain**, where natural hot springs with curative properties were used in religious centres and as sites for pilgrimages since ancient times (Erfurt 2012).

In **Japan**, hot spring sites usually include small shrines and statues of gods (Fig. 6.15). Another religious symbol are Torii gates of varying sizes, which are positioned either at the entrance or near a *Shinto* temple located at hot springs (Fig. 6.16). Because of the close proximity of active volcanoes in many parts of Japan, opportunities for religious observance are everywhere, not just for worship at the site of healing waters but also for prayers to appease local volcano deities.

Fig. 6.15 Small shrines can be found at most Japanese hot springs with public access. Visitors often leave flowers or gifts on the ledge

6.5.1 Water Worship

The Temple and hot Spring of Aquae Sulis were important as a place of worship. Many people came from far and wide to pray at the Temple and make offerings to the goddess Sulis Minerva at the Spring. Priests led the ceremonies in the Temple courtyard.

Bath, UK—Poster in the Hot Spring Museum.

Ancient cultures worldwide considered water to be a sacred natural resource and of divine origin. The earliest deities were gods and spirits, who were worshipped as guardians of the springs (Arvigo and Epstein 2003; Ryrie 1999). Despite regional variations, the cultural significance of water as a connection to the spiritual world was extremely important to many different societies. As the spiritual and the physical environments met in those early times, the result were various forms of *water worship*, where the life-giving water source is attributed to selected divinities. Unable to explain the occurrence of natural hot springs, these were accepted as a gift from the gods with temples erected in their vicinity, a tradition that is still observed in many countries (Erfurt-Cooper 2009; Erfurt 2012; Lee 2004).

In Europe, the discovery of Celtic shrines at prehistoric settlements indicates that the worship of local water gods by tribal communities goes back well before the introduction of Christianity (Fig. 6.17). One of the key factors supporting water worship were the health-giving properties of natural hot springs and their frequent health improvements, which led in ancient Greece to the dedication of important water

Fig. 6.16 Torii gates emerging from the steam of a large hot spring

sources to relevant deities such as Asclepios, Hygieia, Heracles (son of Zeus), Artemis Thermia and the Nymphs (Fytikas et al. 1999). Also closely linked to water worship was the belief in many parts of Europe that the nymphs and guardian spirits of sacred springs and holy wells can be trusted and will bestow blessings and cures (Arvigo and Epstein 2003; Erfurt-Cooper 2009). Water worship further included votive offerings such as small statues, clay tablets and gemstones as well as living creatures. The offering of such sacrificial objects were deposited in the spring water as a gift to the gods, either with the expectation that the water spirits would reward the gift with another, or after a wish had come true or an illness was cured (Nilsson 1992).

6.5.2 Purification and Cleansing

Another reason for the significance of water 'is its universal high regard as an agent for purification and the ritual washing away of sins' (King 1966:186). When combined with prayer, water is thought to have transformational powers, which refresh, calm and balance the body, mind and soul (Arvigo and Epstein 2003; Erfurt-Cooper 2009). As a medium that is constantly moving, water has the ability to carry away impurities and restore life (Emoto 2006). Religions all over the world have considered water as an element of divine origin, and in many cultures and religions water is equated with energy and existence (Arvigo and Epstein 2003). The Japanese *Records of Ancient Matters* (Kojiki 2012) describe sacred rites of purification by water, and Christian denominations in all parts of the world commonly practice baptism by immersion, the most significant use of water in a symbolic religious sense and a sign of being reborn. In **Iceland**, at the time of conversion to Christianity in the year 1000 AD, many people chose to be baptised using hot springs in preference to immersion in cold water (Erfurt 2012; Frazier 2000; Hjálmarsson 1993). Perceived in this way, hot springs gained in spiritual significance and as a symbol for purification and renewal (Bischoff 2001; Bullard 2004).

For the reasons discussed above, water has been an important medium as a cleansing agent and for the purpose of ablution in various religions (Fig. 6.18). While water was the most common way of purification in the ancient world, to this day ritual baths are used in the Islamic and Jewish religions to restore a contaminated person to a state of purity (Attridge 2007; Bremmer 2007).

Fig. 6.17 The Gorgon's head at this temple is a symbol of Sulis Minerva, which the Romans assigned as their goddess of Aquae Sulis, today known as Bath (UK). The sun-like appearance of the head could suggest a combination of the heat of the hot springs and the double role of Sulis as the Celtic sun goddess as well as the relevant deity to protect the waters

Water has always been associated with religious ceremonies, legends of gods and miraculous healing (King 1966), and natural hot and cold springs were used for religious purposes by many cultures alike. The close connection between water and divinity continues to this day through the association of healing springs and patron saints:

- Saint-Thomas-les-Bains, French Pyrenees
- Thermal baths of Saint Gervais, France
- St. Ann's Well, Buxton, UK
- Terme Sant'Elena (Saint Helena natural thermal springs), Tuscany, Italy
- Thermal springs at Santa Rosa de Cabal, Colombia
- San Marcos (Saint Mark), the patron and protector of Aguascaliente, Mexico
- John the Baptist, patron saint of baptism and healing springs, Germany.

Considering the continuing acknowledgment of patron saints in present times it is therefore no surprise that in the past natural springs were considered sacred and honoured with temple buildings and statues dedicated to important deities (Cataldi and Burgassi 1999; Melillo 1995).

6.6 Legends and Mythology

Myths and legends about the origin of hot springs have been passed on from one generation to the next, and for ancient civilisations spirituality and miraculous health improvements after bathing in natural hot springs were closely related to the spiritual world with countless narratives relating cases of mysterious healing events. Accordingly, many countries have a rich mythology surrounding the discovery of healing springs, which has contributed to the development of traditional hot spring use. Springs credited with curative powers were traditionally assigned a guardian and the spring was named in their honour (Erfurt-Cooper 2009).

The Greek mythology is especially rich in legends of natural hot springs connected to deities like *Artemis Thermia*, the patron saint of all springs, and the *Nymphs* (Naiades), who were responsible for the protection of hot

Fig. 6.18 Water basins for cleansing and purification before entering temple grounds are a common sight in Japan

springs, but were also called upon for good health and other favours (Fig. 6.19). According to mythology, *Apollo Thermios* used hot springs to cure people with medical conditions (Erfurt 2012; Katsambas and Antoniou 1996; Ryrie 1999). However, even though Apollo was considered as a 'healer' under the gods, there are variants in the mythology that describe him also as the bringer of disease.

On occasion there has been a darker side associated with natural hot springs. In the mythology of the ancient *Hittites* for example the mystical netherworld played a significant role with natural springs seen as a passageway to the underworld (Wright 2007). One such passage was the Plutonium in the ancient city of Hierapolis (Turkey), which was described by Pliny (the Elder) and Strabo as Pluto's Gate to hell (Fig. 6.20; Özgüler and Kasap 1999; Pfanz et al. 2018). As one of the most important historical settings, the world heritage listed ruins of Hierapolis were left in their original position and the whole area was turned into an open-air museum with the Plutonium one of the major tourist attractions (Chap. 5).

As a matter of fact, gates to hell are not that uncommon. A Māori-owned geothermal reserve in New Zealand called Hell's Gate is a popular tourist destination of major cultural and social importance (Fig. 6.21). Steeped in local legend, the hellish landscape is said to be a gift from *Rūaumoko,* the god of volcanic activity and earthquakes (Hell's Gate 2019).

The city of Beppu (Japan), which is known for its thousands of hot springs, offers several hellish experiences. As a result of volcanic activity Beppu is a city of steaming lakes, boiling mud pools and erupting geysers, which are located in small hydrothermal parks and gardens known as *jigoku*, the Japanese word for hell (Fig. 6.22). As a special attraction visitors can embark on a tour to ten (10) different hells, the *jigoku meguri*. While sometimes referred to as a pilgrimage, the tour of the hells is not of a religious nature for every visitor, although each hell has several corners dedicated to a quiet moment of worship at one of the numerous shrines (Fig. 6.15) and temples (Erfurt-Cooper 2009).

6.6.1 Legends

Many hot springs became known for their curative powers and were regarded as miraculous, often based on legends about original healing events that happened at a particular hot spring. Such legends have added interest to hot spring sites and are a bonus in the promotion of historical spa towns

Fig. 6.19 Porcelain figure of a Water Nymph emptying an amphora (nineteenth century—Italian origin). *Source* Invaluable (2020)

(Sieveking 1913). Outlined below are some legends connected with the original discovery of individual hot springs in different countries.

In **Japanese** legends, priests were repeatedly led to natural hot springs with curative powers by mystical creatures, which materialised as visions in their dreams (Erfurt 2012; Hotta and Ishiguro 1986). Animals including brown bears and white herons were perceived by the Japanese people as messengers from the gods, who guided humans to locations with healing and purifying springs and were therefore credited with the discoveries of many hot springs. The discovery of the healing hot springs at *Takeo* on Kyushu Island are attributed to Empress Jingu, who reigned during the third century AD. According to the legend she discovered a white heron in the warm spring water (Altman 2000; Erfurt-Cooper 2009). This has parallels with various other countries, where hot springs were also discovered by different animals. For example, in Baden (Switzerland) the legend refers to goats and in Cieplice (Poland) it was a deer that led humans to the healing springs (Erfurt 2012). The legend from **Switzerland** tells about a young man who was looking for one of his goats that had gone missing, when he found the animal near a rocky outcrop where hot water welled up from the ground. The man arranged for his paralysed wife to have a bath in the hot water, and as per the legend the woman was miraculously cured (Erfurt-Cooper 2009; Lüscher 1946). The legend from Cieplice in **Poland** recounts a hunting excursion from the year 1175 AD, when Prince Boleslaus the Tall was following a wounded deer to a spring, where the animal had sought refuge and was drinking water. Apparently, the deer regained enough strength from the mineral-rich water to escape its followers. On closer examination of the water it was found smelling of sulphur, but of a pleasant temperature. This accidental discovery led the Prince to build a hunting lodge near the hot spring (Erfurt-Cooper 2009).

A legend from the **Czech Republic** attributes the discovery of Karlovy Vary's hot springs to Charles IV, King of Bohemia, also during a deer hunt (Altman 2000), and although some authors have dismissed this to the realms of fantasy (Burachovič 2009), as a folktale the theme is similar to many others. A different legend from Karlovy Vary refers to the mysterious Ghost of the Springs, who resided where the Castle Spa is located today. This spiritual being is said to rule over all the hot springs by protecting them from depletion and the harmful interference of disobedient and wicked elements like fire, flood and wind, which continuously endangered the hot springs of Karlovy Vary and thereby the existence of the town itself. Rumour has it that the Ghost of the Springs transformed the rebellious elements into a rock under which he resided and from where he released the forces of nature only under strict supervision (Zámecké Lázně 2019).

In **New Zealand**, the Māori recognised hydrothermal phenomena as a valuable resource, and as a treasured gift, which they inherited from their ancestors (Stokes 2000). According to Māori legends, the geothermal activity in the Taupo Volcanic Zone was created when *Ngātoroirangi*, a powerful high priest who had arrived from the far away Hawaiki, was climbing Mount Tongariro to lay claim to new lands, where he encountered extreme weather and nearly died from the severe cold. He called out to his sisters *Kuiwai* and *Haungaroa* in Hawaiki to send him fire to warm him. The sisters came, carrying fire to Ngātoroirangi and everywhere they stopped in search for their brother, the heat of the fire created hot springs, geysers and boiling mud pools that still exist today (Stokes 2000; Te Ara 2019a). A different version of the legend claims that the sisters travelled underground and wherever they surfaced to look for their brother, they left hot spots behind. Yet another account states that Ngātoroirangi summoned his ancestors to send him fire to save him from the cold and certain death, which arrived in the form of volcanic activity (Stokes 2000). A different story again tells of Ngātoroirangi suffering from the cold on the summit of Mount Tongariro and so he ordered three baskets of fire, but only one arrived. In his disappointment he hurled the basket down the mountain where the fire created the *Ketetahi Springs* on the northern slope of Mount Tongariro (Te Ara 2019b).

Fig. 6.20 The Plutonium at Hierapolis (Turkey) with its entrance blocked for safety reasons

In **England** the legend of King Bladud laid the foundation to Bath's reputation for its healing hot springs (Fig. 6.23). Bladud is said to have suffered from leprosy and had left his father's court to herd swine in an isolated swamp. While tending to his herd he noticed that some of the animals, which had a skin disorder, seemed cured after wallowing in the steaming swamp waters. With nothing to lose, Bladud immersed himself in the warm spring-fed ponds and was miraculously healed (Bowman 1998; Bullard 2004; Havins 1976; White 2000), upon which he returned to his family and later became the next king.

A legend from **Georgia** in the Caucasus region recounts a hunting trip of King Vakhtang Gorgasal, during which a pheasant was attacked by one of the king's falcons. The wounded pheasant fell into one of the hot springs nearby and is said to have recovered instantly. The king was surprised by the apparent healing properties and ordered the city of *Tbilisi* to be built on this site (Lerner 2001). Different versions of this legend claim that the pheasant was boiled in the hot water or that it was an entirely different animal such as an injured deer, which recovered miraculously after jumping into the healing springs.

More fact than mythology, indigenous people of **South Africa**, the *Khoisan*, told early European settlers about locations of sacred thermal waters with curative properties. As they were hunter-gatherers, the Khoisan were frequently led to natural hot springs by elephants and baboons; animals, that instinctively sought the healing springs when they were injured or ill (Spicer and Nepgen 2005). Not all hot spring origins relate to animals though. In the north of **India**, hot springs at *Manikaran* in the Kullu district are associated with the god *Shiva* and his wife *Parvati*. The local legend tells that the springs changed their temperature after Parvati lost her earrings in the pool while she was bathing, and Shiva made the water boil out of anger. Since then the hot springs have been used for cooking and when cooled down, the mineral enriched waters are used to treat various medical conditions (Erfurt-Cooper 2009).

6.7 Summary

Since the dawn of time natural hot springs have played a consistent role in human societies as an inseparable symbol of life and a key to physical, emotional and spiritual health. Consequently, hot and cold springs were held in high regard and often presided over by local priests and tribal elders. Native peoples, who lived in close association with those natural water sources, developed a strong belief in gods and other spiritual beings in an attempt to explain the diversity of hydrothermal manifestations. In particular the violence of erupting geysers, the steam rising from boiling lakes and

Fig. 6.21 Hells Gate is one of New Zealand's active hydrothermal parks, but also offers mud baths and hot spring pools for health and recreation (see Fig. 10.15)

bubbling mud pools was seen as a potential connection to the underworld, as a gate to hell, which required the protection of patron saints and powerful deities. As a result, most countries have religious, spiritual and mythological links to natural springs, which had a strong influence on the development style and purpose of hot spring facilities. The cultural significance of water as a connection to the spiritual world is still of great importance in various societies, which has led to water worship with regional distinctions.

Over time and depending on the country, the knowledge reflecting cultural, mythological and religious associations of earlier cultures with natural water sources including hot springs was passed down through generations. For thousands of years different societies used hot spring sites as centres for cultural activities and communal gatherings. Valued as sacred sites and reserved as neutral ground, they were only to be used for peaceful activities. As hot springs normally originate in natural environments, developed sites include not just nature, but also a highly developed infrastructure essential to prestigious and fashionable health resorts and spas all over the world. While cultural and social settings of hot spring destinations vary from one country to another, they share the common goal of seeking health and recreation in a safe and peaceful setting together with other social and cultural factors (Table 6.5).

In ancient times the built environment surrounding hot springs consisted of a diversity of structural features, which usually included temples and shrines adjacent to thermal bathing and treatment facilities. Modern infrastructure includes a wide variety of buildings from Roman style bathhouse complexes to aquatic entertainment centres and waterparks. On the other hand, cultural expectations from visitors of hot spring towns often include a focus on traditional atmosphere together with the physical appeal of an attractive architectural heritage. Also typically expected are cultural events and regional attractions, which offer unique experiences for visitors from other countries with different sociocultural backgrounds (Chap. 9). For this reason, spa towns all over Europe have retained much of their historical ambience, which is an essential part of the experience of 'taking the waters' for health and recreation.

At the other end of the spectrum is the endeavour of individual hot spring spa managers to not only cater for health, wellness and recreation, but also include multicultural events and experiences, if they have the space or a specific setting to do so. In the southern hemisphere in Australia the Peninsula Hot Springs (PHS) are an example of a natural hot spring setting, surrounded by a forest of trees, which was established on degraded grazing land (Appendix 5.7). In only two decades a remarkable hot

Fig. 6.22 The Demon of the Furnace Hell (Kamado Jigoku) resides in a large cauldron, frequently described as an over dimensional rice cooker (Beppu, Japan)

Fig. 6.23 A statue of King Bladud, the discoverer of the Hot Springs of Bath (UK) is still watching over the King's spring, which is part of the old Roman hot spring complex in Bath. The left image is from c. 1910 and the right image was taken at the same location 100 years later (compare Appendix 5.5)

Table 6.5 Natural hot springs in the cultural context

• Traditional connections between natural hot springs, spirituality and religion
• Strong beliefs worldwide in the sacredness of natural springs
• Diversity of regional socio-cultural traditions related to natural hot springs
• Rich legacy of legends and mythology, commonly used to promote hot spring sites and their geoheritage
• Similarities in customs and traditions related to hot spring use in many countries

spring geoheritage has been generated; one that embraces cultural experiences combined with thermal therapies from different countries to further enhance the wellness and recreation component. In addition an immune-boosting program at PHS promotes clean air and sunlight, clean fresh water, organically grown food and herbs, aromatherapy, music and mindful physical activities and therefore offers much more than a relaxing soak in one of the many hot spring pools (Appendix 6.4). Saunas, hammams and ice caves are all part of a contrasting therapy at PHS that aims to enhance the immune system as well as the overall wellbeing and point toward the future of hot spring development.

In contrast to the contemporary way of using hot springs for health, wellness, and recreation for thousands of years human civilisations have searched for divine intervention from individual deities with the main aim of obtaining and maintaining physical and spiritual health. To this day purification and cleansing rituals in combination with prayer are still a tradition to refresh and balance body, mind, and soul. Another reason for the significance of water is its use for baptism, which is commonly practiced by Christian denominations in all parts of the world. Countless myths and legends related to the origin of hot springs have been conveyed from one generation to the next: from the ancient Greek mythology to the Māori legends of New Zealand and the Australian Dreamtime stories, from the Caucasus to South Africa and from Japan to the Americas. And they all tell about the apparent health benefits gained from the use of hot spring water and have created a wealth of regional cultural traditions around individual hot spring settings. The additional appeal for recreational purposes has ensured that most developed hot springs have retained their attractiveness as health and wellness destinations.

Reflection of the Bath Abbey in the pool of the Roman hot spring complex of Bath, UK

6.8 Appendices

Appendix 6.1 Old Customs—Coins in Springs

Throwing coins in hot and cold springs or fountains (wishing wells) is an old worldwide custom. In exchange for a wish to come true or for good health and good luck people do not mind parting with some small coins—just in case

Appendix 6.2 Roman Thermal Baths—Floor Plans

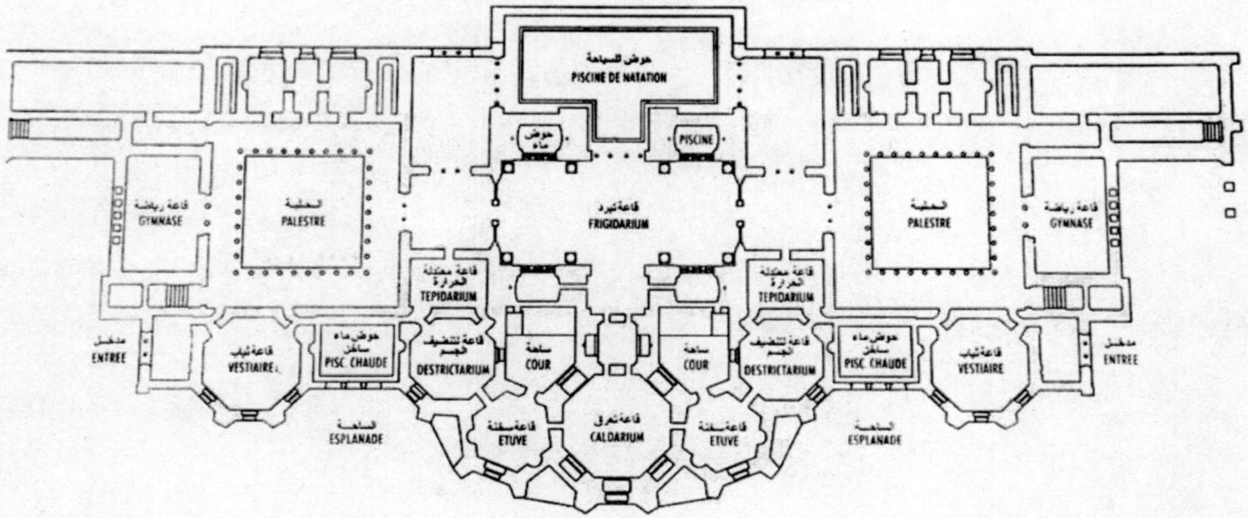

1—One of the Roman baths (Bücheltherme) in Aachen, Germany; Drawing by Lersch (1878); 2—In contrast the floor plan of the Antonine Thermal Baths in Carthage, Tunisia, displays a much more generous layout

Appendix 6.3 Peninsula Hot Springs, Australia, Site Map

The modern version of a large hot spring bathing facility is represented in this site plan of the Peninsula Hot Springs (Australia). An area of 42 acres of former grazing land devoid of trees has been transformed into a hot spring paradise with multicultural hot spring bathing and wellness experiences. Since the land was bought in 1997 by Charles and Richard Davidson, over 100,000 trees have been planted to create a forest of native vegetation to provide a natural setting that surrounds the many hot spring pools

Appendix 6.4 Peninsula Hot Springs—Health and Wellness Experience

1—Hot springs, fresh air, sunshine, and a natural peaceful environment help to relax and contribute to a sense of wellbeing; 2—While this hot spring looks like a natural pool in the wilderness, it is part of the highly developed Peninsula Hot Springs complex and one example of many different hot spring experiences; 3—Healthy activities at PHS include hot spring yoga classes; 4—Even the drinking water fountains completely blend into the natural vegetation and remind guests to stay hydrated; 5, 6—Fresh health food organically grown at PHS—from the garden to the table; 7, 8—Day and night view of the hot spring pools in the Bathing Gully, which was built with local granite and limestone

References

Ah Chee D (2014) Indigenous people's connection with Kwatye (water) in the Great Artesian Basin. Presented at the Australasian spa conference 2014

Altman N (2000) Healing Springs: the ultimate guide to taking the waters—from hidden springs to the world's greatest spas. Healing Arts Press, Rochester

Arvigo R, Epstein N (2003) Spiritual bathing: healing rituals and traditions from around the world. Celestial Arts, Berkeley

Attridge HW (2007) Pollution, sin, atonement, salvation. In: Johnston SI (ed) Ancient religions. The Belknap Press of Harvard University Press, London, pp 71–83

Bacon W (1997) The rise of the German and the demise of the English spa. Leisure Stud 16(3):173–187. https://doi.org/10.1080/026143697375386

Baden Baden (2020) History of Baden-Baden. Retrieved from https://visit.baden-baden.de/en/history-of-baden-baden

Birkby J (2002) Touring Washington and Oregon Hot Springs, a Falcon guide. Globe Pequot Press, Guildford

Bischoff M (2001) Touring New Mexico Hot Springs, a Falcon guide. Globe Pequot Press, Guilford

Boulangé M (1982) Thermal treatment: current therapeutic status. Phlébologie 35(2):571–579

Bowman M (1998) Belief, legend and perceptions of the sacred in contemporary bath. Folklore 109:25–31

Bremmer J (2007) Ritual. In: Johnston SI (ed) Ancient religions. The Belknap Press of Harvard University Press, London, pp 32–44

Burachovič S (2009) Chapters in the history of Karlovy Vary. Retrieved from https://lavernaromana.com/app-content/files/karlovy-vary/history-of-karlsbad-en.pdf

Bullard L (2004) Healing waters—Missouri's historical mineral springs and spas. University of Missouri Press, Columbia

Cartwright M (2013) Roman baths. Ancient history encyclopaedia. Retrieved from https://www.ancient.eu/Roman_Baths/

Cataldi R, Burgassi PD (1999) Flowering and decline of thermal bathing and other uses of natural heat in the Mediterranean area, from the birth of Rome to the end of the first Millennium. In: Cataldi R, Hodgson SF, Lund JW (eds) Stories from a heated Earth. Geothermal Resources Council, International Geothermal Association, Sacramento, pp 147–164

Der Spiegel (1987) Warm woget die Welle, Wandel der Badesitten: Spaßbäder und Badetempel in der Bundesrepublik. Der Spiegel 24:204–211. Retrieved from https://www.spiegel.de/spiegel/print/d-13523724.html

Emoto M (2005) The true power of water—healing and discovering ourselves. Atria Books, New York

Emoto M (2006) The secret life of water. Beyond Words Publishing, Oregon

Erfurt-Cooper P (2009) Cultural and Religious Use of Water. In: Erfurt-Cooper P, Cooper M (eds) Health and wellness tourism—spas and Hot Springs. ChannelView, Bristol, pp 110–129

Erfurt-Cooper P (2010) The importance of natural geothermal resources in tourism. In: Proceedings world geothermal congress. Bali, Indonesia, 25–29 Apr 2010

Erfurt P (2012) An assessment of the role of natural hot and mineral springs in health, wellness and recreational tourism. UnpublishedDoctoral Dissertation, School of Business, James Cook University

Frazier D (2000) Colorado's Hot Springs, 2nd edn. Pruett Publishing Company, Boulder

Fytikas M, Leonidopoulou, Cataldi R (1999) Geothermal energy in ancient Greece: from mythology to late antiquity (3rd century AD). In: Cataldi R, Hodgson SF, Lund JW (eds) Stories from a heated Earth. Geothermal Resources Council, International Geothermal Association, Sacramento, pp 68–101

Gamulin S (2001) The two millennia of Varaždinske Toplice (Varazdin's Spa). Croat Med J 42(1):1–3

Haley AJ, Snaith T, Miller G (2005) The social impacts of tourism—a case study of Bath, UK. Ann Tourism Res 29(3):429–438. https://doi.org/10.1016/j.annals.2004.10.009

Havins PJN (1976) The Spas of England. Robert Hale & Company, London

Hell's Gate (2019) Where the centre of the earth meets the sky—where myth and legend live on. Retrieved from https://www.hellsgate.co.nz/about

Hjálmarsson JR (1993) History of Iceland—from the settlement to the present day. Iceland Review/Edda Publishing, Reykjavík

Hotta A, Ishiguro Y (1986) A guide to Japanese Hot Springs. Kodansha International, Tokyo

Invaluable (2020) Figur einer Quellnymphe—Italien, wohl Mailand, 19. Jh. Retrieved from https://tinyurl.com/tpas32d

Ivanov L (2020) File: deception-tourists (GNU free documentation license, version 1.2). Retrieved from https://tinyurl.com/t4636yt

Katsambas A, Antoniou C (1996) Mineral water and spas in Greece. Clin Dermatol 14(6):615–618. https://doi.org/10.1016/S0738-081X(96)00093-4

King T (1966) Water: miracle of nature. Collier Books, New York

Kojiki (2012) Records of ancient matters (Hall Chamberlain B transl.), 2nd edn. Originally published in 712 AD. Tuttle Publishing, Tokyo

Lee G (2004) Spa & Wellness in Europa: Hotels, Anwendungen, Rezepte. Christian Verlag, München, De

Lerner CB (2001) The "river of paradaise" and the legend about the city of Tbilisi: a literary source of the legend. Folklore 16. https://doi.org/10.7592/FEJF2001.16.tbilisi

Lersch (1878) Aachen Roman bath drawing. In: Festschrift der Stadt Aachen zu einem Kongreβ der Ärzte und Naturforscher, 1900

Lund JW (1993) Spas and balneology in the United States. GHC Quart Bull 1–3

Lund JW (2002) Balneological use of geothermal waters. Geo-Heat Center, Oregon

Lund JW (2005) Basic principles of balneology and examples in the United States. In: Proceedings of the world geothermal congress 2005. Antalya, Turkey, 24–29 Apr 2005

Lüscher G (1946) Die Thermen von Baden und die Mineral und Heilquellen der Schweiz. Buchdruckerei Neue Aargauer Zeitung, Aargau

Melillo L (1995) Thermalism in the ancient world. Med nei Secoli 7 (3):461–483

Ministerio Bienes Nacionales (2014) Bienes Nacionales entrega a comunidades atacameñas concesión de uso gratuito de Geysers del Tatio. Retrieved from tinyurl.com/spln3eew

Monti S (2003) Thermalism between past and future. In: Proceedings of the conference 'the cultural turn in geography'. 18–20th of september 2003—Gorizia Campus. Part VI: tourism, sustainable development and culture turn. Retrieved from https://www.openstarts.units.it/bitstream/10077/862/1/f1monti.pdf

Nelson J (1999) Glenwood Springs: the history of a Rocky Mountain resort. Western Reflections Inc., Ouray

Nickell J (2005) Healing waters—investigative files, sceptical briefs, June 2005. Retrieved from https://skepticalinquirer.org/newsletter/healing_waters_-_part_i_spas/

Nilsson MP (1992) Geschichte der griechischen Religion: die Religion Griechenlands bis auf die griechische Weltherrschaft, Dritte Auflage. Verlag C.H. Beck, München

Onsen Tipster (2015) Are Tattoos really Taboo at onsen in Japan? Retrieved from https://tinyurl.com/u6349k6

Ottaway P, Cyprien M (1987) A traveller's guide to Roman Britain. Routledge & Kegan Paul, London

Özgüler ME, Kasap A (1999) The geothermal history of Anatolia, Turkey. In: Cataldi R, Hodgson SF, Lund JW (eds) Stories from a heated Earth. Geothermal Resources Council, International Geothermal Association, Sacramento, pp 51–67

Pfanz H, Yüce G, Gulbay AH, Gökgöz A (2018) Deadly CO_2 gases in the Plutonium of Hierapolis (Denizli, Turkey). Archaeol Anthropol Sci 1–13. https://doi.org/10.1007/s12520-018-0599-5

Ryrie C (1999) The healing energies of water. Journey Editions, Boston, Tokyo

Sanner B (2000) Baden-Baden—a famous thermal spa with a long history. GHC Bull 21:16–21. Retrieved from https://tinyurl.com/y2nnxejo

Schafer EH (1956) The development of bathing customs in ancient and medieval China and the history of the floriate clear palace. J Am Orient Soc 76(2):57–82. https://doi.org/10.2307/595074

Shiva V (2002) Water wars: privatisation, pollution, and profit. South End Press, Cambridge

Sieveking H (1913) Die Radioaktivität der Heilquellen. Die Naturwiss 21:497–499

Spicer S, Nepgen J (2005) Holistic holidays in South Africa: health spas, hot springs, magical places and sacred spaces. Human & Rousseau, Cape Town

Stokes E (2000) The legacy of Ngatoroirangi; Maori customary use of geothermal resources. Retrieved from https://tinyurl.com/r65nwz7

Te Ara (2019a) Story: Ngāti Tūwharetoa—the journeys of Ngātoroirangi and Tia. Retrieved from https://teara.govt.nz/en/ngati-tuwharetoa/page-2

Te Ara (2019b) Story: Volcanic Plateau region—Ketetahi Springs. Retrieved from https://teara.govt.nz/en/photograph/15143/ketetahi-springs

UNESCO (2019) World heritage list—city of Bath. Retrieved from https://whc.unesco.org/en/list/428

Valenza JM (2000) Taking the waters in Texas: springs, spas and fountains of youth. University of Texas Press, Austin

Walsh R, Allom T (1836) The Bath. Illustration from Constantinople and the Scenery of the Seven Churches of Asia Minor illustrated. Fisher, Son & Co. (1836–38), London/Paris. Retrieved from https://commons.wikimedia.org/wiki/File:The_Bath_-_Walsh_Robert_%26_Allom_Thomas_-_1836.jpg

Wettstein R (2010) Römische Bäder (Badruinen). Retrieved from www.bad-bad.de/sehen/badruin.htm

White G (2000) Bath: a world heritage site—the Bath Spa Project. GHC Bull 21(3):12–15

White J (2020) Maruia Hot Springs, New Zealand. Images used with permission

Wright DP (2007) Anatolia: Hittites. In: Johnston SI (ed) Ancient religions. The Belknap Press of Harvard University Press, London, pp 189–196

Zámecké Lázně (2019) History of Castle Spa—the legend of the Springs. Retrieved from https://www.zamecke-lazne.com/en/history#legend

Steam rising from the red 'Blood Pond' (Chinoike Jigoku) in Beppu, Japan

7 Hot Springs and Their Natural Heritage

Contents

P. Erfurt, *The Geoheritage of Hot Springs*, Geoheritage, Geoparks and Geotourism,
https://doi.org/10.1007/978-3-030-60463-9_7

A landscape of salt and sulphur formations shaped by volcanic activity at Dallol in the Danakil Depression, Afar Region of Ethiopia. *Source* A. Savin (2018) (Public Domain)

The natural geoheritage of hot springs includes their unique geodiversity based on distinct geophysical environments and geographic settings, their outstanding value from a geological standpoint, as well as their intrinsic value for health and recreation as a natural resource with curative properties.

7.1 Introduction

The cultural and natural heritage of hot springs is closely linked with many intersecting margins, and it can be challenging to separate some of their individual characteristics. While the main objective of this book is to explore and examine the wide-reaching geoheritage of hot springs in different settings, this chapter is looking primarily at aspects of the natural heritage related to hydrothermal areas.

Hot springs have always encouraged people to use them for a variety of purposes including tourism. Especially their more extreme features found in active hydrothermal areas present unusual attractions for visitors and are located on every continent, even Antarctica (Fig. 6.6), and in many countries along the *Ring of Fire* where countless hot springs, geysers, fumaroles and other phenomena emerge from the ground (Table 7.1).

In Europe, the visit of hot spring environments has been common for centuries and was part of the *Grand Tour,* an extensive journey on which well-to-do individuals embarked during the eighteenth and nineteenth centuries to broaden their intellectual horizon. Exploring hydrothermal sites such as the *Campi Flegrei* in Italy or Iceland's extreme hot springs and geyser fields was an important element of inspirational journeys that could last from several months to a few years (Erfurt-Cooper and Cooper 2009, 2010; Hróarsson and Jónsson 1992).

7.1.1 Hot Springs—Close Connections Between Natural and Cultural Geoheritage

Natural hot springs are traditionally used for health and wellness tourism and are often successfully combined with related hydrothermal attractions, which indicates that in many cases aspects of the natural and cultural geoheritage of hot springs are presented side by side. People visiting hot springs for recreational purposes often find that natural and cultural features have a close relationship and show some form of reciprocal integration. This implies that the interface between the cultural and the natural geoheritage of hydrothermal resources is connected to the point, where the margins between the two subjects may have become indistinct and where cultural and natural elements have merged into a geoheritage component of their own.

Table 7.1 Examples of hydrothermal features that are used as popular tourist attractions worldwide, with the inclusion of geothermal power stations, as some of them welcome visitors for educational tours

Natural hydrothermal tourist attractions	
Fumaroles (*steam vents*)	Boiling mud pots
Geysers	Explosion craters
Hot lakes	Sinter (*travertine*) terraces
Hot rivers and streams	Hot springs (*for bathing*)
Hot waterfalls	Hot springs (*for cooking*)
Mud volcanoes	Geothermal power stations

Source Author

The close connection between the natural and cultural geoheritage of hot springs is supported by various shared elements. A comparative framework (Fig. 7.1) clearly indicates that a separation of the natural and cultural values pertaining to the overall geoheritage of hot springs is not always possible. Especially as hot springs and their natural settings are a vital part of *any* hot spring experience and are key attractions at many destinations.

Active hydrothermal environments and their distinct manifestations are part of a unique natural heritage that requires effective management strategies, which need to focus on the conservation of the landscape as well as on visitor safety (Migoń and Pijet-Migoń 2016). This leads to the potential hazards and risks for tourism in active hydrothermal areas, a topic that is discussed in detail in Chap. 9. For now a brief overview of common hazards in hydrothermal environments includes (a) unexpected hydrothermal eruptions or earthquakes, (b) increased steam discharge at fumaroles or geysers causing severe and life-threatening thermal burns, (c) toxic gas emissions including H_2S, CO_2, SO_2, and (d) unstable ground with only a thin crust covering hydrothermal systems lying beneath (Chap. 3; Erfurt-Cooper 2018). Although active hydrothermal geosites can harbour potential dangers, their geoheritage is also quite fragile (Newhall 2014) and requires protection from visitor pressure (Chap. 4). Therefore it is essential to find the right balance between sustainable tourism development and site conservation, and at the same time establishing suitable management practices for visitor safety.

7.1.2 Natural Geoheritage—Some Examples

With reference to best practice management, in the case of New Zealand's *Pink and White Terraces* (Example 7.1) nothing could have prevented the loss of the natural and cultural geoheritage of this once spectacular natural wonder.

Example 7.1 Pink and White Terraces, New Zealand

The unique beauty of the Pink and White Terraces was created by the constant flow of hot spring water rich in dissolved minerals. Once termed the eighth wonder of the world (Fig. 7.2; Appendices 7.1 and 7.2), the extensive travertine terraces attracted visitors from as far as Europe to see this remarkable natural phenomenon until their disappearance in 1886 during an eruption of

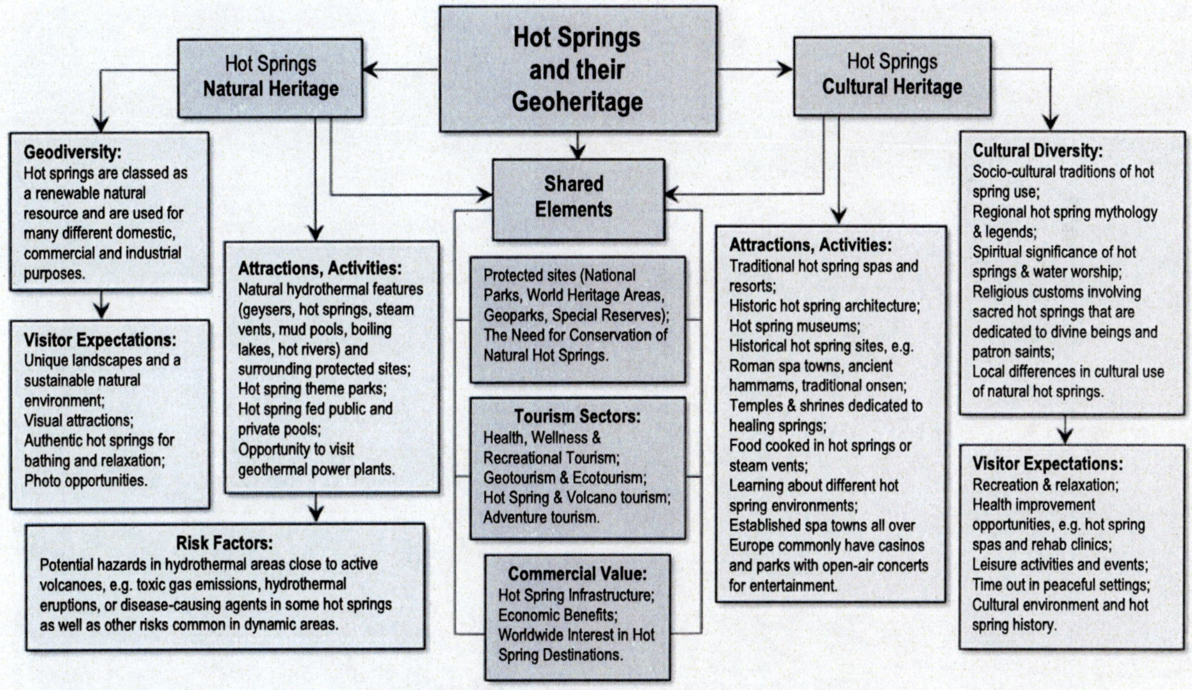

Fig. 7.1 A conceptual representation of hot spring environments and their geodiversity. *Source* Author

the nearby volcano Mt. Tarawera. Located at Lake Rotomahana near Rotorua, the White Terraces were known as *Te Tarata* or the tattooed rock, and the smaller Pink Terraces as *Ō-tū-kapua-rangi* (fountain of the clouded sky). Tourists were taken to this famous natural treasure in canoes by local tour guides until the terraces vanished from sight (thought to be at the bottom of Lake Rotomahana). From a cultural perspective the terraces were revered by the local Māori as *taonga* (treasure), in part because of the curative properties of the mineral-rich waters, but also for their unusual landform.

Source De Ronde et al. (2016), Erfurt (2012), Erfurt-Cooper and Cooper (2009), Erfurt-Cooper (2010), Lee and King (2008), Stafford (1986), Te Ara (2019)

While other examples of outstanding natural geoheritage are located all over the world, worth mentioning here are two specific hot spring locations in Japan and Turkey (Examples 7.2 and 7.3). Each in their own way represent remarkable natural features while at the same time they are easy to access and receive millions of visitors every year.

Example 7.2 Tour of Hells (Jigoku Meguri), Beppu, Japan

In the Japanese city of Beppu the visually attractive hydrothermal features of the *jigoku* (Japanese for hell) are of particular interest for tourists. Located in ten small geoparks gushing geysers, boiling crater lakes, bubbling mud pools, hissing steam vents and differently coloured hot springs embody a remarkable natural geoheritage that has attracted visitors for centuries, who embark on a 'pilgrimage of the hells' (Fig. 7.3, Appendix 9.1). Adhering to cultural traditions all jigoku offer purification basins (Fig. 6.15) and small shrines (Fig. 6.12) for religious observance. A popular treat, especially in winter, are hot spring foot baths (Fig. 7.4), and as a culinary experience not to be missed, food items such as eggs, sweet potatoes and dumplings are cooked in boiling ponds or over hissing steam vents (Fig. 7.8), a delicious experience. In addition, onsen baths at varying temperatures are available for visitors at several jigoku as well.

Source Erfurt (2012).

Fig. 7.2 The Pink Terraces prior to their complete disappearance in 1886 during Mt Tarawera's catastrophic eruption. *Source* Glimpses of New Zealand (1896) (compare Appendix 7.1)

Fig. 7.3 Visitors observing a hydrothermal feature on their Jigoku Meguri (Pilgrimage of the Hells) early last century. Uniformed guards are standing by to watch over the safety of the visitors. *Source* Photo taken at an exhibition of historic pictures at one of the Jigoku in Beppu (Japan)

Fig. 7.4 Jigoku visitors in Beppu are taking advantage of the warm and relaxing foot spas, which can be found in hot spring areas all over Japan

Fig. 7.5 The ruins of Hierapolis and the travertine terraces of Pamukkale—one of the closest connections between cultural and natural geoheritage

Similar to Japan's spiritual relationship with their hot springs, many natural hot spring sites and other water bodies have a strong association with the divine and the mythical (Chap. 6), which further confirms the cultural dimension added to the natural geoheritage (Kiernan 2015).

Example 7.3 Hierapolis—Pamukkale, Turkey

The famous white travertine terraces of Pamukkale together with the ruins of the ancient city Hierapolis represent an outstanding example of the close connection between cultural and natural heritage and have been UNESCO World Heritage listed since 1988. The unique blend of natural hot springs and architectural ruins attracts approximately 2.5 million people every year (Chap. 4). While the hot springs continue to create an impression of shining white petrified waterfalls, above the travertine terraces the historic architecture of Hierapolis, dating back as far as the second century BC when Hierapolis was first established (Chap. 5), continues to fall into disrepair (Fig. 7.5). Other remarkable features at this unique geoheritage site are the petrified water channels, built from layers of minerals deposited by the flowing hot springs. Not more than 30–40 cm wide, these ancient channels (Fig. 7.6) are now dry and do not carry hot spring water any longer. However, further down from the plateau, on which the remains of Hierapolis stand, new petrified water channels (Fig. 7.7) are work in progress, fed by the constant stream of warm spring water enriched with mineral elements that flows over the travertine terraces.

Source Erfurt (2012).

7.2 Natural Hot Springs, Health and Recreation

A review of the historical use of hot springs (Chap. 5) has verified the integration of natural springs into health treatments and recreational activities and goes back to ancient civilisations all over the world. To appeal to visitors of all age groups, genders and backgrounds, at the present time hot spring tourist destinations typically aim for a balance

Fig. 7.6 One of the ancient petrified water channels on the site where the ruins of Hierapolis are located. All that remains is the travertine build-up and the clearly defined channel

between natural and cultural attractions, with an abundance of additional leisure activities such as hiking, skiing or water sports.

In addition to natural hot springs used for health and recreation, the visual impact of hydrothermal phenomena (e.g. geysers, boiling lakes and bubbling mud pools) is a major tourist attraction and draws visitors to the most remote and distant locations to experience unique natural environments in action. Without this priceless natural resource, various tourism industry sectors would lack one of their most significant natural infrastructure components that play an essential role in attracting higher visitor numbers at many destinations.

One of the objectives of this chapter is to demonstrate the close ties of natural hot springs with health resorts and spa towns and their large supporting service industry. These close ties are evident at many traditional and newly developed hot spring destinations, where transport, accommodation, restaurants and retail opportunities as well as cultural attractions contribute to the overall hot spring experience (Erfurt 2012). The appeal of individual hot spring destinations is often further enriched with a complementary fusion of local food (Fig. 7.8), idyllic scenery and interesting cultural activities. A prime example for the clear and distinct interface between the natural and cultural geoheritage of hot springs is represented by the English city of Bath, where the ancient Roman cultural heritage blends with the contemporary use of natural hot springs. This outstanding universal value has led in 1987 to the inscription of Bath on the UNESCO World Heritage list.

7.2.1 Hot Spring Infrastructure—Built Environment

Long-established hot spring spa towns are proud of their traditional architecture and ambience, which developed around the local natural mineral waters, the key element of their ongoing success. The built environment is a significant component of any spa town and consists of hot spring resorts and spas with special treatment facilities for bathing, drinking, inhaling as well as various forms of physical therapy; all designed with high quality standards in mind. The built infrastructure is essential for destinations where hot springs are a major tourist attraction, whether they are used

Fig. 7.7 A present-day petrified water channel at Pamukkale's travertine terraces in the process of accumulating mineral deposits, which gives it a similar appearance compared to the ancient channels

for medical purposes or as a natural geosite in a protected area. Part of the supporting infrastructure are public access roads, visitor centres, walking tracks and interpretive sign boards. Communication services such as phone reception and internet access unquestionably are modern-day fixtures, which everybody relies upon. Further components are shops, cafes, restaurants and of course accommodation, which differentiate the infrastructure of the built environment from the essential infrastructure of natural areas; a topic discussed later in this chapter in Sect. 7.5.3. With the aim to illustrate the spa town infrastructure of different countries, two short case studies (Examples 7.4 and 7.5) of Germany and Japan may shed some light on the diversity related to the origin of hot springs, their history and their past and current use.

Example 7.4 Baden Baden, Germany

The traditional spa town Baden Baden developed from a luxurious thermal spa centre under Roman occupation circa two thousand years ago into an international destination for health and recreation based on the use of natural hot springs. Since the beginning the infrastructure of the resort town has been improved continuously. The built environment includes state-of-the-art conference centres, a festival hall for 2650 visitors and the large Caracalla Spa. The original hot springs, as well as two wells drilled in the 1960s, directly supply the spa facilities with thermal water. Spring water (both hot and cold) is also delivered to several public drinking fountains and other users such as hotels and hospitals. The essential infrastructure of the spa town Baden Baden includes not only the exceptional quality of the thermal spring water, an extremely mild climate and pleasant surrounding landscapes, but other important criteria such as excellent services, good public access and transport, as well as traditional architecture and other historic facilities.

Source Erfurt (2012), Sanner (2000), Wettstein (2010), Wohnlich (1996).

In Europe and Asia specialised clinics with balneotherapy centres for rehabilitation as well as health education centres are also fundamental elements of the built spa infrastructure.

Fig. 7.8 Fresh vegetables, eggs and buns filled with bean paste are cooked over steam vents and sold at street corners and hot spring parks in Japan

Natural surrounds with parks, gardens and walking areas provide relaxation and a connection with nature and are usually complemented by fitness and sports facilities. Depending on the destination, accommodation buildings may be directly connected to natural hot springs for personal use. And as a final point, entertainment for resort guests and outpatients plays an important social role and is typically catered for and promoted in programs for social activities and cultural events for all age groups (Chap. 6).

Example 7.5 Hot Spring Infrastructure in Japan
The overall number of natural hot springs in Japan is exceptional (more than 27,000) and has made it possible to cater for the large demand for traditional onsen bathing facilities in spa towns in every prefecture. The built infrastructure is generally excellent and most national parks and other protected areas with hydrothermal attractions are relatively easy to access. To accommodate the vast numbers of hot spring tourists (over 150 million p/a), a well-established local tourism infrastructure includes visitor information centres, historical architecture and cultural sites including shrines and temples, as well as museums with exhibitions themed around the local natural hot springs. A large supporting service industry for hot spring tourism is the backbone of many local economies in Japan with abundant retail opportunities, restaurants, public transport (Fig. 7.9), traditional and western accommodation, which together contribute to the overall hot spring tourist experience. Natural settings surrounding the hot springs combined with cultural attractions are evident at most traditional Japanese onsen destinations and, like other historical spa towns elsewhere, present a tangible interface between the natural and the cultural geoheritage that ranges from the perceptible to the physical and observable.

Source Erfurt (2012), Erfurt-Cooper and Cooper (2009).

On the European continent, one hundred years ago the French Office National du Tourisme (1921) published a book

Fig. 7.9 The small hot spring town Yufuin in Kyushu is a popular destination for mass tourism. Visitors arrive by bus for a sightseeing and shopping tour and return to their nearby accommodation. The infrastructure of this town is completely dedicated to hot spring tourism, combined with other local attractions such as forests, lakes and the volcano Yufudake towering over the township

to promote the hot spring destinations of France. Special consideration was given to (a) the medicinal properties, mineral content and temperature of the springs, (b) the qualified medical teams at the balneotherapy centres who administered the treatments, (c) the unspoilt healthy natural surroundings, (d) the attractive entertainment opportunities available to visitors, and most importantly, e) the local infrastructure and its advanced state compared to other areas (Chap. 8, Sect. 8.2.1). Figure 7.10 gives an impression of the type of architectural infrastructure at the turn of the previous century common at French thermal resorts and spa towns.

7.2.2 Endorsement of Hot Springs

Hydrothermal phenomena such as geysers or boiling mud pools were already used to attract visitors to Iceland during the eighteenth century. Indeed, all countries rich in hydrothermal resources have a history of promoting their active environments to increase visitor numbers, e.g. China, Italy, Japan, New Zealand, Turkey and the United States. And while active geysers and boiling mud pools do not seem to lose their appeal, the spa industry had its ups and downs over time. By good fortune though, when the global wellness movement started in the 1990s, countless hot spring health resorts and spas were experiencing a revival, just as they were about to close their doors after a lengthy decline. The novel trend also encouraged the owners and administrators of many traditional and long-established hot spring facilities to renovate and modernise to compete with a new generation of health resorts and spas (Erfurt-Cooper and Cooper 2009). On the other hand, historical spa towns in many European regions that were under Roman occupation 2000 years ago, managed to sustain their viability and are still flourishing for several major reasons:

- Continued sponsorship and support from local and state governments
- Continued sponsorship from medical and/or health associations (varies by country)
- Continued promotion through the local travel industry as attractive destination packages including nature, hot springs, climate and landscape
- Ongoing support from the host communities to protect their cultural identity, the natural geoheritage as well as opportunities for ongoing employment

Fig. 7.10 Celestin's Spring in Vichy (France) c. 1920. *Source* Office National du Tourisme (1921)

- Historical, cultural and natural significance, e.g. traditional architecture and/or official heritage listing
- Awareness of the curative value of the mineral-rich hot springs for health and recreation
- Reputation for the successful treatment and rehabilitation of certain medical conditions
- Located in a country or region that has always been a trendy destination
- Located in a country that is known as a safe destination or
- The natural hot springs are easy to access.

Today, hot spring-based tourism remains a growing segment of the health and wellness tourism sector on a global scale, as well as being integrated in geotourism, ecotourism and adventure tourism (Erfurt 2012).

7.3 Protected Hot Spring Sites

Travellers for health and recreation often set their sights on natural environments including national parks and other protected sites, with many hydrothermal areas developed and promoted as tourist attractions. In fact, these regions are powerful points of interest (POI) and genuine hot spots when it comes to unique travel destinations. In addition, the lasting trend 'back to nature' has given rise to the term *geo-wellness*, which relates to the natural heritage of our environment as a significant component in the search for different levels of health and wellbeing for body, mind and soul (Erfurt-Cooper and Cooper 2009; Fig. 7.1).

Natural heritage sites containing hydrothermal features are usually protected in nature reserves with the aim to avoid endangering unique environments and preserve them for future generations. Many countries have national heritage lists, which places particular natural heritage sites under government protection. Hydrothermal landscapes across the world are possessing such unique features that it has led to their protection, either through the application of environmental regulations and government laws, or they have been declared as natural and/or cultural geoheritage of national interest to stop unsuitable urban or industrial development in their immediate neighbourhood (Erfurt 2012). Often co-located with volcanic environments, hot springs can exhibit forceful displays of steam and water erupting at extreme temperatures, which essentially serves to increase the appeal of nature-based tourism destinations, especially if they

Fig. 7.11 The old sulphur mines at Krýsuvík are depicted in a painting by J. Wright from 1879. *Source* Hróarsson and Jónsson (1992). While the Krýsuvík-Seltún Geothermal Area once was used to extract sulphur, today it is a major tourist attraction with steaming fumaroles, mud pots and boiling springs

are located within city limits (Fig. 9.12). Here, the natural heritage of hot spring sites presents itself in the form of their remarkable geodiversity, which offers the unusual experience of witnessing the intense power of hydrothermal activity.

National parks, world heritage sites, geoparks, and areas designated as national heritage are major tourist attractions, which contribute significantly to the revenue derived from hot spring related tourism. The common denominator of all protected hot spring sites lies in the unique appeal of their environment on one side and the impressive hydrothermal activity on the other side. Especially extreme hot springs like geysers (Appendix 7.3), fumaroles and colourful mineral deposits in boiling pools attract travellers to hydrothermal areas as famous as *Yellowstone* (US) and as remote as the *Valley of Geysers* on the Kamchatka Peninsula in far east Russia. Also of rare beauty are the travertine terraces of *Badab-e Soort* in Iran (Appendix 7.4), although they are not as well-known as Pamukkale and do not have the infrastructure to cater for large visitor numbers. Other hot spring sites can be privately owned and managed like the unique hydrothermal parks in New Zealand (Example 7.6). In Iceland the *Reykjanes Nature Reserve* with the *Krýsuvík Geothermal Area* (Fig. 7.11) is a natural setting where visitors can experience hydrothermal activity including boiling springs, mud springs and steam vents. The Krýsuvík area is one of the 55 official geosites of the Reykjanes Global Geopark, which was established in 2015 (Reykjanes Geopark 2020; UNESCO 2020, 2021).

Example 7.6 Rotorua, New Zealand
Rotorua is well-known for its hydrothermal manifestations and the sulphur smell that lingers in certain areas of the town. The natural infrastructure of the surrounding areas includes attractions such as hot spring-fed streams, geysers, sinter terraces, and mud pools. Most of these unique features are protected in parks or privately owned areas that can be accessed after paying an entrance fee.

7.3.1 National Parks

National parks are areas reserved by national governments for the preservation of the natural environment. A national park may be set aside for the purpose of public recreation and enjoyment or because of its historical or scientific interest, with most of the landscapes and their accompanying plants and animals in a national park kept in their natural state (Encyclopedia Britannica 2019). A well-known case is Yellowstone (Example 7.7), the first national park of the world, which was established in 1872 to protect its cultural and natural heritage for the benefit and enjoyment of future generations (NPS 2020).

Example 7.7 Yellowstone National Park (USA)
Not only a National Park, but also a World Heritage area, Yellowstone is a well-known destination that contains the largest concentration of hydrothermal features worldwide with about 100 different hot spring groups adding up to more than 10,000 individual hydrothermal features. Some of the most popular tourist attractions at Yellowstone are the travertine terraces at Mammoth Hot Springs, the reliable eruptions of the geyser Old Faithful (Fig. 9.5) and the colourful boiling springs (Fig. 7.12). The combined attractions of this National Park and World Heritage site, apart from the hydrothermal activity, also include unique wildlife and vegetation and attract over four million visitors on average every year.

Source Bargar and Muffler (1975), Bryan (1986), Erfurt-Cooper and Cooper (2009), Rhinehart (1980), White et al. (1988).

National Parks in Japan
Japan's hot spring or *onsen* tourism sector greatly benefits from the abundance of volcanic hot springs. Apart from the numerous hot spring spa towns or onsen resorts, which attract visitors in pursuit of health and recreation, national parks and other protected areas throughout the country (Table 7.2) comprise a wide range of areas with opportunities to observe hydrothermal features from erupting geysers to thermal lakes (Fig. 7.13). In fact, the Japanese people are known for travelling far and wide to discover and experience as many different hot springs in their country (and elsewhere) as possible.

The conservation of natural treasures such as volcanic hot springs goes back to the 1930s, when the first national parks were established in Japan in recognition of the need to preserve natural resources for the future. Currently there are 34 National Parks in Japan; however, an additional 56 Quasi National Parks as well as 314 Prefectural Parks have been established with further divisions between terrestrial and marine park environments. A specific Natural Parks Law from 1957 is in place to oversee the different categories and their status of protection.

7.3.2 World Heritage Areas

What makes the concept of World Heritage exceptional is its universal application. World Heritage sites belong to all the peoples of the world, irrespective of the territory on which they are located. (UNESCO 2020, 2021)

While not everybody may be able to physically visit World Heritage sites, the UNESCO philosophy of ownership gives the impression of inclusiveness. At the same time, the world heritage brand sells destinations with currently 1121 properties inscribed on the UNESCO World Heritage list. This list contains 869 sites based on their cultural criteria, while less than a quarter (213) are natural sites and 39 are mixed sites. The World Heritage list also includes over fifty endangered sites, of which many are striving to retain their prestigious world heritage brand.

Three already discussed examples of hot spring locations are world heritage listed and include Hierapolis/Pamukkale (Turkey), Yellowstone National Park (north America) and the city of Bath (England). While Hierapolis/Pamukkale is represented as a mixed UNESCO World Heritage area, Yellowstone is listed as a natural site and, despite its natural hot springs, Bath is listed as a cultural site only (Fig. 7.14). The World Heritage *Volcanoes of Kamchatka* is an extended site that includes the Valley of Geysers as one of the hydrothermal features of the overall natural geoheritage. Other listed sites also include natural hot springs, but they are not mentioned as a key feature.

7.3.3 Geoparks

UNESCO Global Geoparks are single, unified geographical areas where sites and landscapes of international geological significance are managed with a holistic concept of protection, education and sustainable development. UNESCO (2020, 2021)

Regional, national and global geoparks are a more recent option to promote the conservation of geosites and their

Fig. 7.12 Aerial view of the Grand Prismatic Spring, one of more than 10,000 hydrothermal features in the Yellowstone National Park. *Source* Jim Peaco (2006) (Public Domain)

Table 7.2 Protected sites in Japan include diverse hydrothermal features of volcanic origin

Examples of National Parks in Japan with hydrothermal attractions		
Akan National Park Hokkaido	est. 1934	Hot springs and boiling mud ponds, onsen resorts
Daisetsuzan National Park, Hokkaido	est. 1934	Hot spring resorts Asahidake Onsen, Fukiage Onsen, Sounkyo Onsen and Tenninkyo Onsen—c. 6 million visitors per year
Shikotsu-Toya National Park, Hokkaido	est. 1949	Includes the popular hot spring resort town of Noboribetsu
Towada-Hachimantai National Park, North Honshu	est. 1936	Hot springs, boiling mud pools, steam vents and fumaroles, onsen facilities
Fuji-Hakone-Izu National Park, Honshu	est. 1936	Countless hot springs around Mt Fuji, c.150 million visitors every year, onsen resorts onsen resorts
Chūbu-Sangaku National Park, Honshu	est. 1934	Many hot springs are located at the foot of the mountains, onsen resorts and spas
Daisen-Oki National Park, Honshu	est. 1936	Several hot spring spas in the park, e.g. Kaike, Tamatsukuri and Sanbe
Jōshin'etsu-kōgen National Park, Honshu	est. 1949	Onsen resorts, e.g. Yudanaka, Kusatsu, Karuizawa, Yamanouchi
Aso-Kuju National Park, Kyushu	est. 1934	Popular onsen resort areas, e.g. Kokonoe, Yufuin, Beppu
Kirishima-Kinkowan National Park, Kyushu	est. 1934	Popular hot spring resorts, active volcanoes
Unzen-Amakusa National Park, Kyushu	est. 1934	Several hot spring spa towns, e.g. Shimabara Onsen

Source Various

Fig. 7.13 Steam rises from Lake Kinrin in Yufuin, a small hot spring resort town in Kyushu (Japan)

diversity including natural hot springs and other hydrothermal manifestations. Also, geoparks represent an important part of the tourism infrastructure of a region and focus on the promotion of sustainable development. For example, the diversity of natural and cultural geoheritage side by side is strongly reflected throughout the *Azores Geopark* (Portugal). Here, a network of geosites of scientific and educational significance includes countless hydrothermal features (Nunes 2014) and represents the geoheritage of nine individual islands. The Azores archipelago joined the Global Geoparks Network (GGP) in 2013 with a revalidation in 2017 after the first four years. Apart from the rich natural and cultural geodiversity of the nine volcanic islands, several active hydrothermal areas (Fig. 7.15) and a *Thermal Spa Route* for health and recreation add to the remarkable geosites of the Azores archipelago.

At the time of writing in 2020, 161 UNESCO Global Geoparks have been established in 44 countries with more being added every year. Like some of the UNESCO World Heritage areas, national and global geoparks contain natural hot springs and are visited for health and recreation.

7.4 Hot Spring Waterparks

Hot spring tourism can take place in any region where hydrothermal resources are accessible, although the long-term sustainability of this unpolluted and cost-effective resource depends on the responsible management and preservation of all water sources (Erfurt-Cooper 2010). National parks and other protected sites with hot springs and related hydrothermal manifestations are generally controlled through appropriate management strategies with the aim to preserve their integrity as a natural environment for the future. Protected areas in different countries have their own individual site-specific laws and regulations for a best practice management approach (Chap. 4). Some regions may have stricter policies regarding conservation of natural assets such as hot springs than others; however, officially managed hot spring sites can be largely classed as a sustainable resource in a protected natural environment.

On the other side of the fence we find artificial waterscapes using hot water from hydrothermal reservoirs, which

Fig. 7.14 The original hot spring in Bath (UK), where the water is still flowing as it has done for more than 2000 years, simultaneously represents the natural and cultural geoheritage of the only genuine hot spring in England

have become very popular and are an important component of tourism in many countries. The following examples demonstrate the colossal size of some of these facilities, built with a mindset of unlimited access to hot spring water as well as cold water resources, but perhaps not all of them with enough consideration for their long-term sustainability.

7.4.1 Aquatic Entertainment on a Grandiose Scale

The fact that natural hot spring water is available at considerable volumes and at low or negligible cost, has led many municipalities worldwide to develop large-scale water entertainment facilities. Natural thermal water is still a relatively inexpensive resource, which presents substantial savings in the running of such extensive waterscapes. Consequently, waterparks have taken the way of exploiting natural hot springs to an entirely different level and come in many different sizes, depending on the following conditions:

- A large parcel of land available for development
- An abundant supply of hydrothermal resources
- Local government permits
- Encouraging feasibility studies
- Favourable stakeholder support (tourism industry, service industry)
- Extensive financial backing
- Access to public infrastructure (transport, communications)
- Impact assessments (environmental, social, economic)
- A demand for such a facility; and (hopefully)
- Appropriate management strategies for wastewater re-use or disposal.

Waterparks are built for recreation, fun and entertainment and have become popular in recent times as an endeavour to attract large numbers of visitors all year round. Many of these ambitious projects can only be described as spacious and extravagant waterscapes with countless bathing and swimming pools, giant waterslides and children's water

Fig. 7.15 The 'Caldeira Grande', one of the boiling springs at the Caldeiras das Furnas in São Miguel (Azores)

playgrounds. Waterparks are typically developed as aquatic theme parks, often with a touch of adventure, but also catering for health and wellness combined with cultural attractions including local cuisine. While aquatic theme parks are particularly attractive because of their thermal facilities, many of their pools are using cold groundwater resources as well.

To cater for all types of visitors from small children to more mature aged people, most waterparks using hot springs are based on the same infrastructure with region dependent variations:

- Water sourced from local hydrothermal systems
- Pools of many different sizes, depths, and other features such as geysers, mushroom fountains and waterfalls
- Steam rooms, salt caves and saunas
- Therapy centres for health treatment with mineral-rich hot spring water
- Water slides with enormous towers and several ways down as well as small slides in children's play pools
- Interactive water playgrounds for different age groups
- Amusement and entertainment centres with various gaming zones
- Retail opportunities including souvenirs shops and/or shopping malls
- Restaurants, cafes, food courts, bars
- Natural environments and adventure activities, sports including abseiling, skiing, hiking and
- All year-round events schedules to attract visitors to the thermal water action.

The following examples discussed here highlight the popularity of themed waterparks or aquatic entertainment centres in several different countries. While some facilities focus more on recreational activities and aquatic entertainment, others have incorporated hot spring-based health and wellness spa facilities. Large aquatic centres constructed around natural hot springs are frequently promoted as *Thermal Waterparks* or *Hydrothermal Parks.* Mixed sites depending on hot as well as cold water resources may be termed *Waterparks* and *Aqua Parks* or *Amusement Waterparks*, *Aquatic Theme Parks*, *Aquatic Entertainment Parks* and *Aquatic Adventure Parks*. Sometimes the hot spring component is only small and other times it is the principal foundation of a thermal waterpark. Nonetheless, the contrast to protected natural hot spring environments is quite extreme.

7.4.2 Hydrothermal Waterparks—Examples from South America and Asia

Example Brazil—Rio Quente

In South America, in the Brazilian state of Goiás, *Rio Quente* is a large-scale waterpark, which is advertised as the 'largest hydrothermal park in the world' together with the nearby city of *Caldas Novas.* Over one million people visit Caldas Novas every year, which has the largest hot spring tourism industry of the region with many hotels having built individual waterparks for their guests. Additional recreational options include ecotourism and adventure tourism activities, as well as natural pools and an artificial beach with thermal water (Caldas Novas 2019).

Example Uruguay—Acuamanía

Acuamanía is a thermal waterpark in Uruguay and obtains its natural hot water from the large transboundary *Guarani Aquifer System* (GAS), which underlies four South American countries, Argentina, Brazil, Paraguay and Uruguay (Chap. 2; Example 2.4; Chap. 4). A land parcel of 15,000 m^2, together with a concession to access the hot water reservoirs for recreational purposes, was initially granted for 25 years by the municipality of Salto (Uruguay). The attractions of Acuamanía's large waterpark complex include thermal pools for relaxation, swimming pools, waterslides, an artificial river and whirlpools. For sports minded visitors, basketball and volleyball courts offer additional activities and kiosks, boutiques and a food court complete the range of essential features (Acuamanía 2019).

However, since the first thermal spa developments in the mid-1990s on the border between Uruguay and Argentina, concerns have been raised about the increasing exploitation and resource depletion of certain parts of the GAS (Chaps. 2 and 4). But this has not stopped the tourism industry to continue to take advantage of this natural resource and capitalise on the opportunity to extract the hot water needed to supply thermal waterparks like Acuamanía.

Example Argentina—Termas Cacheuta

The thermal waterpark *Termas Cacheuta* in Argentina emphasises their close connection with nature and the mineral-rich hot water. Thermal pools with different features and temperatures invite visitors to spend time immersed in the warm water and absorb some of the health-giving minerals. Surrounded by indoor and outdoor picnic areas the waterpark Termas Cacheuta is a place where people gather to embark on guided adventure tours in the surrounding mountains and other sports activities before their return to the thermal pools to relax (Termas Cacheuta 2019).

Example Japan—Ooedo-Onsen Monogatari

In contrast to the artesian hot water reservoirs of South America, Japan's waterparks are based on the abundance of volcanic hot springs. One of the largest entertainment parks is the *Ooedo-Onsen Monogatari,* a hot spring theme park in Tokyo, which is marketed as an authentic onsen experience from the era of 'Good Old Japan'. This theme park offers large common baths, foot spas, and many other different baths to suit individual needs. In addition to open air hot spring pools, steam baths and whirlpools, a special facility for therapeutic spa treatments and a rock salt sauna invite visitors to spend as much time as they like. The entire centre is developed in the traditional style of the Edo Period with nostalgic overtones throughout this Onsen Theme Park that allows visitors to feel like travelling 400 years back in time. Further adding to the entertainment are restaurants and traditional street food, gaming arcades for entertainment, and the opportunity to see a fortune teller. Speciality shops are selling onsen merchandise and souvenirs, food courts are lining the streets of the re-enacted Edo township and natural garden settings are complementing the cultural and social attractions to provide a special experience (Monogatari 2019).

Example Japan—Spa Resort Hawaiians

The *Spa Resort Hawaiians* is another large-scale recreational complex in Japan based on aquatic entertainment (Chap. 4; Figs. 4.9, 4.10). This hydrothermal waterpark revolves around Hawaiian and Polynesian culture, which is combined with traditional Japanese onsen bathing. The water attractions include several large water slides, a flowing river pool (Fig. 7.16), public swimming areas and pools for children of all ages, with the majority under the roof of a large domed complex with an artificially created tropical indoor environment that is kept at a constant temperature of 28° Celsius. All pools and onsen facilities are supplied with hot spring water rich in mineral elements, which is sourced from the *Iwaki Yumoto* thermal reservoir. Around three million litres every day are used at the facilities of Spa Hawaiians with a total potential of up to five million litres (Spa Resort Hawaiians 2019).

The Resort complex is divided into several coloured zones with different themes, including a large open-air hot spring bath, which was designed to recreate the ambiance of a public bathhouse from the popular Edo period, surrounded by a townscape from the same era. Similar to the Ooedo-Onsen Monogatari, the Spa Resort Hawaiians has effectively conjured up an atmosphere of times gone by, setting apart individual cultural and natural zones of the complex. Featured zones include a *European Hot Spring Oasis* reminiscent of a Roman thermae (Fig. 7.17), typical Japanese communal hot spring bathing areas, and a zone that is devoted exclusively to health, wellness and beauty.

Fig. 7.16 The 'endless' flowing river pool at the Spa Resort Hawaiians, Fukushima Prefecture, Japan

In adherence to the Hawaiian theme, which in reality has no connection with Japanese hot springs, one of the Gourmet and Retail Zones was recreated to appear like a street scene from the 1960s in Hawaii's *Kauai Island*. Other zones contain food courts, shopping decks and an amusement pavilion, while daily Hawaiian dance performances take place in a large stage area. The whole waterpark complex is surrounded by hotels, which are also supplied with natural hot spring water. To remain an attractive destination and continue to attract large visitor numbers, a 27-hole golf course is part of the overall Spa Resort Hawaiians experience (Erfurt 2012; Erfurt-Cooper and Cooper 2009).

Visitor numbers at the Spa Resort Hawaiians before the Great East Japan Earthquake in 2011 were high with circa 3200 visitors daily and up to 10,000 at weekends; including the holiday seasons this adds up to between 1.3 and 1.5 million people each year. Although it is not known whether these figures have remained the same since the re-opening in 2012, the whole complex has undergone substantial remodelling and repairs after the structural damage caused by the magnitude 9 earthquake. But most likely the popularity of this facility has remained the same or has increased further.

7.4.3 Hydrothermal Waterparks in Europe

The next two examples of thermal waterparks are from Europe, where subterranean thermal reservoirs are tapped with wells to extract the hot water needed to supply large water entertainment facilities. These destinations are also examples for the linkage between natural resources and regional culture to offer visitors a wider range of attractions.

Example Slovakia—Tatralandia

The waterpark *Tatralandia* in Slovakia is essentially an amusement complex with many opportunities for fun and adventure for every age group and all year round. Tatralandia offers over a dozen individual pools for bathing, swimming and relaxing as well as waterslides and wave surfing. Apart from warm mineral water and freshwater, some pools contain saltwater and are promoted as a Tropical Paradise with 'seawater healing', a somewhat dubious claim, as Tatralandia is nowhere near the sea. It is more likely that the complex has access to saline groundwater, similar to many other spa towns in northern and central Europe. The wellness facilities, restaurants and visitor accommodation are surrounded by mountains and enriched with local cultural attractions (Tatralandia 2019).

Fig. 7.17 A 'Roman Style' bathing zone at the Spa Resort Hawaiians, Fukushima Prefecture, Japan

Example Poland—Zakopane Aqua Park

The pools at the thermal entertainment complex *Zakopane Aqua Park* in Poland include basins for health and wellness, filled with mineral-rich water with health-promoting properties, massage nozzles and underwater geysers between 26 and 38 °C. The recreational pools (29 and 32 °C) offer swimming lanes as well as fun for the whole family with water jets and cascades, designed to improve the feeling of wellbeing. Thermal pools and therapeutical basins emphasise the medicinal properties of the mineral water and its curative value for a variety of health conditions. Recreational areas include aqua fitness, wading pools for children, waterslides, whirlpools, a sauna zone, a salt cave, plus the inevitable non-aquatic pleasures such as restaurants, bowling alleys and gaming areas. Surrounded by nature and the Tatra mountains, Zakopane offers health improvements based on natural resources such as thermal water and landscape, combined with some cultural tourist attractions (Aqua Park Zakopane 2019).

While the descriptions of various waterparks may sound like a promotion for this type of hot spring use, this is not the case; the intention is rather to demonstrate the extravagance of exploiting vast amounts of natural resources to provide large waterparks with thermal water.

7.4.4 Now and Then—A Comparison with the Past

Somewhere within the impressive waterparks of our time are striking parallels to the Roman bathhouse history; another example where the natural and cultural geoheritage of hot spring settings is interrelated. While not all Roman bathing facilities, left behind in so many countries, may have had access to natural hot springs, the principle of these monumental establishments was the same throughout the provinces of the former Roman empire. Lavish interior design and a diversity of available services ranging from opulence to sheer decadence were signs of the time and eventually followed by the inevitable demise. The enormous domed and vaulted Caracalla Baths (Fig. 7.18) for instance were constructed in typical Roman fashion from brick, concrete and marble and adorned with floor mosaics and wall murals. Exercise areas, hot and cold pools and steam rooms took up more than 100,000 m^2 with a capacity for 1600 people all at once. These large-scale facilities were used for more than bathing or workouts; social activities and business transactions were common in such a relaxed environment (Chap. 6,Social Settings). This culture of aquatic entertainment from the past is certainly comparable to the contemporary waterpark establishments, which often incorporate Roman features (Fig. 7.17) as a reminder of the grandiose thermal bathhouses of 2000 years ago.

7.4.5 How Sustainable Are Thermal Waterparks Versus Protected Natural Hot Spring Sites?

Visitor Pressure at Thermal Waterparks

Regrettably, there is a considerable potential for certain negative effects. The common denominator of all extensive aquatic theme parks is their excessive use of thermal water sourced from hydrothermal systems as well as cold freshwater from subterranean reservoirs. Apart from valid legitimate environmental concerns, contrasting these large-scale waterscapes to areas so dry that people do not have enough water for cooking and washing on a daily basis, there certainly are some moral and ethical concerns as well regarding the potentially uneconomical use of water for recreational developments. Unless a significant amount of the wastewater is recycled and re-injected back into the ground, the sustainability of waterparks with areas covering several tens of thousands of square metres should be seriously called into question. Also, a significant percentage of the water used in these parks is continuously evaporating and must be replenished, which further

Fig. 7.18 The ruins of the Caracalla Baths. Drawing by Giovanni Battista de' Cavalieri, Rome, 1569 AD

strains the sustainability of what otherwise would be a renewable resource. Another adverse effect of large-scale waterparks is the extensive use of chemicals necessary to avoid the risk of anthropogenic pollution of the water and the possible contamination with disease causing agents due to the immense crowds of people using these types of facilities. To put this into perspective, some Chinese aquatic entertainment parks are large enough to cater for up to 30,000 people on a daily basis. Therefore it can be assumed that the chemical treatment of large volumes of water and surrounds represents an unprecedented environmental impact with no results from long-term studies available yet.

As noted earlier in this chapter, the objective of purpose-built themed aquatic parks with large waterscapes is to attract as many visitors as possible, preferably 365 days a year. To achieve this, exciting water features and aquatic amusement are the key attractions, surrounded by entertainment and service facilities to cater for thousands of visitors. While waterparks can be structured either as outdoor or indoor facilities, as well as a combination of both, the more than generous way of utilising a major natural resource such as hot and cold spring water might prove unsustainable in the long term. But at this point in time, research appears to be more focused on the prevention of injuries and waterborne infections (Davis 2009) than on the lasting sustainability of large waterparks, although this is of course another serious concern. Unfortunately, the size of thermal waterparks and the volume of visitors presents an inherent risk for the spread of infectious diseases (Chap. 9) if not maintained correctly.

7.5 Tourism Opportunities in Hot Spring Environments

Several tourism sectors have incorporated natural hot springs and other hydrothermal attractions in their destination development (Table 7.3). Countries such as Iceland, Japan, China and New Zealand have a long history of combining visits to natural hot spring sites with other regional tourist attractions. In more recent times Brazil, India, Malaysia, Peru, the Philippines and Singapore are also increasingly using natural hot springs to develop tourism destinations for health and recreation (Erfurt 2012).

Natural heritage sites including hydrothermal attractions are often part of a sightseeing agenda, with tours specialising in visits to active hydrothermal areas to respond to an increased demand for interesting geosites. Apart from being an adjunct to all forms of sustainable nature-based tourism, the geoheritage of hot springs is successfully utilised to supplement other tourism sectors such as health and spa tourism, food and wine tourism, ecotourism, adventure tourism, and even industrial tourism (Table 7.4). In the case of natural hot springs, industrial tourism may include visiting geothermally heated greenhouses or geothermal power stations. Countries with an abundance of high temperature hydrothermal resources like Iceland use geothermal power for the generation of clean energy, but also offer educational tours at their facilities, which is unique and awareness-raising. Several of Iceland's geothermal power plants include visitor centres: *Nesjavellir, Hellisheiði* and the *Krafla Geothermal Energy Exhibition.* Also of interest is the *Hveragerði Greenhouse Village*, where tropical temperatures allow bananas to grow along with other exotic vegetation uncommon in Iceland's climate. Greenhouse heating is also common in Japan where they are used for agricultural purposes and as regional attractions (Fig. 11.1a, b).

Visiting active hydrothermal areas falls into the category of adventure tourism due to the potential risk associated with such environments (Erfurt-Cooper and Cooper 2009). For adventure-minded tourists with a curiosity for unusual or extreme natural experiences, visually exciting hydrothermal features present some of the most interesting geosites, that are frequently located close to active volcanism (Erfurt-Cooper 2018, Chap. 11; Fig. 11.9, 11.10).

Table 7.3 Short typology of important elements of hot mineral springs in relation to their natural geoheritage with an emphasis on the sustainable use of natural hot spring water

Attributes of natural hot springs	
• Valuable tourism infrastructure in the form of natural and cultural geoheritage • Renewable natural resource of authentic and chemical-free hot spring water • Visually attractive and dynamic natural environments • Common occurrence worldwide and often located in or near national parks and other protected sites • Unique natural destination drawcard—hot spring access is a unique selling point (USP) • Independent from seasons and climate	• Hot spring-based waterparks with potential for sustainable development if following principles of conservation • Good prospects for future expansion of recreational facilities • Benefits for health and recreation—integrated into health tourism • Value adding thermal by-products (e.g. cosmetics, therapeutic mud, mineral bath powders

Source Author

Table 7.4 Hydrothermal features play an important role in the development and the promotion of nature-based tourist destinations

Natural hot springs and their role in tourism	
Hot Spring Spa Tourism	Use of mineral-rich hot spring water with curative properties—alternative health resource—hot springs contribute actively to the improvement of health without any known side effects Recreational Value, health and wellness oriented
Geotourism	Opportunity to learn about the geological background of hot springs Natural environment, interesting landscapes Promotion of conservation
Ecotourism	Opportunity to enjoy the natural and cultural heritage of hot springs Sense of place, ambience, cultural heritage Promotion of conservation
Adventure Tourism and Nature-based Tourism	Visits of extreme natural environments—seasonally independent Exploration and observation of active hydrothermal areas highly regarded for their remarkable and unusual features
Food and Wine Tourism	Co-branding with other attractions, e.g. hot spring wellness spas or visits of hot spring geosites
Industrial Tourism	Educational tours at geothermal power stations, greenhouses or food drying plants
Sustainable Tourism	Development based on natural hot springs as a renewable resource if managed responsibly Promotion of conservation
Economic Benefits of Natural Hot Springs	Value adding when combined with different tourism sectors USP for destination marketing Capacity building for host communities—especially in rural and/or remote regions Employment opportunities

Source Author

7.5.1 Geotourism and Hot Springs

Geotourism is a sustainable form of tourism based on the natural and cultural geodiversity of our environment and encourages a responsible use of natural resources. Geotourism provides a positive and optimistic response to the demand for the conservation of natural environments and their recreational value and aims for a low impact on the environment. Not only highly educational with regards to the natural environment, geotourism also embraces the cultural significance of the overall geodiversity of a place or region as well as recreational activities such as the opportunity to experience hot spring bathing. Learning about natural landforms helps to raise awareness about the sustainability of geosites and the need for the protection of their unique features for future generations, while at the same time the long-term conservation goals of protected areas are promoted (Erfurt 2013, 2014). Bound by the local environment, geotourism has the potential to enhance the natural and cultural diversity and distinctiveness of a place for visitors through elements of discovery and adventure.

Initially geotourism was viewed as a form of nature-based tourism and was frequently associated with a main focus on the promotion of the geological heritage of a region and the development of geosites (Dowling and Newsome 2006; Gray 2011; Hose 2010; Newsome and Dowling 2010; Newsome et al. 2012). As a rather new tourism sector, geotourism has taken hold since the early 2000s and subsequently was integrated in the tourism policies of many countries. The first country that made geotourism their national tourism strategy was Honduras in 2004, followed by Norway, Romania and Mexico in 2005. Since these early days many other countries have recognised that geoheritage in its various forms is a way to promote not only geosites but also raise awareness about the need for conservation and sustainable development (Erfurt 2012; Ruban 2015). But is it enough to stop the degradation of the natural environment through educational measures? Time will tell.

Gradually a broader focus was applied that embraced the overall geodiversity of natural heritage by including sites that are significant for their cultural and historical, as well as for their educational value (Brocx and Semeniuk 2007), and because geosites represent landforms of a high scientific value that are attracting people with their uniqueness and their rare beauty. Taking into account the sustainability of geoheritage areas, Professor José Brilha, a geologist at the University of Minho (Portugal) has developed a detailed set of criteria, indicators, and parameters for the assessment of geosites and their diversity. The criteria include the evaluation of (a) the scientific value, (b) the potential for educational use, (c) the potential for touristic use and (d) the risk of degradation, e.g. the deterioration of geological elements (Brilha 2016).

To protect the natural geoheritage, geoparks are developed all over the world with geotourism as a sustainable way to educate about conservation of natural assets as well as preserving the cultural identity of individual regions (Brilha 2016; Farsani et al. 2012). China for example has already over 200 national geoparks with many sites including natural hot springs either as a major natural feature or as a valuable geoheritage resource (Luan et al. 2016). In a similar way Germany's natural springs are considered as a major geosite type at historical sites and are included in the state of Bavaria's Geosite Register (Lagally and Loth 2016).

7.5.2 Community Development and Capacity Building in Remote and Rural Areas

Tourism based on natural hot springs represents a substantial part of a region's economy, with the key focus on income generation and employment (Fig. 7.19). As a labour-intensive sector, tourism not only supports economic diversity, but it is also a valuable source of foreign exchange and therefore a potential vehicle for community development in regions without any other industry prospects. The opportunities for regional communities to develop geotourism around local hot springs and their natural geoheritage include a priceless asset—the natural resource itself. Moreover, hot springs are often located in already protected areas with some essential infrastructure in place. Based on the existing demand for authentic experiences and nature-based education, the sustainable use of undeveloped hot spring areas for geotourism presents a possibility with great potential for stakeholder partnerships (Erfurt 2013).

Among the main stakeholders taking advantage of natural hot spring areas are health resorts and spas together with the overall tourism and service industry of a region. The unique characteristics and advantages of natural springs denote a considerable commercial benefit for host communities in hot spring areas, especially in more remote or rural regions. The economic value may consist of one or both of the following: (a) seasonally independent unique settings in which tourism can take place, and (b) a long established direct relationship between hot springs, health and recreation. In combination with natural hot springs these factors are a substantial contribution to local communities with the potential for commercial tourism development (Erfurt 2012).

In countries such as Iceland, Japan, China, Portugal (Azores) or New Zealand the diversity of geosites with active hydrothermal manifestations is used by the tourism industry to promote their natural geoheritage in conjunction with hot spring health resorts and spas. This synergy establishes natural hot springs as a primary tourism resource and therefore as a key element for health and recreational tourism, which has developed into a global movement and caused the redevelopment of existing hot spring facilities as well as opening up new hot spring areas (Erfurt 2012; Foster et al. 2011). To attract visitors of different interest groups, hot spring destinations offer a diversity of outdoor activities including hiking, trekking, climbing, canoeing and fishing. After a long trek cross country or skiing and snowboarding, a relaxing soak in a hot spring is one of the main incentives to look forward to following an action-packed day (Erfurt-Cooper and Cooper 2009). For example, to integrate rural and remote areas into travel itineraries for several days, guided bus tours on New Zealand's north island offer a broad range of attractions that always include the natural hot spring geoheritage. Online promotions are focusing specifically on exploring *Rotorua's geothermal wonders* or viewing the famous *Pohutu Geyser* and boiling mud pools.

Another representative case from the German health resort and spa industry demonstrates the economic relevance of spa tourism with an annual turnover of around 25 billion Euros and 500,000 people either directly or indirectly employed. The latest available statistics for annual visitor numbers at German hot and mineral spring-based spa towns

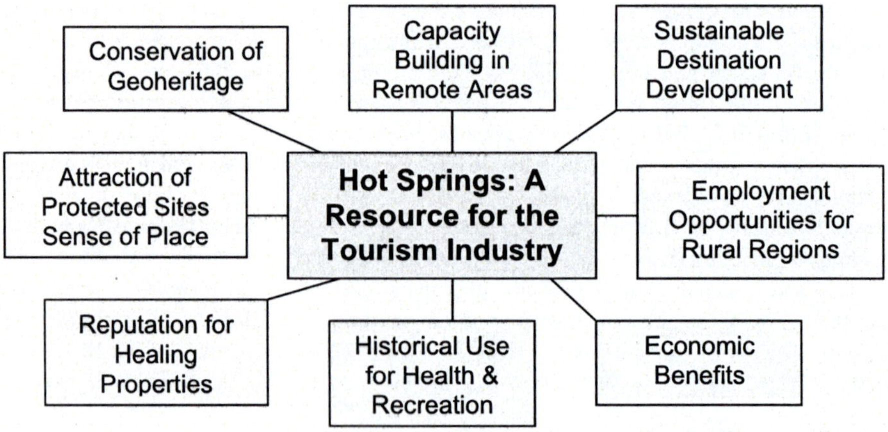

Fig. 7.19 Natural hot springs and some of the potential benefits for their local communities. *Source* Author

from 2018 to 2019 are reporting 9,978,803 visitor arrivals with 45.25 million overnight stays (average length of stay 4.5 nights; Fig. 9.1). Considering that this part of the health tourism sector represents only about 35% of the overall spa industry, this signifies a substantial economic contribution at an increase of 2.2% compared to the previous year (Deutscher Heilbäderverband 2020). Although not all hot spring areas can be as highly developed as in Germany or Japan, regions with an otherwise low economic status can certainly benefit from sustainable developments that contribute to progress and reduce financial hardship in rural and remote regions with access to natural hot springs. Especially if the experience can be combined with other local attractions including cultural activities, retail therapy and sampling the regional cuisine (Erfurt 2012).

7.5.3 Essential Infrastructure—Natural Environment

The infrastructure of more sophisticated hot spring spa towns includes not only natural resources such as the mineral-rich hot spring water, but other elements such as pristine landscapes, a healthy seaside climate or unpolluted mountain air, which further enhance a health and wellness destination. In areas more oriented towards natural geoheritage, the supporting infrastructure at designated hot spring sites depends on the location and size of the area but should preferably include:

- Access roads
- Carparks and rest rooms
- Visitor or interpretive centres
- Guided tours
- Accessible and well-maintained walking tracks
- Wheelchair access
- Viewing areas/decks/outlooks
- Shaded sitting areas
- Well-designed information boards at suitable locations
- Emergency shelters in active hydrothermal areas, as well as
- Warning signs and boardwalks with safety barriers where required.

Infrastructure can further consist of geotrails, which may be short walking circuits through an area of interest, or they can be many kilometres long, following a map with POIs over long distances and are therefore relying in part on existing infrastructure such as roads, transport, public services and telecommunications. Most hot spring destinations also take advantage of their seasonal independence, which adds another dimension to the natural tourist attractions.

Hot springs located in the 'middle of nowhere' are often favoured for their anticipated remoteness and therefore not overrun by crowds of tourists; at least that is the expectation until arriving and finding tour buses already there. *Landmannalaugur* in the south of Iceland is a well-known example for a natural hot spring setting in a somewhat desolate and bare countryside. Despite the minimalist infrastructure, people always gather here, rain or shine, to explore the unique volcanic landforms and to take a relaxing soak in the natural hot springs (Erfurt-Cooper 2010; Figs. 2.21, 7.20).

At the other end of the spectrum concerning hot spring tourism infrastructure, Japan's Ministry of the Environment in 2006 recorded 3157 onsen resorts as well as circa 6400 public hot spring bathhouses. These are frequented on average by over 150 million, mainly domestic overnight visitors every year, which is only made possible by more than 27,200 hot springs all over the country. Still, this does not even include other adjunct hot spring tourism infrastructure or the integration of public infrastructure that facilitates, promotes and encourages Japan's passion for onsen bathing (Erfurt-Cooper 2010; Erfurt 2012; Statista 2015).

7.5.4 Geo-certification to Support Conservation

Geotourism is widely recognised as a sustainable and responsible form of tourism with the objective to contribute to a much needed reduction of the global ecological footprint. This is especially important as many natural areas are facing increasing degradation from growing visitor numbers, ignorant and irresponsible actions, anthropogenic pollution, as well as tour guides not sufficiently qualified to lead by example and provide essential geoheritage related educational talks. From a conservation point of view, geotourism aims for the lowest impact on land use, reinforced by sustainable destination management and the responsible use of natural resources. To identify and confirm natural and cultural geoheritage sites, including hydrothermal areas that provide a genuine experience for geotourists, destinations could introduce a geo-certificate to authenticate the regional geodiversity and at the same time endorse qualified professional geotourism operators and/or provide tour guides with reliable credentials. Geo-certification could be in a similar format as already established eco-certification programs but with a focus on the natural and cultural geoheritage and its conservation and protection. Geo-certificates should be an official endorsement of authentic geotourism destinations and globally recognised.

Fig. 7.20 When you think nobody will be there because it is a cold and rainy day—Landmannalaugar in Iceland's remote highlands with tour buses and other vehicles

7.6 Final Comments

The appeal of hot spring areas is based on a range of attractions, from warm springs used for bathing and soaking in the wilderness to other open-air activities such as hiking and taking photos, collecting information and general sightseeing. Following the concept of geotourism, hot spring environments offer an opportunity to learn about their natural geoheritage and their geophysical background. Because natural hot spring areas are highly valued POIs, this means that these landscapes are at risk due to their growing popularity and large visitor numbers, a combination, which has an increasingly damaging effect on their fragile environment (Erfurt 2012).

From undeveloped to developed hot springs, their natural origin and their surrounding areas must be protected from any harmful impact caused by overuse and/or pollution of hot and cold groundwater reservoirs (Chap. 4). While new destination developments provide employment for local communities in many countries, and national parks, geoparks or world heritage areas have been successful in protecting endangered geoheritage in the past, new developments in untapped markets must adhere to the principles of sustainable management. In particular, the trend towards large-scale thermal waterparks should be closely monitored to avoid negative environmental impacts that affect the long-term availability of what could be a renewable water resource under sustainable management.

The crater of Old Faithful between eruptions. *Source* W. H. Jackson, 1872 (Public Domain)

7.7 Appendices

Appendix 7.1 The Pink Terraces, New Zealand

1, 2—The Pink Terraces (New Zealand) were painted several times by Charles Blomfield, some of them after their disappearance (image 2 c. 1890). In the middle background of image 1 the steam indicates the location of the White Terraces; 3, 4—Two rare photographs of the Pink Terraces before they vanished during the eruption of Mount Tarawera in 1886

Appendix 7.2 The White Terraces, New Zealand

1—The White Terraces painted by Charles Blomfield in 1913 after they had disappeared; 2—One of the few remaining images that were taken of the White Terraces using early photographic technology

Appendix 7.3 Unique Hot Spring Geoheritage—The Great Geysir, Iceland

The Great Geysir, South Iceland. *Source* Painting by T. K. Fairless, 1849. Unfortunately, this remarkable hydrothermal feature only erupts occasionally now

Appendix 7.4 Unique Hot Spring Geoheritage of Badab-e Surt, Iran

The red travertine terraces of Badab-e Surt in the north-west of Iran are formed from calcium carbonate deposits, which are enriched with iron carbonate and responsible for the different shades from yellow to dark orange. *Source* M. Samaee, 2010 (Creative Commons)

References

Acuamanía (2019) Atracciones. Retrieved from https://www.acuamania.com.uy/

Aqua Park Zakopane (2019) Attractions (translated from Polish). Retrieved from https://www.aquapark.zakopane.pl/pl/atrakcje

Bargar ML, Muffler LJP (1975) Geologic map of the travertine deposits, Mammoth Hot Springs, Yellowstone National Park. Wyoming. In: Geological survey miscellaneous field studies Map MF-659 1:48000 and 1:2400 scale. USGS, Washington DC

Brilha J (2016) Inventory and quantitative assessment of geosites and geodiversity sites: a review. Geoheritage 8:119–134. https://doi.org/10.1007/s12371-014-0139-3

Brocx M, Semeniuk V (2007) Geoheritage and geoconservation—history, definition, scope and scale. J Roy Soc West Aust 90:53–87. oai:researchrepository.murdoch.edu.au:10140

Bryan TS (1986) The Geysers of Yellowstone. Colorado Associated University Press, Boulder, Colorado

Caldas Novas (2019) Rio Quente. Retrieved from https://www.visitbrasil.com/attractions/rio-quente.html

Davis T (2009) Water quality of modern water parks. J Environ Health 71(9):14–19. https://www.jstor.org/stable/26327917

De Ronde CEJ, Fornari DJ, Ferrini VL, Walker SL, Davy BW, LeBlanc C et al (2016) The Pink and White Terraces of Lake Rotomahana: what was their fate after the 1886 Tarawera Rift eruption? J Volcanol Geotherm Res 314:126–141. https://doi.org/10.1016/j.jvolgeores.2016.02.003

Deutscher Heilbäderverband (2020) Jahresbericht 2019 – Oktober 2018 bis Dezember 2019. Berlin: Deutscher Heilbäderverband e.V. (DHV). Retrieved from https://tinyurl.com/y4ztpwtr

Dowling R, Newsome D (2006) Geotourism. Elsevier, London, UK

Encyclopedia Britannica (2019) National park. Retrieved from https://www.britannica.com/science/national-park

Erfurt-Cooper P, Cooper MJ (2009) Health and wellness tourism: spas and hot springs. Channel View Publications, Bristol, UK

Erfurt-Cooper P (2010) The importance of natural geothermal resources in tourism. In: Proceedings world geothermal congress, Bali, Indonesia, 25–29 Apr 2010

Erfurt P (2012) An assessment of the role of natural hot and mineral springs in health, wellness and recreational tourism. Unpublished Doctoral Dissertation, School of Business, James Cook University

Erfurt P (2013) Geotourism in Australia—mega-potential for sustainable development. Presented at Global Eco Asia-Pacific Tourism Conference, Noosa 18–20 Nov 2013

Erfurt P (2014) Geotourism: protection and sustainable use of geodiversity—achieving conservation goals through educational encounters with nature and culture. Presented at IUCN World Parks Congress, Sydney Nov 2014

Erfurt-Cooper P (2018) Active hydrothermal features as tourist attractions. In: Fearnley CJ, Bird DK, Haynes K, McGuire WJ, Jolly G (eds) Observing the Volcano World: volcano crisis communication. Advances in volcanology series. chap 27. Springer, Berlin, Heidelberg. https://doi.org/10.1007/11157_2016_33

Farsani NT, Coelho COA, Costa CMM (2012) Analysis of network activities in geoparks as geotourism destinations. Int J Tour Res 16:1–10. https://doi.org/10.1002/jtr.1879

Foster LT, Keller CP, McKee B, Ostry A (2011) British Columbia Atlas of wellness, 2nd edn. Retrieved from https://dspace.library.uvic.ca/handle/1828/3838

Glimpses of New Zealand (1896) Beautiful Britain of the South. New Zealand Publishing Company Ltd., Christchurch, NZ

Gray M (2011) Other nature: geodiversity and geosystem services. Environ Conserv 38(3):271–274. https://doi.org/10.1017/S0376892911000117

Hose TA (2010) The significance of aesthetic landscape appreciation to modern geotourism provision. In: Newsome D, Dowling R (eds) Geotourism: the tourism of geology and landscape. Goodfellow Publishers, Oxford, UK, pp 13–26

Hróarsson B, Jónsson SS (1992) Geysers and hot springs in Iceland. Mál og Menning, Reykjavik, Iceland

Jackson WH (1872) Crater of old faithful geyser, taken during Hayden Geological Survey of 1872 by W. H. Jackson. Retrieved from https://commons.wikimedia.org/wiki/File:CraterofOldFaithful1872.jpg

Kiernan K (2015) Landforms as sacred places: implications for geodiversity and geoheritage. Geoheritage 7:177–193. https://doi.org/10.1007/s12371-014-0128-6

Lagally U, Loth G (2016) Experiencing Bavaria's geological heritage —the project "hundred masterpieces." Geoheritage 9:519–531. https://doi.org/10.1007/s12371-016-0209-9

Lee CF, King BE (2008) Using the Delphi method to assess the potential of Taiwan's hot springs tourism sector. Int J Tour Res 10 (4):341–352. https://doi.org/10.1002/jtr.661

Luan F, Wang F, Xiong H, Wang Z, Li B (2016) A study on classification and zoning of Chinese geoheritage resources in National Geoparks. Geoheritage 8:247–261. https://doi.org/10.1007/s12371-015-0157-9

Migoń P, Pijet-Migoń E (2016) Geoconservation and tourism at geothermal sites—lessons learnt from the Taupo Volcanic Zone, New Zealand. Proc Geol Assoc 127:413–421. https://doi.org/10.1016/j.pgeola.2016.04.002

Newhall C (2014) A dilemma for tourists and land managers alike: what risks to take? In: Erfurt-Cooper P (ed) Volcanic tourist destinations. Springer, Berlin, Heidelberg, pp 351–353

Newsome D, Dowling RK (eds) (2010) Geotourism: the tourism of geology and landscape. Goodfellow Publishers, Oxford, UK

Newsome D, Dowling R, Leung Y-F (2012) The nature and management of geotourism: a case study of two established iconic geotourism destinations. Tour Manag Perspect 2(3):19–27. https://doi.org/10.1016/j.tmp.2011.12.009

NPS Stats (2020) Yellowstone NP. Recreation Visitors (1904–Last Calendar Year). Retrieved from https://tinyurl.com/kgk5xh8

Nunes JC (2014) The Azores Archipelago: islands of geodiversity. In: Erfurt-Cooper P (ed) Volcanic tourist destinations. Springer, Berlin, Heidelberg, pp 57–67

Office National du Tourisme (1921) The Spas of France. Ministère de Traveaux Publics, Paris

Ooedo-Onsen Monogatari (2019) Enjoy the Good Old Japan—Onsen Entertainment. Retrieved from https://daiba.ooedoonsen.jp/en/

Peaco J (2006) Aerial view of Grand Prismatic Spring. Yellowstone NP. Public Domain. Retrieved from https://tinyurl.com/yjmv8u9m

Reykjanes Geopark (2020) Upplifðu Reykjanes UNESCO Global Geopark. Retrieved from https://www.reykjanesgeopark.is/

Rhinehart JS (1980) Geysers and geothermal energy. Springer, New York

Ruban DA (2015) Geotourism—a geographical review of the literature. Tour Manag Perspect 15:1–15. https://doi.org/10.1016/j.tmp.2015.03.005

Sanner B (2000) Baden-Baden—a famous thermal spa with a long history. GHC Bulletin, PDF. Retrieved from https://tinyurl.com/y2nnxejo

Savin A (2018) Landscape at Dallol volcano, Afar Region, Ethiopia. Retrieved from https://commons.wikimedia.org/wiki/File:ET_Afar_asv2018-01_img36_Dallol.jpg

Spa Resort Hawaiians (2019) Theme parks. Retrieved from https://www.hawaiians.co.jp/english/tp/wp.html

Stafford DM (1986) The Founding years in Rotorua: a history of events to 1900. Ray Richards, Rotorua

Statista (2015) Total number of hot springs (onsen) in Japan from 2006 to 2015. Retrieved from https://www.statista.com/statistics/682636/japan-hot-spring-numbers/

Tatralandia (2019) Tatralandia. Retrieved from https://www.tatralandia.sk/en/welcome/tatralandia/about-us

Te Ara (2019) Bathing at the Pink Terraces. The Encyclopedia of New Zealand. Retrieved from https://teara.govt.nz/en/photograph/15155/bathing-at-the-pink-terraces

Termas Cacheuta (2019) Thermal waterpark—welcome to nature. Retrieved from https://www.termascacheuta.co/en/thermal-water-park/

UNESCO (2020) World heritage. Retrieved from https://whc.unesco.org/en/about/

UNESCO (2021) UNESCO Global Geoparks. Retrieved from https://tinyurl.com/hyjbx29

Wettstein R (2010) Römische Bäder (Badruinen). Retrieved from www.bad-bad.de/sehen/badruin.htm

White DE, Hutchinson RA, Keith TEC (1988) The geology and remarkable thermal activity of Norris Geyser Basin, Yellowstone National Park, Wyoming, U.S. Geological Survey Professional Paper 1456, USGS, Washington DC Wikimedia Commons (2009a) Map Wyoming

Wohnlich S (1996) The spa of Baden Baden, Germany. Environ Geol 27(2):108–109. https://doi.org/10.1007/BF01061680

Termas de Ferraria, a hot spring spa next to the Atlantic Ocean. São Miguel, Azores

8

Hot Springs, Health and Wellbeing

Contents

P. Erfurt, *The Geoheritage of Hot Springs*, Geoheritage, Geoparks and Geotourism,
https://doi.org/10.1007/978-3-030-60463-9_8

The colonnades and pump house (Trinkhalle) are part of the spa complex in Baden-Baden. *Source* Alexander Savin (2015b) (Public Domain)

People searched for the healing powers of nature and found them in mineral springs. They sought companionship, leisure and alleviation of bodily pain in an inspiring and comforting environment. (Valenza 2000:6)

8.1 Introduction

In ancient times places of healing were repeatedly established near hot springs and a belief in their divine curative powers developed (Arvigo and Epstein 2003; Bishoff 2001; Bullard 2004; Frazier 2000; Havins 1976; King 1966; Lund 2005; Ryrie 1999). Many springs were assigned a patron saint or a guardian deity, who was expected to watch over the healing waters and protect them (Chap. 6). Natural sources of water were usually associated with divinities of the Earth and Roman temples that offered thermal baths were dedicated to Aesculapius, the Roman god of medicine[1] (Mellilo 1995; Chap. 5). To this day, water is one of the key elements in health and recreation with many facilities relying on hot springs and their mineral content as a powerful visitor attraction (Erfurt-Cooper and Cooper 2009, 2010). But even without hot springs, many recreational parks use water features such as fountains (Fig. 8.1) or lakes as a focal point.

Hot springs are traditionally considered as a natural option to improve health conditions, while at the same time offering recreational opportunities at tourist destinations centered around natural springs. Based on the demand for health treatments such as balneotherapy or *taking the cure*, hot spring resorts, clinics and spas are promoted as destinations for better health, wellbeing and recreation (Erfurt 2012).

Numerous European countries and other nations including Japan, New Zealand or Tunisia have a longstanding reputation for health treatments based on the use of hot spring water (Table 8.1; Erfurt-Cooper and Cooper 2009, 2010).

In France and Germany, the hot spring health and wellness industry has always been supported by the medical profession, which continues to back the existence and survival of hot spring-based health resorts, clinics and spas (Erfurt 2012; Weisz 2001). Hundred years ago, in 1921, French spa towns were strongly promoted by the National Tourism Office in an endeavour to once again attract foreign visitors, who remained largely absent after the First World War. A book titled *The Spas of France* listed and described the most important spa towns and provides interesting insights into the marketing of French hot spring spa destinations. The following is an excerpt from the Preface of this book.

> *Spa treatment is one of the most invaluable aids to modern therapeutics, and it is in no sense to be regarded as the outcome of a passing fashion. From the earliest times our waters have been sought by sick persons of all races. That the Romans attached extreme importance to this method of natural medication, is shown by the many architectural remains in the neighbourhood of our springs.*
>
> *Albert Robin, 1921, France*
> *Honorary Professor and Member of the Academy of Medicine*

[1] Aesculapius was the Roman god of medicine and healing.

Fig. 8.1 The large 'Sultan Ahmad Maidan' fountain is a central point in the park between the cathedral Aya Sofya and the Sultan Ahmed mosque in Istanbul (Turkey). People gather here during their break from work or meet in the afternoon for a picnic

Table 8.1 Some of the countries with a longstanding history of hot spring use for medical benefits based on the mineral content of the water

European countries		Non-European countries	
Austria	Italy	Algeria	Korea
Bulgaria	Poland	Australia	New Zealand
Czech Republic	Portugal	Canada	Russia
France	Slovakia	China	Taiwan
Germany	Slovenia	Iran	The Americas
Greece	Spain	Israel	Tunisia
Hungary	Switzerland	Japan	Turkey
Iceland	United Kingdom	Jordan	

This list is not exhaustive; many other countries in Asia, Africa and the Americas also have a history of hot spring use for health and recreation

Over one hundred spa facilities throughout France are using four hundred hot and mineral springs that are officially recognised as beneficial in providing thermal cures (Tabone et al. 2012). In Germany health resort and spa medicine plays a vital role in the national health care system with its effectiveness supported by evidence based on scientific investigations carried out over more than a century (Erfurt 2012). In France, various public and private organisations, some linked to university institutes, are carrying out scientific research into medical balneology (Tabone et al. 2012). And in the southern hemisphere, although scientific research into the use of natural hot springs for health, wellness and recreation in Australia is more recent, several PhD students are currently involved in clinical trials and studies of the benefits of balneotherapy at Melbourne's RMIT and Victoria Universities (Clark-Kennedy and Cohen 2017).

Even so, the use of hot springs for medicinal purposes predates the more recent centuries of European and Asian developments and goes back to earlier societies on a global scale. Described in poems from the Homeric Period (seventh century BC), hot spring bathing at different temperatures was used in ancient Greece to treat various health conditions. During the pre-Hippocratic period (fifth–sixth centuries BC) hot spring spa medicine was practiced at the temples of Asklēpiós[2] with treatments carried out by priests at purpose-built facilities near natural hot springs (Erfurt 2012; Katsambas and Antoniou 1996). In the fifth century BC Hippocrates of Kos was using balneotherapy to treat patients from Mediterranean countries at his Asclepian Centre (Cataldi et al. 1999; Erfurt-Cooper and Cooper 2009; Monti 2003), where he wrote a medical encyclopaedia that included a systematic classification of hot spring waters together with their specific healing properties (Fytikas et al. 1999; Katsambas and Antoniou 1996). Several other classical authors including Plato (427–347 BC), Aristotle (384–322 BC), and Pliny the Elder (23–79AD) also described the

[2]Asklēpiós (Asclepius) was the god of medicine in ancient Greek religion and mythology.

beneficial use of natural hot spring waters (Cataldi et al. 1999; Erfurt 2012; Chap. 5).

Hot spring water was not only used for bathing in the past. In fact, on occasion it has been employed as a multi-purpose resource (Example 8.1) to assist in the battle against disease.

Example 8.1 Baden Baden, Germany

The inhabitants of Baden Baden had so much confidence in their health-giving natural water, that they used it as a defensive measure against the raging plague in 1349 AD by flooding the streets with hot spring water. A few centuries later in 1625 AD the physician *Johan Küffer* from Straßburg (Strasbourg) was commissioned by Margrave Wilhelm von Baden to write a book about the healing springs of Baden Baden. With the objective to promote them worldwide, the published book describes twelve (12) hot mineral springs supplying thermal water at temperatures between 52 and 67 °C to treatment facilities (Fig. 8.2).

Source Heilbad Baden Baden (2020), Küffer (1625), Sanner (2000).

Even though earlier civilisations were aware of the healing benefits of natural hot springs (Chap. 5), it was mainly the Romans who promoted thermal bathing throughout their occupied territories by establishing a vast bathhouse architecture, of which parts can still be observed all across Europe, in northern Africa and in parts of the Middle East. This chapter explores the traditional use of hot springs in spa medicine with examples from different countries to demonstrate the enduring popularity of this natural resource for health, wellbeing and recreation on a global scale.

Fig. 8.2 The title on the front cover of Johan Küffer's book about the healing springs of Baden Baden (published in 1625 AD) was written in New High German, a version that was used from 1600 AD onward

8.1.1 The Attraction of Hot Springs

Granting the fact that natural hot springs across the world are highly regarded for their curative value, it is needless to say that with such a reputation come certain expectations for health improvements when using hot spring-based treatments to recover from illness (Table 8.2). Although one of the main appeals of hot springs may be their natural origin, equally attractive is the absence of negative side effects of balneotherapy. In addition, hot springs are a renewable resource when used in a sustainable manner (Chap. 4) and are a priceless natural asset for health resorts, clinics and spas.

8.1.2 Hot Spring Health Clinics, Resorts and Spas

A few words in advance regarding hot spring facilities: for the purpose of this chapter the terms health resorts, therapy centres, clinics and spas based on natural hot springs are used where appropriate. Different types of hot spring facilities can be further defined within their own context, as they often depend on translations from other languages. For example, the Japanese onsen hospital (温泉病院) is similar to a German *Thermalklinik* or a French *Clinique Thermale*, and a hot spring spa town is the same as an Austrian or Swiss *Thermalbad*, a French *Ville Thermale* or a Japanese *Onsen Machi* (温泉街). Even so, European spa towns do not rely on hot springs alone, but usually promote health and recreation based on the beneficial climate of a destination as well. Exceptional environmental conditions of either gentle climates (pure air, lower mountain regions) or stimulating climates (seaside, higher mountain regions) add restorative properties to health and wellness centres with or without natural hot springs (Erfurt 2012; Vereinigung für Bäder-und Klimakunde 2016).

8.1.3 The Purpose of Hot Spring Health Facilities

Health resorts, clinics and spas are designed as special facilities offering spa treatments (the cure) in a health inducing environment combined with local ambience. To

Table 8.2 The abundance of hot springs worldwide has contributed significantly to their popularity as attractions for health, wellness and recreation

Medical use as non-invasive treatment	Respected throughout Europe, Asia, Middle East, northern Africa	Health
Healing therapies (Thermalism or Balneotherapy)	Offered at hot spring health resorts, clinics and spas	Health
Mineral-rich hot spring waters	Expected health benefits from mineral content	Health, wellness
Health improvements	Rehabilitation and convalescence at thermal health clinics and spas	Health, wellness
Benefits for health and wellbeing	Taking the waters—prevention of illness	Health, wellness
Warm and hot spring bathing	Assisting in the maintenance of good health—especially in colder climates	Health, wellness, recreation
Authentic, unpolluted hot spring water with no added chemicals	Expected by hot spring connoisseurs	Health, wellness, recreation
Peaceful and natural idyllic settings	Parks and gardens *(Kurparks)* in Germany, France and other European countries (Figs. 8.3, 8.4 and 8.6)—outdoor onsen in Japan (Fig. 11.12), wilderness hot springs (Fig. 10.21)	Health, wellness, recreation
Change of scenery	Taking time-out from several days to several weeks to rejuvenate	Health, wellness, recreation, relaxation
Friendly socialising in comfort, luxury and safety	E.g. German *Kur* environment or Japanese Onsen settings	Health, wellness, recreation
Traditional atmosphere, historical significance	Expected at many hot spring destinations with architectural heritage sites	Recreation, relaxation
National Parks and World Heritage Sites	Visual appeal of natural hot springs and unique hydrothermal environments	Recreation
Aquatic Theme Parks	Waterparks based on natural hot spring water resources	Recreation, family entertainment
Undeveloped hot springs, active hydrothermal sites	Sought after by people who enjoy the outdoors	Recreation

provide relief from medical conditions and to prevent future health problems, they (a) serve as therapy centres with treatment programs for chronic diseases using localised natural remedies, (b) provide rehabilitative therapies after accidents, injuries or surgery together with (c) preventive measures including health education (Erfurt 2012). Health resorts, clinics and spas present opportunities to spend time in a health-oriented setting where medical treatments can be combined with a holiday, while spa towns and their residents are united in providing services in support of health and recreation (Example 8.2). For traditional spa towns the use of classical forms of preventive and rehabilitative treatment options provides economic incentives for regional health care industries and the health tourism sector (Erfurt 2012; Pratzel 2001b).

Example 8.2 Wiesbaden, Germany

After the First World War the historic spa town Wiesbaden lost its reputation as a leading health spa. To revive its status the town started to focus strictly on clinical treatments in highly specialised medical areas. Apart from various hospitals catering for general medicine and cosmetic enhancements, large health clinics (*Kurkliniken*) offer rehabilitation and recreation and are associated with the thermal centre in *Aukammtal*, which is supplied with natural hot spring water. Aside from health clinics and spas one of Wiesbaden's most important landmarks is the 'Kurhaus', a witness of former times and a centre for national and international conventions, art exhibitions, lectures, seminars and cultural events (Fig. 8.4). Following extensive restorations, the magnificent, historical architecture includes ballrooms, restaurants and recreation facilities with optimal infrastructure and up-to-date technology. Surrounded by a large 'Kurpark' with quiet places to relax, the Kurhaus is an important feature of Wiesbaden's cultural scene.

Source Erfurt (2012), Wiesbaden (2019).

The infrastructure of health resorts and spas in most European countries generally consists of a number of basic elements (Table 8.2; Erfurt 2012; Größchen 1998), which are combined with different types of entertainment for leisure and recreation (Table 8.3). This serves the purpose of adding to the successful treatment at health resorts, clinics and spas (with or without hot spring access) by achieving a

Fig. 8.3 'Kurpark' scene from Bad Pyrmont (Germany) in the year 1886. In the background are the main fountain and the pump room, a place where people met and took the waters. *Source* Public Domain

Fig. 8.4 Illustration of the original Thermal Establishment (Kurhaus) in Wiesbaden, Germany. Steel engraving from 'Views of the Rhine' by William Tombleson (1795–1846). The main architectural features have remained the same to this day. *Source* Tombleson (1840) (Public Domain)

Table 8.3 Classic European spa towns aim to offer their visitors a variety of options how to spend their time. Quality plays an important role as this is an extremely competitive part of the health and wellness tourism industry

Spa Town infrastructure—expected standards
• Natural resources and remedies certified for treatment, e.g. hot mineral-enriched spring water
• Highly specialised physicians and medical teams
• Health resorts with special facilities for thermal bathing, mineral water fountains (Fig. 8.5), graduation towers for brine vapour inhalation (Fig. 8.6; Example 8.3) and physical hydrotherapy
• Relaxing and therapeutic environments with parks, gardens and walking areas reminiscent of the 'old days' (Figs. 8.3 and 8.4)
• Amenities for fitness, sports, and relaxation
• Special accommodation catering for different categories of patients, visitors and/or guests
• Cultural events, festivals, entertainment (e.g. casinos, restaurants, sightseeing, retail opportunities; Table 6.4)
• High quality public service industry
• Health education centres

Source Erfurt (2012), Größchen (1998)

Fig. 8.5 Mineral water fountains such as this one in São Miguel (Azores Islands) are used as a natural water source for drinking and cooking by many residents and are appreciated for their health benefits

natural balance of body, mind and soul (Deutscher Heilbäderverband 2019a; Erfurt 2012).

Example 8.3 Brine Graduation Towers

Brine or salt graduation towers are unique constructions consisting of a timber framework that is tightly packed with blackthorn twigs (Fig. 8.6; Appendix 8.1). These towers can be up to 10 metres high and several hundred metres long and are located in parks, which are surrounding health facilities in European spa towns with access to thermal brine springs. Some thermal facilities also have smaller indoor graduation structures. Brine graduation towers saturate the air with a wide range of beneficial microelements including bromine, iodine, magnesium, potassium and sodium. The healthful effects are achieved by pumping the brine to the top of the graduation tower from where the mineral and salt laden water flows down through the blackthorn. Assisted by air flow the dripping brine evaporates and creates a microclimate similar to climatic conditions near the sea. The effect of the brine evaporating and thereby saturating the area around the tower with airborne mineral elements could be compared to a large-scale air purifier. While the natural brine typically has a salinity between 8 and 10%, the evaporation process increases the salt content to 25%, resulting in the distribution of highly concentrated saline vapours. In fact, the sign in Fig. 8.6 refers to the elevated salt content as 'white gold' due to its health benefits. Structures such as these graduation towers exist in various countries, where they are used for the prevention and treatment of a number of medical conditions affecting the upper respiratory tract, e.g. allergic respiratory diseases, bronchial asthma, chronic bronchitis, chronic laryngitis and eustachitis, chronic rhinitis, sinusitis, emphysema, thyroid gland diseases and a number of other diseases including allergic skin disorders.

Source Solanki (2019), Tarita Salt Spa and Wellness (2019), Therme-Thermen.de (2019).

Health resorts, clinics and spas may not always have access to genuine hot spring water, but they all use water as a key element, either (artificially heated) for health and

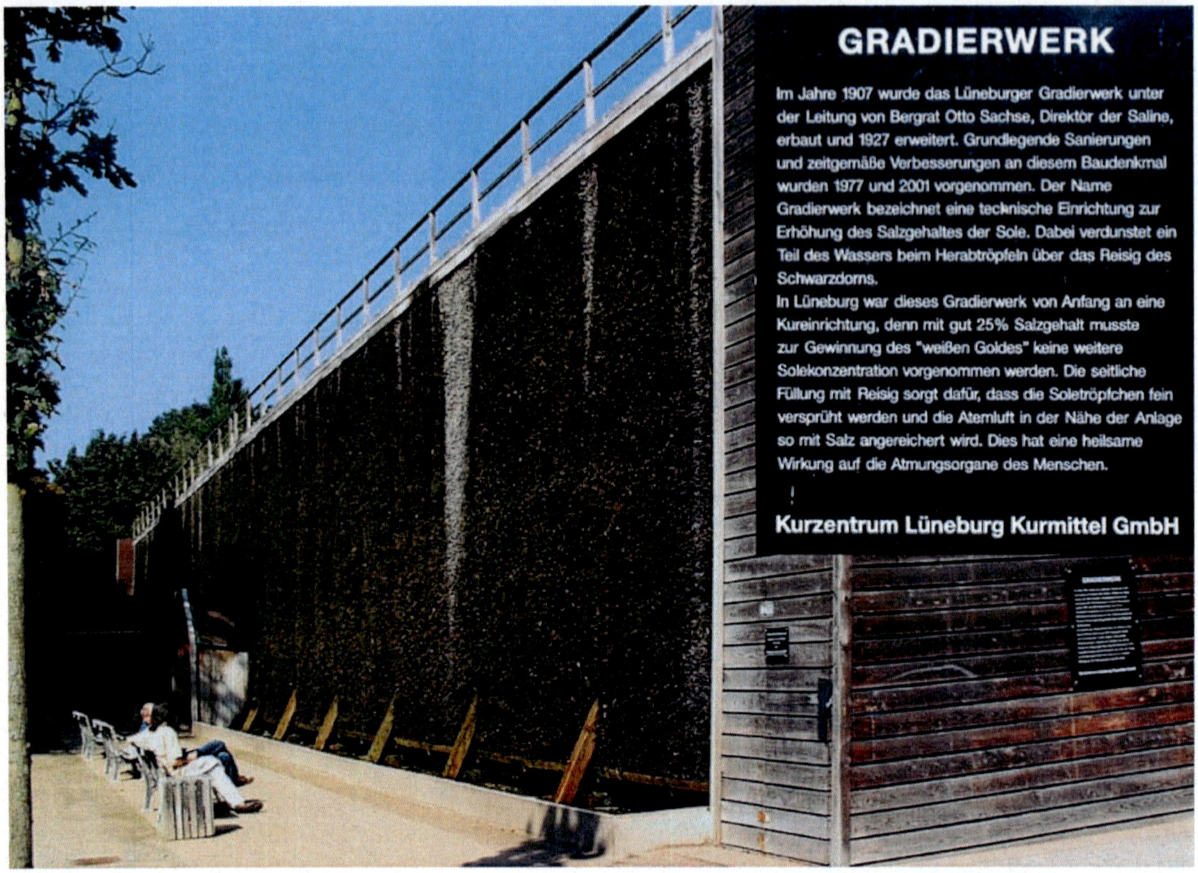

Fig. 8.6 Graduation tower (Gradierwerk) for brine vapour inhalation in the German spa town Lüneburg (Example 8.3; Appendix 8.1). The graduation tower pictured here was constructed in 1907, followed by improvements and enlargements. The term 'graduation' refers to the increase of the salt content during the evaporation process

recreational purposes or as decorative landscaping features. Given that natural hot springs are one of the main foundations for facilities offering spa medicine or holistic wellness therapies, the ever growing health and recreational tourism sector will exploit natural resources such as hot springs wherever these are available to attract health and wellness seeking visitors (Erfurt-Cooper 2010; Erfurt 2012). Natural hot springs have acquired a permanent place in various treatment environments where they play a significant role with some examples listed in Table 8.4 highlighting the diversity of hot spring facilities.

The following examples from different countries demonstrate the widespread use of natural hot springs for health and recreation. In **Japan** hot spring-based medical treatments (commonly translated into spa medicine) are fundamental to balneotherapy (Agishi 2001; Agishi and Ohtsuka 1998) and are practiced at national onsen hospitals with specialised medical practitioners conducting studies and research into health resort medicine (Mitsanobu et al. 2003). However, for insurance purposes, balneotherapy is not officially recognised and although medical specialists are certified by the *Japanese Society of Balneology, Climatology and Physical Medicine* (2019) to perform balneotherapy, they do this more in an advisory role (Ohtsuka 2012).

In Rotorua, **New Zealand**, the *Queen Elizabeth Hospital* (QE Health) has traditionally used natural hot springs for health treatments. In recognition of the proven curative effects by the New Zealand Ministry of Health these treatments are generally covered by national health insurance. As an approved rehabilitation centre QE Health offers spa medicine, including thermal baths using natural hot spring water, as well as volcanic mud treatments under the supervision of a qualified health care team (Erfurt-Cooper and Cooper 2009; QE Health 2019).

Table 8.4 Apart from health-related facilities, most contemporary wellness and recreation centred spas embrace hot springs as an important natural resource

Health, wellness and recreational hot spring settings	
Traditional Health Clinics	Hot spring hospitals where inpatients 'take the waters' to cure various medical conditions
Health/Medical/Thermal Clinics	Specialised in rehabilitation for inpatients after accidents, injuries and post-operative conditions
Health Resorts	Medical thermalism as well as holistic health elements and recreational facilities including fitness centres and entertainment
Medical Spas	Traditional and complementary medical services supervised and/or administered by medical professionals
Health Spas	Health-enhancing programs and spa treatments under medical supervision with a focus on healing body and mind
Wellness Spas	Holistic treatment options to maintain good health and prevent illness with a focus on rejuvenating the body, boosting the immune system and energising the spirit
Wellness Resorts	Option for people who want to relax in comfort, but do not require specialised forms of medical treatment—the main objective is regeneration, stress relief and enhanced wellbeing
Hot Spring Guesthouses, Inns, Motels	Common in Iceland, New Zealand and Japan—their main attraction is the access to genuine hot spring pools including private hot spring baths
Mineral Spas	This type of spa often depends on cold spring water, which is heated and then used in the same way as hot spring water—natural springs are a valuable source for drinking cures due to their content of mineral elements
Holistic Spas	Apart from thermal hydrotherapies, treatment programs may include herbal medicine, acupuncture, iridology, reiki, naturopathy, Ayurveda, sound healing, aroma therapy and detoxification

In **Iceland** the *Blue Lagoon* has developed special skin care products and therapy programs based on the chemical composition of the silica-rich geothermal seawater, the source that created this unique lagoon several decades ago. Empirical observation and scientific research are acknowledging the beneficial effects of bathing in the water of the Blue Lagoon, especially for individuals suffering from skin afflictions including psoriasis (Eysteinsdóttir et al. 2014; Grether-Beck et al. 2008; Ólafsson 1996; Sigurgeirsson and Ólafsson 2003).

8.2 Hot Springs: Are the Expectations Justified?

There is hardly a similar healing remedy associated with comfort and wellbeing like a thermal bath
(Pratzel and Schnizer 1992)

Most people who have used hot springs will agree that it has been a positive experience; one that may have left them feeling more relaxed and reinvigorated. Because mineral-rich hot spring waters allow for a natural and holistic approach to better health, they can be used for many different medical conditions (Erfurt-Cooper and Cooper 2009). For example, patients suffering from various joint and muscle afflictions are generally rewarded at a minimum with some improved movement after a few weeks of applied balneotherapy.

8.2.1 Taking the Cure

An inscription at the *Friedrichsbad* in Baden Baden from the German writer Johann Wolfgang von Goethe refers to 'the miraculous flow of the springs and the hot steam', which is making people more courageous as their blood is renewed and ends with a salute to the hot spring water (Heilbad Baden Baden 2020).

Wunderwirkend strömt die Welle – Strömt der heiße Dampf der Quelle

Mut wird freier – Blut wird neuer
Heil dem Wasser – Heil dem Feuer

In the past it was common to refer to 'taking the waters' or 'taking the cure' when embarking on a journey to a fashionable hot spring spa. For centuries these health destinations were established in many countries at locations where access to natural hot springs was a guarantee to attract visitors from near and far. Whereas illness and chronic diseases were supposedly the main cause for a trip to a spa, apart from seeking health improvements, spa guests were also pursuing leisure and recreational activities. French and German spa destinations for instance were highly developed in this aspect and offered their visitors entertainment of all kinds to pass the time between undergoing hydrothermal treatments. Casinos, theatres, golf courses and tennis courts were standard options as visitors often stayed several months at trendy places like Vichy, Dax, Chaudes Aigues or Bourboule (Example 8.4). In German spa towns the picture was similar. Baden Baden or Wiesbaden catered for the social upper class, who would stay for longer terms and thoroughly enjoy themselves at the same time. This is still a European tradition, as in Germany alone over eight million health tourists are visiting thermal spas every year with high expectations not just for improved health and wellbeing but also for a certain standard of recreation and entertainment (Table 8.3).

Example 8.4 Spa Medicine in France

In France the popularity of spa medicine (*thermalisme*) based on natural mineral-enriched hot springs has been historically sustained by the positive attitude of the French medical profession towards thermalism and is a highly valued sector of traditional medicine. Supporting the medical elite were government policies and favourable regulations that have continued to back the survival and further the development of thermal facilities. To establish the medical significance of spa medicine, numerous studies and clinical trials were and still are undertaken, a process that mirrors the German spa development. Spa therapy remains as a strong sector in prescriptive medicine supported by the State because thermalism is firmly established in the health structure of France with most medical spa treatments covered by the national health system. In more recent times, public and private organisations are involved in the scientific investigation of spa medicine with research activities supported with an annual budget disbursed by the *Association Française pour la Recherche Thermale* (AFRETH).

Well known examples of traditional French health spas are La Bourboule and Vichy, which were established in response to an increasing demand for balneotherapy in a sophisticated environment, together with quality entertainment and first-class accommodation. Similar to German spa towns the construction of casinos, theatres, opera houses and racecourses in the vicinity of health spas was common in France at the beginning of the twentieth century (Figs. 8.7 and 8.8a, b). At this stage the French see their spa towns and thermal bathing facilities as medically respected, of historical importance and contemporary significance, all reasons to continue their use as annual health holiday destinations.

Source Erfurt (2012), Mackaman (2007), Office National du Tourisme (1921), Roques (2012), Roques et al. (2012), Tilton (1981), Weisz (2001, 2002).

8.2.2 Hot Springs and Their Curative Value

Although the importance of drinking mineral water and bathing in hot springs for health purposes was already emphasised by early Greek writers (e.g. Homer, Hippocrates and Asklepios), clinical studies under the control of specialised researchers and qualified medical professionals have proven beyond doubt the physical benefits of treatments with hot spring water (Altman 2000; Bullard 2004; Mitsanobu et al. 2003; Nagasawa et al. 2001; Ohtsuka et al. 2003; Pratzel and Schnizer 1992; Vu and Mitsanobu 2004). Aside from the known benefits of balneotherapy, a further important factor is the time frame during which treatments are received (Example 8.5), although this might vary depending on the medical condition, the location of the treatment centre and the therapy options available.

Example 8.5 The German *Kur* (cure)

Despite an immediate effect a thermal bath can have on the human body, based on practical experience spa therapies are more successful using series of treatments over several weeks. To heal or cure, an intensive treatment plan including at least 15 applications over three, preferably four weeks duration is considered necessary. Two weeks can alleviate a medical condition, while one week assists in prevention of illness and one to several days are the minimum interventions for regeneration and recovery. The sustainability of a correctly dosed and applied treatment program carries a measurable effect over a term of nine months, which can be slightly extended with follow-up therapies.

Source Erfurt (2012), Bäderland Deutschland (2018), Pratzel and Schnizer (1992).

The connection between mineral components and other trace elements contained in hot spring water and their potential health benefits has a long history of investigation by scientists, especially in Europe and Japan (Barabanov and Disler 1968; Erfurt 2012; Ghersetich and Lotti 1996; Jorden 1631; Parish and Lotti 1996). As a result, hot springs are considered as an important natural resource for health and wellness destinations, a fact that has been recognised by medical professionals and visitors alike (Erfurt 2012), and is reinforced by Example 8.6, which provides a brief insight into Italian spa medicine.

Fig. 8.7 The Casino in Plombières-les-Bains (France) around 1921. *Source* Office National du Tourisme (1921)

Example 8.6 Balneotherapy in Italy
In Italy the traditional practice of balneotherapy has been applied for centuries, based on the many natural hot springs that occur throughout the country. Medical spa therapies are researched on an ongoing basis and include fields such as cardiovascular diseases, gastroenterology, otorhino-laryngology, pneumology, rheumatology, skin diseases and urology. For the purposes of advancing the knowledge related to preventive medicine and rehabilitation the Foundation for the Spa Scientific Research (*La Fondazione per la Ricerca Scientifica Termale*—FoRST) provides grants for clinical and scientific research into balneology, e.g. from 2003 to 2011 nearly five million euro were disbursed for research activities.

Source Vitale and Vaccarezza (2012).

8.2.3 Mineral Content and Temperature

On account of their content of mineral elements natural hot springs have traditionally been used as an essential foundation for spa medicine (Andreassi and Flori 1996; Fabiani et al. 1996; Hann 1996; Ledo 1996; Omulecki et al. 1996; Picoto 1996; Routh and Bhowmik 1996; Vassileva 1996; Wolf 1996). In other words, the mineral composition is more significant than a pleasant water temperature and a natural environment alone, as the desired therapeutic effects are directly linked to the variety of dissolved minerals and metallic trace elements contained in the water and which are assisting in the improvement of various illnesses (Barabanov and Disler 1968; Erfurt 2012; McGeary et al. 2001; Pesce 2002). This is confirmed by clinical research and controlled studies, which indicate that the curative effect of thermal bathing is directly associated with the absorption of mineral elements through the skin, and thereby supporting the healing process for a large number of medical conditions (Tables 8.5, 8.6, 8.7, 8.8 and 8.9) (Abels et al. 1996; Bernstein 1996; Boer 1996; Erfurt 2012; Ghersetich and Lotti 1996; Karam 1996; Katsambas and Antoniou 1996; Ólafsson 1996; Omulecki et al. 1996; Oumeish 1996; Tsankov and Kamarashev 1996; Vassileva 1996; Wolf 1996). The mineral elements acknowledged for their therapeutic benefits include Calcium (Ca), Chloride (Cl), Fluorine (F), Iron (Fe), Magnesium (Mg), Potassium (K), Silica (SiO_2), Sodium (Na) and Sulphate (SO_4) (Erfurt 2012). An overview of the most common chemical elements contained in hot spring water is provided in Table 8.10 (introduced in Chap. 3 as Table 3.1).

a

b

CLINICAL INDICATIONS

GENERAL INDICATIONS. — Taken by the mouth the waters of Cauterets are strongly indicated in the management of all *chronic affections of the respiratory passages*, whether associated with lymphatic or arthritic conditions, or whether arising from inflammation due to occupational abuse or strain (singers, barristers, stock-brokers, lecturers, etc.). The local application of sulphur waters has an aseptic action and reduces mucous congestion and these results are reinforced by the general application of the treatment.

Nose. — Purulent naso-pharyngial catarrh, atrophic rhinitis, the spasmodic rhinitis of arthritics, operative treatment of the sinuses (nasal douches, inhalations).

Ear. — Catarrh of the tubes, infiltration of the mucosa with accompanying deafness (tubo-tympanic vapourisation, inhalation by the nose).

Pharynx. — Recurring tonsillitis and pharyngitis, the hypertrophic angina of children, the sequelae of adenoid growths (gargles, sprays).

Larynx. — Occupational catarrhs and purulent conditions, complete or nodular hypertrophy of the vocal cords (inhalation, inhalation of the natural gases of the springs, mixed with vapourised water, throat douches).

Lungs. — Tracheal-bronchitis, both recurring and chronic, catarrhal asthma, emphysema, chronic pleurisy, sequelae of influenza, broncho-pneumonia, empyema, tracheabronchial glandular disease, inert pretuberculosis.

General nutrition. — Pathological fatigue, over-work, convalescence. Arthritic signs : articular (rheumatism) respiratory (asthma), cutaneous (eczema, acné, seborrhoea), nervous (debility, neurasthenia, the algias), digestive (loss of appetite, atonic dyspepsia), gynaecological (dysmenorrhoeae, metritis, growths). These conditions respond to baths, local and general douches and foot baths.

Wounds. — Old wounds, fistulae, sequestra, muscular atrophy, painful scars. These are treated by re-education and physical exercises.

Syphilis. — The cachexic stage which follows the active period of the trypanosome. The treatment aims at remineralisation and elimination of toxins.

CONTRA-INDICATIONS. — Acute disease, febrile tuberculosis, haemoptysis, cardiac conditions with broken compensation, albuminuria.

MEDICAL MEN. — Drs. Anglade, Armengaud, G. Bernard, Bouyer, Castay, Corone, Ducos, Flurin, Fourcade, Got, Julia, Labayle, Lassalette, Mary, A.-E. Meillon, Pradal, Thierry.

The following waters are exported : La Raillière, César, Mauhourat. For particulars application should be made to La Société Thermale.

LIFE AT CAUTERETS

The season begins at the end of May. Immediately after the Paris Derby, society leaves the capital and Cauterets becomes crowded with jaded and debilitated members of the fashionable world. Each is a devotee of a particular form of treatment, and each is intent upon the speedy recovery which will enable him once more to enjoy the pleasures and distractions of life, for which Cauterets offers many facilities.

Simple excursions. — Mountain walks (there are numerous shelters put up by the T. C. F. and the C. A. F.) mountain climbing, botanising, camping out, excursions on horse-back or with mules, carriage driving, motor excursions.

Sports. — Tennis and golf tournaments, pigeon-shooting competitions ; shooting, trout-fishing in the lakes and torrents of the Hautes Vallées ; dog-shows ; gymkhanas ; donkey-polo ; battles of flowers, balls, fairs, etc., etc.

Musical and artistic attractions. — Two theatres, two casinos, popular concerts, classical concerts, cinema, conferences, exhibitions.

The "*Nature Theatre*". — Artistic performances are given by well-know actors from Paris in the magnificent setting where Marguerite de Valois, sister of Francis I., once held her famous Courts of Love.

Motor tours. — *Cauterets is an excellent centre for mountain tours* in the grandest part of the Hautes-Pyrénées. It is a halting station for the service of motor cars belonging to the company of the *Grande Route des Pyrénées*, which connects Biarritz on the Atlantic with the Mediterranean. Electric funicular railways facilitate excursions to the Gorges de Pierrefitte, to Luz-Saint-Sauveur, to the cirques of Gavarnie and Troumouse, to the Abbey of Saint-Savin and to Notre-Dame-de-Lourdes, the goal of pilgrims from all over the world.

Winter sports. — From Christmas onwards, the winter sports at Cauterets are in full swing. There are sleighing excursions, ski-ing competitions, bob-sleighing, luging and skating.

ACCOMMODATION. — There are 4 first-class hotels with hot and cold water, electric light, central heating, bath-rooms, etc. Between them they have 1.400 bedrooms of which 300 have private bath-rooms.

There are 9 second-class hotels. These are comfortable and have 800 bed-rooms.

There are 10 third-class hotels, unpretentious in style and with 600 bed-rooms.

There are 50 furnished houses. Villas and apartments of every kind are available.

INFORMATION concerning the treatment and the export of the waters may be obtained from La Société Thermale des Pyrénées, 8, rue Lamennais, Paris (8e).

All enquiries concerning the journey, details of residence, villas, boarding-houses, excursions, shooting, fishing, touring, etc., should be addressed to the Syndicat d'Initiative de Cauterets-Attractions, Cauterets (Hautes-Pyrénées).

Fig. 8.8 **a** The 'Thermes du Petit-Saint-Sauveur' at Cauterets, a thermal spa town in the Pyrenees, a mountain range on the border between France and Spain. *Source* Office National du Tourisme (1921). **b** The clinical indications present a number of health conditions treatable at the spa town Cauterets together with a list of the medical practitioners available. The text on the right describes various entertainment options as well as different types of accommodation. *Source* Office National du Tourisme (1921)

Table 8.5 Some of the mineral elements contained in natural hot springs and their purported function in balneotherapy

Mineral components	Function
Arsenic	Arsenic in larger doses is toxic to the human body but minute amounts may assist with plasma and tissue growth. Foot baths in thermal water with a high content of arsenic are used to address fungal infections of the feet
Bicarbonate	Bathing in bicarbonate waters is used to address hypertension and mild atherosclerosis
Boron	Boron is considered to build muscle mass, increase brain activity and strengthen bones
Chloride	Saline hot springs contain sodium chloride. Based on research, amounts between 0.5 and 3% are considered to be beneficial in the treatment of rheumatic conditions, arthritis, central nervous system conditions, post traumatic and post-operative disorders, as well as orthopaedic and gynaecological diseases
Magnesium	Magnesium converts blood sugar to energy and promotes healthy skin
Potassium	Potassium assists in the normalisation of heart rhythm, assists in reducing high blood pressure, helps to eliminate body toxins and promotes healthy skin
Sodium (compare chloride)	Sodium and natural salts assist with the alleviation of arthritic symptoms, and may stimulate the body's lymphatic system when used in baths
Sulphur and Sulphates	Hot springs rich in sulphur elements are used to address a wide variety of conditions, including skin infections, respiratory problems, and skin inflammations

Source Compiled from various sources

Table 8.6 The various states of ill health are subject to interpretation and can overlap in their definition

Different states of illness	
Affliction	Impairment, physically unfit, affected by disease, suffering from persistent pain
Ailment	Minor illness or physical disorder (e.g. suffering from a cold) or may be a chronic condition
Disability	Physical or mental restriction that limits a person's movements and activities (e.g. blindness, paralysis, depression, anxiety)
Disease	Abnormal condition, has a pathological cause (e.g. virus), abnormalities in systemic functions
Disorder	Functional disturbance, can be physical, mental or emotional
Illness	A disease, a period of sickness or a condition
Medical Condition	Abnormal state of health, unhealthy state, includes diseases and disorders
Sickness	The state of being ill, being temporarily affected by disease
Unwell	Feeling unwell can range from not being in good shape, feeling sick, being in poor health to incurable

In addition to their benefits in balneotherapy, mineral elements have traditionally been included in solutions, crèmes or ointments, which are used to address and cure individual health conditions. Also, taking mineral spring water internally by drinking is an unabated practice world-wide and supported by the demand for alternative medicine and a holistic approach to wellbeing (Erfurt-Cooper and Cooper 2009). For people who prefer to take mineral baths at home, their favourite bath powders can be obtained from local chemists or ordered online from many different hot spring destinations (Figs. 8.9 and 11.6a, b, c).

Research from the year 1936 describes the particular benefits of spring water containing iodine, which has helped to improve the condition of patients suffering from thyroid problems, who found relief through the internal (Fig. 8.10) and external use of iodine-rich spring water. Also mentioned for their curative value at the time were brine baths based on their high salt content as well as sulphur baths (Erfurt 2012; Selke 1936).

Table 8.7 Exercise therapies in thermal water (After Schnizer 2002)

Thermalism (balneotherapy)—aims and applications
• Weight support, joint relief, facilitation of movement, muscle relaxation and pain relief
• Increasing the peripheral circulation, draining and stimulating effect of sensorimotor functions
• Exercise and training of muscular strength, muscular endurance, general aerobic endurance, motor coordination and agility
• Improvement of muscle-joint function in orthopaedic, rheumatological and neurological disease and damage patterns
• Promotion of rehabilitation after surgery and trauma of the musculoskeletal system
• Supporting care for cardiovascular disease, metabolic and psychosomatic illness and immobilisation syndromes
• Promotion of physical fitness and wellbeing

Table 8.8 Overview of health conditions treatable with balneotherapy based on natural hot spring water

Health Conditions treated with Balneotherapy in Germany	
Allergies	Neurological diseases *(including chronic pain)*
Ankylosing Spondylitis *(chronic, inflammatory arthritis, can cause fusion of the spine)*	Neurological rehabilitation
Blood circulation, blood pressure	Oncological diseases *(tumours, cancer)*
Cardiovascular diseases	Osteoporosis *(loss of bone density)*
Childhood diseases	Postoperative therapies
Chronic pain	Postural and locomotor diseases *(e.g. Parkinson's Disease)*
Convalescence	Prevention of illness
Degenerative joint disease	Psoriasis vulgaris *(chronic, non-contagious disease characterized by inflamed lesions)*
Depression	Psychosomatic disorders *(physical symptom caused by mental or emotional disturbance)*
Diseases of the digestive tract	Rehabilitation after accident or injury
Eye disorders	Rehabilitation after operations
Fibromyalgia *(pain in muscles and connective tissue)*	Renal and urinary tract diseases
General exhaustion	Respiratory diseases, Asthma
Gynaecological diseases	Rheumatoid Arthritis, rheumatism
Metabolic symptoms and disorders	Skin diseases
Musculo-skeletal system *(motion impairment, muscle malfunction)*	Stress management
Nervous system disorders *(from headaches to Alzheimer's)*	Tinnitus *(hearing noises which do not exist)*
Neuro endocrine diseases *(immune system and rheumatic diseases)*	Weakness, fatigue and convalescence

Source After Pratzel (2001a)

Table 8.9 Health conditions treated and cared for at special onsen facilities with trained medical staff in Japan

Health conditions treated with balneotherapy in Japan		
Aches and Pains	Digestive Disorders	Obesity
Allergies	Eye problems	Palsy, paralysis and Poliomyelitis
Anaemia	Fatigue	Recovery and Rehabilitation
Atherosclerosis	Female disorders	Relaxation and rejuvenation
Arthritis	Gout	Respiratory system
Athlete's Foot	Haemorrhoids	Rheumatism
Beautification	Heart problems	Rest and recuperation
Beriberi	Hypertension	Sciatica
Blood circulation	Insect bites	Skin problems
Brain Haemorrhage	Insomnia	Sprains and breaks
Burns	Internal disorders	Joint problems
Chilblains	Longevity	Stomach disorders
Common Cold	Muscular problems	Surface wounds
Detoxification	Nervous disorders, stress	Venereal diseases
Diabetes	Neuralgia	Whiplash

Source Compiled and translated from Japanese Literature

Table 8.10 Some of the common components found in natural hot springs (Erfurt 2012)

Common components	Other substances present		Main gases in solution
Calcium (Ca)	Aluminium (Al)	Iodine (I)	Carbon Dioxide (CO_2)
Carbonate (CO_3^2)	Ammonium (NH_4)	Lithium (Li_3)	Hydrogen Sulphide (H_2S)
Chloride (Cl^-)	Arsenic (As)	Manganese (Mn)	Nitrogen (N)
Fluoride (F^-)	Bicarbonate (HCO_3)	Potassium (K)	Oxygen (O)
Iron (Fe)	Bromine (Br)	Radium (Ra)	Radon (Rn)
Magnesium (Mg)	Caesium (Cs)	Rubidium (Rb)	
Sodium (Na)	Cobalt (Co)	Silica (SiO_2)	Argon (Ar) and Helium (He) can occur in some sulphur springs
Sulphate (SO_4^{2-})	Copper (Cu)	Zinc (Zn)	
Sulphide (S^{2-})	Fluorine (F)		

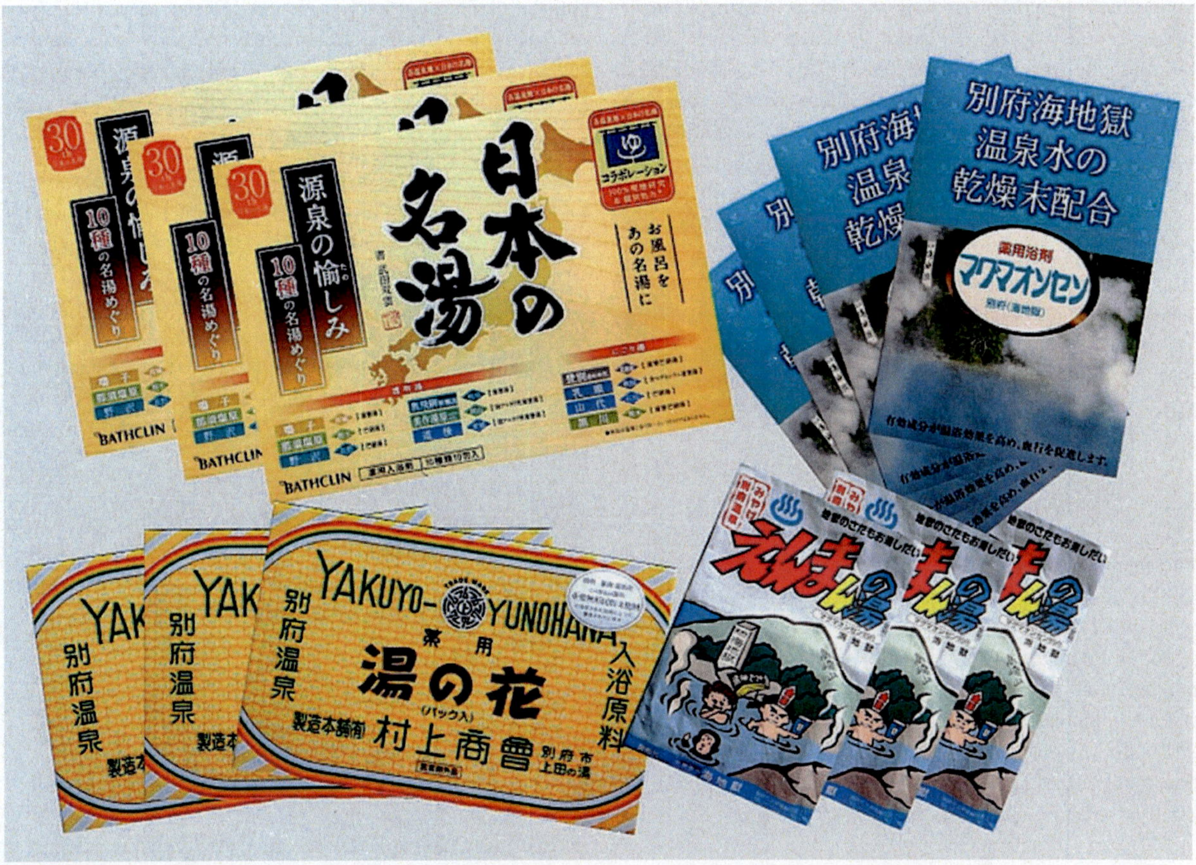

Fig. 8.9 Bath powders containing minerals from different Japanese onsen are used at home to replicate the effects of natural hot spring water

8.3 Hot Springs and Spa Medicine

Spa medicine plays an important role in the areas of treatment, rehabilitation and prevention of a wide range of medical conditions. In spa medicine, also known as medical hydrology, hydrotherapy, balneotherapy and thermalism, the preferred medium is water, which is used for medicinal bathing to achieve health benefits through prevention or cure (Pratzel and Schnizer 1992). Despite the fact that the terminology varies from one country to another, there are also differences related to the recognition of spa medicine, the clinical practice, public health cover and related research (Cantista 2012).

Medicinal baths are used therapeutically as a supporting agent in the treatment of many diseases as partial or full immersion baths, and at temperatures ranging from cold to hot in combination with other substances or by using mineral-rich natural spring water (Pratzel and Schnizer 1992). Therapy-relevant factors are influenced by mechanical aspects such as hydrostatic pressure, buoyancy and viscosity of the water, while at the same time thermal aspects affect the temperature regulation of the human body. A third relevant factor is the effect of chemical substances dissolved in water such as minerals or the addition of herbal supplements (Agishi 2001; Pratzel and Schnizer 1992).

Incorporated into the complexities of spa medicine are exercise through thermal bathing, drinking cures and inhalation of mineralised vapour (Deutscher Heilbäderverband 2019b), complemented by steam baths, hydromassages and supervised water gymnastic (Erfurt 2012). To assess the state of health or illness (Table 8.6), individual patients who have been recommended by their personal physician for a course of spa medicine, are evaluated by medical professionals at a specialised health clinic or treatment centre. After this examination, a suitable treatment plan based on the availability of site-specific therapy resources is developed (Deutscher Heilbäderverband 2019c; Erfurt 2012). A key part of the course of the treatment involves thermal hydrotherapy, which is supporting physical mobility, helps to reduce inflammation, promotes shedding of the skin and has antibiotic qualities while improving the penetration of UV radiation (Titzmann and Balda 1996).

8.3.1 Therapy Options

The medical literature covers different methods of treating health conditions with spa medicine or balneotherapy (Fabiani et al. 1996; Lund 2005; Parish and Witkowski 1994; Williams et al. 1996; Witt 1990); however, researching the topic in more depth, shows that even some Earth science publications include research articles, frequently under the subject of *geomedicine* (Erfurt 2012). The terminology of the more commonly known treatment options is briefly explained and compared below.

Thermalism: This term is common in Europe and Latin America (*termalismo*) and refers to the medical use of hot springs and their by-products (e.g. mud, minerals, algae) to assist with physical rehabilitation after illness (Bar-On 1989; Nahrstedt 2004; Pratzel 2001a). Thermalism has experienced a global resurgence due to a consumer demand for natural and ethical health and wellness treatment options and may also be considered as a socio-cultural resource (Stevens et al. 2018). In health care, thermalism refers more specifically to the therapeutic application of warm to hot [spring] water and is used to stimulate circulation with underwater massage to improve muscle and joint movement. Therapeutic thermalism combines new scientific evidence with traditional treatments for different diseases as well as assisting in the prevention of illness and the maintenance of good health.

Balneotherapy: Therapeutic treatment programs that are based on the recognition that water, especially hot spring water rich in dissolved naturally occurring minerals, has medicinal benefits and is effective in the treatment of a wide range of medical conditions. The use of balneotherapy goes back to the earliest civilisations who treated illness with bathing in hot spring water and mud baths, and which is widely used in modern clinical practice (Agishi 2001; Baysal et al. 2018; Fioravanti et al. 2017; Huang et al. 2018). As balneotherapy or spa medicine are increasingly recognised as a specialised medical field, this has strengthened the network of balneological associations in the in European scientific community, which involves numerous countries (Karagülle 2012).

Balneology: The science/scientific study of bathing in medicinal springs and the therapeutic benefits. Balneology also includes the research of applications of internal use of spring water rich in mineral elements (see Hydropinic Therapy).

Crenotherapy: A term that describes healing with natural hot spring water rich in essential mineral and trace elements. Treatments can incorporate the internal use of mineral water (drinking, inhalation—Figs. 8.10 and 8.11; Appendix 8.2) and external applications such as hot spring baths and *Pelotherapy*.

Hydrotherapy: An essential part of balneotherapy, where thermal water therapies are applied for different medical conditions and which include baths, showers and exercise therapies at varying water temperatures (Table 8.7).

> **Note**: The practice of hydrotherapy also uses mineral water from cold springs, which is artificially heated and enriched with supplements (Bender et al. 2005).

Hydropinic Therapy (Drinking Cure): Also referred to in the past as 'taking the waters', which involves the consumption of mineral water usually directly from the source of the spring. To benefit from the drinking cure, the water should be taken under the supervision and guidance of a professional health care provider, who determines the quantity of spring water consumed and the timeframe of the treatment.

Pelotherapy, Fangotherapy or Mud Therapy: Treatments using either thermal mud baths or mud packs, also known as *Fangotherapy* in some countries (e.g. Italy, Germany, France). Fango is a special type of clay that is rich in minerals such as sulphur of volcanic origin, and which is derived

Fig. 8.10 This fountain is dispensing mineral spring water for all visitors who are willing to sample the King's Spring in the Pump Room, Bath (UK)

from hot springs in Italy. Fangotherapy involves either heated mud packs to cover certain parts of the body to relieve pain and heal skin conditions, or the complete immersion in warm mud baths to relieve joint and muscle pain. These mud baths consist of hot spring water combined with volcanic ash or clay and were already known and applied by the Celts and the Romans. This type of treatment can also be used as a detox therapy and is relaxing and rejuvenating.

Spa Therapy: Spa therapy uses various treatment options including balneotherapy and hydrotherapy (Bender et al. 2005). It is also another definition of *spa medicine* with more emphasis on the inclusion of complementary and alternative methods in the treatment program. Spa therapy further represents an opportunity to retreat from everyday life to enjoy some rest and recreation for a certain period of time to improve health and wellbeing (Baysal et al. 2018).

Hot spring-based spa medicine can include some rather interesting methods, which are nevertheless acknowledged as successful treatments. One of these therapies is the fish cure (Example 8.7), which has become increasingly trendy and accepted especially by people suffering from psoriasis.

Example 8.7 Ichthyotherapy—Hot Spring Fish Treatment

A unique alternative treatment option for skin afflictions comes with the help of a rare breed of small fish (*garra rufa* or commonly known as Doctor Fish of Kangal) that exist in thermal spring water at temperatures up to 43 °C. The unusual feeding activities (they thrive on dead skin) of these fish and its effects has made them a somewhat atypical therapy in addressing dermatological problems including *psoriasis*. Studies carried out by dermatologists on patients with psoriasis appear to verify the effectiveness of this unique spa therapy. Hot spring fish spas are now a popular attraction in many countries (Figs. 8.12 and 10.3), although reports point to their origin in Turkey (Kangal Fish Spring), Iran and Iraq. Some sources suggest a treatment course of 21 days with up to eight hours daily immersed in a fish spa to successfully combat psoriasis, reportedly with excellent results and no side effects.

Source Grassberger and Hoch (2006), Kangal Fish Balikli Kaplica Treatment Center (2019), Kangal Balikli Kaplicasi (2019), Kelly (2011), Özçelik et al. (2015), Sayili et al. (2007), Undar et al. (1990).

Because of the uniqueness and growing popularity, the fish spa or doctor fish treatment has been exported from its origin in Turkey to many other countries, also promoted as 'fish pedicure'. While protected in Turkey, fish spa suppliers in Thailand, Pakistan, and India breed the *garra rufa* for export in response to the increasing demand. Concerns have been raised though over the potentially negative effects of transporting these small fish to environments where they are unlikely to find an ecological habitat that is suitable for their species and required for their survival (Kelly 2011).

8.3.2 Treatable Medical Conditions

Spa medicine is used to treat chronic health conditions including back pain, rheumatism and arthritis, and bathing in hot spring water has long been suggested as an effective method in treating skin conditions as well (Erfurt-Cooper and Cooper 2009; Erfurt 2012; Parish and Lotti 1996). Table 8.8 provides examples of health conditions treated with hydrotherapy based on natural thermal springs in Germany.

When comparing health conditions that are treated with spa medicine in different countries, several diseases are frequently mentioned: skin diseases, musculo-skeletal impairment, arthritis and rheumatic diseases (Example 8.8; Table 8.9). Depending on their thermal resources, treatment

Fig. 8.11 Hot spring steam emitting at 100 °C can be used by the public to help with nasal congestion and other respiratory complaints (see also Appendix 8.2). A real treat on a cold winter day. Kamado Jigoku, Beppu (Japan)

Fig. 8.12 'Doctor fish' at work in a foot bath at the Spa Resort Hawaiians (Japan)

centres usually specialise in therapies that are most relevant to certain conditions and generally are expected to provide some form of relief.

Example 8.8 Karahayıt, Turkey
The town of Karahayıt is located northwest of the famous world heritage site Pamukkale. The Karahayıt thermal spa centre provides services to people who are searching for a cure for their personal medical conditions with Fig. 8.13 depicting the more casual side of the spa centre. The health giving springs emerge at 53–58 °C from fissures in the ground without being affected by seasonal changes. Despite the reddish coloured water containing radioactive elements, it is sought after for the treatment of various illnesses including arteriosclerosis, heart diseases, high blood pressure, lumbago, rheumatism, sciatica and skin diseases.

Source Alçiçek et al. (2016), Ökçesiz et al. (2014), Özkul et al. (2013).

Looking back over the last two centuries, the literature has left us with some detailed chronicles about specific spa towns in Europe. A medical handbook titled *The Hot Springs of Bath* by John Hatton, published circa 1910, provides a fountain of knowledge covering the healing springs of Bath (UK) and the forms of hydrotherapy that were available to patients then.

> *'The waters appear to have a soothing effect and greatly alleviate pain'*
>
> *John Hatton, Director of the Baths, c. 1910:51*

With the intention to provide a comparison to health conditions treated in countries such as Germany (Table 8.8) or Japan (Table 8.9), the list of most frequently treated diseases in Bath (England) around the beginning of the last century (Table 8.11) includes some medical conditions that may have been a sign of this period, while others are still common. For the purpose of stepping back over a hundred years, the original word choice was retained as far as possible together with some direct quotations; however, some definitions were added for clarification.

Another example comes from **India**, where hot springs are primarily used by local residents, who are aware of the healing benefits of the water. In the Khurda District in western India the hot springs of Odisha are locally considered as healing springs, which are reportedly used for a

Fig. 8.13 People are dropping in during the day or after work for a foot spa in the health-giving thermal springs of Karahayıt (Turkey)

Table 8.11 Even tropical diseases were treated at the healing springs of Bath because 'the waters rapidly benefit cases saturated with the poison of malaria and improve the cachetic [refers to muscle loss] condition'

Diseases most frequently treated in Bath, UK (c. late nineteenth—early twentieth century)	
Anaemia	'Taking the waters' internally combined with thermal bathing
Arterio-sclerosis	Administration of the Bourbon-Lancy treatment [a special form of douching]
Chronic rheumatism	Thermal water baths, underwater massage and vapour baths
Diseases of the nervous system	Reclining thermal baths, massage and douches, vapour baths and spinal sprays
Diseases of the respiratory system	The water is used in pulverised [vaporised?] form administered through sprays
Diseases of the skin	Reclining thermal baths with the water temperature and time in the water being gradually increased in combination with vapour baths
Diseases of women	Satisfactory results with reclining baths and local hot under-current douches
Disorders of the digestive organs	Relief is partially due to thermal bathing combined with massage, dietetic regulations and '*washing out the stomach by drinking the waters'*
Gonorrhoeal rheumatism	Thermal water baths, Aix douche and copious doses of mineral water taken internally—*'The patients, being usually young male adults, stand the baths well'*
Gout	A course of treatment with thermal waters including reclining and deep baths, vapour baths, douches and massage
Muco-membranous colitis	Intestinal douches (lavage, irrigation) with hot mineral water
Rheumatoid arthritis	Based on statistics of the Mineral Water Hospital in Bath (Fig. 8.14) 80.7% of the patients experienced considerable improvement and pain relief
Sciatica	Relief is obtained from immersion in deep thermal baths, vapour baths, Aix massage and undercurrent douche
Results of injury	Relief through thermal douches and massage
Syphilis	A course of thermal baths to *'promote the actions of the skin and help the elimination of the poison'* in combination with the administration of mercury to increase the therapeutic action

Source The Hot Springs of Bath—Medical Handbook by J. Hatton, published c. 1910

Fig. 8.14 The Royal Mineral Water Hospital in Bath (UK) was established in 1738/39 AD. It was recently sold to a hotel development group from Singapore with plans to transform the 'Min' into a luxury lifestyle hotel

number of different health conditions. A recent research article has included a list of diseases (Table 8.12) that benefit from the therapeutic hot spring waters.

The final example is an account from the late 1890s from north America, which lists various health conditions that were considered to benefit from the application of hot spring water

Table 8.12 India has some noteworthy hot spring areas, although not many are well-known outside India. Local people are usually familiar with the benefits of hot spring bathing to improve their health

Odisha Hot Springs, India	
Autonomic nervous system dysfunction Chronic gastrointestinal diseases Chronic gynaecological diseases Chronic mild respiratory diseases (asthma) Circulatory diseases (moderate to mild hypertension) Conditions resulting from trauma	Metabolic diseases (diabetes, obesity and gout) Peripheral circulatory diseases (affected hands and feet) Psychosomatic and stress-related diseases Rheumatic diseases Skin diseases

Source Pattanaik et al. (2019)

Table 8.13 This list of medical conditions thought to be treatable with the natural mineral-rich springs of Missouri was compiled by the chemist Paul Schweitzer from the Missouri Geological Survey in consultation with local physicians (Bullard 2004)

Diseases treated with natural spring water, Missouri, USA		
Bladder diseases Bowel diseases Bright's disease [nephritis] Cholera Chronic diarrhoea Dropsy [oedema] Dyspepsia, stomach problems	General disability Kidney diseases Liver diseases Malaria Neuralgia Old sores Paralysis	Piles Rheumatism Sciatica Skin diseases Sore eyes

Source Schweitzer (1892)

(Table 8.13). However, this attitude towards natural healing was never developed to the same level as in other countries.

8.3.3 Effects of Thermal Bathing or Spa Therapy

The basic effects of a thermal bath take place predominantly in the neuromuscular, haemodynamic, hormonal and renal systems. Further effects influence the cardiovascular system as well as the respiration and lung function, whereby several individual effects can cause a synergistic impact (Pratzel and Schnizer 1992). Physical exercise in thermal water offers several significant advantages in the treatment of certain health conditions (Table 8.7), as the buoyancy of the water provides relief for the support and movement system. This refers mainly to patients with muscle weakness, movement pain, restriction of joint movement and poor coordination (Schnizer 2002; Fig. 8.15). Functional mobility for movement impaired patients can improve following thermal treatments including buoyancy assisted exercise in warm water (34 °C) combined with other physical activities to increase the range of motion and muscle force (Peterson 2001).

Clinical studies have documented and acknowledged that the desired therapeutic effect of mineral-rich hot spring water is supporting the healing process of a variety of health conditions due to the absorption of mineral ingredients and metallic trace elements through the skin (Abels et al. 1996; Bernstein 1996; Boer 1996; Ghersetich and Lotti 1996; Karam 1996; Katsambas and Antoniou 1996; Ólafsson 1996; Omulecki et al. 1996; Oumeish 1996; Pratzel and Schnizer 1992; Tsankov and Kamarashev 1996; Vassileva 1996; Wolf 1996). Immersion in a thermal bath leads to useful therapeutical effects, which depend on temperature and composition of the water as well as the duration and the frequency. Due to the heat of the water, thermal bathing has a muscle-relaxing effect, as well as increasing metabolic processes and blood circulation. In addition, the heat leads to increased elasticity of connective tissue structures such as tendons, ligaments and fascia and provides pain relief in the sense of thermal analgesia and a sedative heat effect favoured by muscle tension (Karagülle et al. 2007; Pratzel and Schnizer 1992). In Germany's renowned spa town of Baden Baden for example, the healing effects of the local hot spring water are used to treat chronic disorders including arthritis and rheumatism, vegetative circulatory disorders, bronchitis, paranasal sinus problems and obesity. Also beneficially influenced can be various glandular disorders, where the most important applications of balneotherapy are thermal baths, drinking cures and inhalations (Gutenbrunner and Hildebrandt 1998; Sanner 2000).

As an illustration, physical exercise in thermal water increases the body heat, which in turn affects the ability of movement at different water temperatures. Therefore, minimal physical activity allows for higher water temperatures around 35–36 °C, while more intense activities (e.g. swimming, water aerobics) necessitate a lower temperature, preferably below 30 °C. Temperatures above 36 °C are considered too high for therapies based on physical activity (Schnizer 2002). However, hyperthermal bathing at even higher temperatures has a long tradition in

Fig. 8.15 Patients with muscle and joint weakness, movement impairment or poor coordination can use special chairs that can be lifted into the water. Left image: onsen facility in Japan (Photo of promotional literature). Right image: The Cross Bath in Bath (UK)

Kusatsu, Japan (Example 8.9), where the immersion in hot spring water (up to 48 °C) is used to treat certain health conditions.

Example 8.9 Kusatsu, Japan
Kusatsu is one of the most popular hot spring spa towns in Japan and famous for a unique therapeutic bathing method called *jikan-yu*. The volcanic hot springs of Kusatsu were first promoted to the world during the Meiji era (1868–1912) by Dr. Erwin Bälz, who was visiting from Germany in 1878 and immediately recognised the health benefits of hot spring (onsen) bathing. Kusatsu's strongly acidic hot spring waters with pH levels as low as 1.6 are favoured for their strong sterilising power to destroy germs and bacteria, and which are used to treat various health conditions including neuralgia, muscle pain, dermatological conditions, diabetes, hypertension and arteriosclerosis. The treatment method jikan-yu (timed bathing) is a process recognised for its severity due to the high temperatures of the water and is a traditional way of onsen bathing at this location. The treatment starts with very hot spring water poured several times over the patient's head to increase the blood circulation, followed by bathing, timed at 3 min in water at temperatures between 46 and 48 °C. This procedure can be repeated four times a day but only under strict medical supervision. There is a possibility that some patients may feel lightheaded or unwell afterwards as the water temperature is significantly above that of normal bath water.

Source Kubota et al. (1992), Kusatsu Onsen (2016), Shirakura and Sugai (1986).

This type of treatment practiced at Kusatsu is unusual and hyperthermal bathing may not suit everybody, but the therapy is known to have beneficial effects when carried out in a controlled environment.

8.3.4 Potential Risks Related to Hot Spring Bathing

Spa medicine is generally considered a safe practice, although there are some contraindications where health conditions make it necessary to avoid thermal bathing. This includes acute skin diseases and/or large skin injuries as well as cases of high fever and infectious diseases. Patients suffering from cardiac diseases (class III and IV), increased cardiac arrhythmia and angina pectoris and hypertension stage 3 may also be at risk if participating in thermal bathing. Complications may also arise when leaving the bath and the hydrostatic pressure ceases, which can lead to a tendency for circulatory collapse (Pratzel and Schnizer 1992 *{translated from German}*).

In addressing the potential risks of hot spring (onsen) bathing, the Japanese Hot Springs Law specifies in detail (Article 18, Paragraph 1) the criteria for notifications on contraindications for bathing and drinking (Example 8.10). These are based on (a) general adverse reactions to hot spring water, (b) on the spring quality, e.g. highly acidic or sulphur springs, and (c) reactions to certain mineral components (Ministry of the Environment—Japan 2016).

Example 8.10 Risks of Thermal Bathing and Drinking Cures, Japan
Contraindications are diseases and states of illness with a risk of adverse events by even one occasion of bathing in or drinking of hot springs. However, it does

not preclude, even with contraindications, hot spring cures under the guidance of a doctor with expertise.

Contraindications include the active stage of diseases (especially when accompanied by fever); active tuberculosis, advanced malignant tumours or the case of significant debility involving severe anaemia; severe cardiac or lung diseases involving suffocation feelings with little movement; severe kidney diseases involving oedema; gastrointestinal bleeding and when bleeding is visible; and acute exacerbation stage of chronic diseases.

Source Ministry of the Environment—Japan (2016) *[translated from Japanese]*.

Additional risk factors related to hot springs used for recreational purposes are discussed in detail in Chap. 9.

8.4 Hot Springs and Their Role in Rehabilitative Medicine

The provision of rehabilitative care facilities is very important for patients who are recovering from surgery, accidents or injuries as well as from a large number of other health conditions. Using Germany as an example, approximately a third of the spa towns throughout the country are based on natural thermal springs and focus on rehabilitation as well as the prevention of illness using spa medicine (Erfurt 2012). Specific therapies and rehabilitative treatment programs have been developed by medical professionals qualified in all aspects of spa medicine *(Kurmedizin)*, including Physical and Rehabilitation Medicine (PRM), and which are applied based on the diagnosis of either a specific illness or a combination of several health complaints. The treatment of individual patients is monitored continually, documented and evaluated for its effectiveness and whether a person with a medical disorder is achieving an optimal improvement (Gutenbrunner 2014). In the handbook of medicinal bathing *(Handbuch der Medizinischen Bäder)* H. Pratzel and W. Schnizer explain in detail how thermal water therapies are beneficial for many health conditions.

Treatment programs in the rehabilitative process based on thermal bathing are an important part of the overall therapy, especially in cases where restriction of movement can only be overcome underwater. It also prepares patients for other physical activities outside the water environment that are necessary for the rehabilitation process (Schnizer 2002). Therefore, movement or exercise therapy in thermal water has become increasingly important within the field of rehabilitation medicine as a physiotherapeutic treatment method. Favoured for the physical properties of water such as buoyancy, heat, viscosity and hydrostatic pressure, this therapy is preferential over other forms of exercise and is particularly beneficial for patients with movement restrictions (Pratzel and Schnizer 1992). The weightlessness simulated by the buoyancy of the water has a positive effect on neuromuscular functions, which allows the patient to perform movements otherwise not possible. The weight reduction on the musculoskeletal system relieves joints and ligaments and changes the functional conditions of the muscular perception and the reception of the joints. Feedback to the central nervous system then causes a re-adjustment of the muscle tone, while the energy expenditure decreases. However, exercise baths preferably should not deviate from the thermo-physiological neutral temperature ($\sim$35 °C) to avoid movement activities under hyperthermic stress (Pratzel and Schnizer 1992).

8.5 Hot Springs in Complementary and Alternative Medicine

Integrative and alternative healing practices are not generally included in conventional medicine and are therefore not part of standard medical care, but frequently take health and wellbeing from the natural environment (Erfurt 2012; Lazzerini 2013; National Center for Complementary and Alternative Medicine 2010). Complementary and alternative medicine (CAM) is part of a holistic approach to health and wellbeing and consists of medical products and practices that are either used in conjunction with conventional medicine or on their own. CAM approaches include traditional treatments such as acupuncture, reflexology, herbal medicine as well as hot spring bathing, often used to help lessen discomfort or pain although alternative treatments and non-invasive therapies are not always supported by the medical profession. But the growing interest in alternative (new age) remedies and health treatments has led to a consumer focus on lifestyle management and health improvements without mainstream medical intervention. This has encouraged many people to reconsider the traditional use of hot springs as an alternative treatment option for a wide range of health conditions. Supported by medical research reports, the role of natural hot springs in CAM is recognised as a valuable treatment option that has no side effects. In fact, the immersion in hot spring water provides not only physical health improvements together with a relaxing time but is an uplifting experience, which creates a sense of mental wellbeing (Erfurt-Cooper and Cooper 2009; Erfurt-Cooper 2010; Erfurt 2012).

Hot spring destinations offer a wide range of health and wellness treatment options, whereby thermalism is complemented by additions from the field of alternative medicine. For example, the implementation of local remedies such as ‘mud’ or ‘brine’ add further value to the range of specific treatment options at health resorts, clinics and spas

(Deutscher Heilbäderverband 2019b; Erfurt 2012). As a non-invasive therapy, mineral-enriched hot spring waters have long been integrated in the field of alternative medicine practiced worldwide.

8.5.1 The Concept of Prevention

With a worldwide focus on health and wellbeing, supported by wholesome lifestyle choices, people are increasingly looking at prevention instead of cure, which has resulted in a renewed demand for 'old fashioned' natural remedies and healing therapies. Together with the desire to remain well while growing older, the demand for CAM is expected to increase in the future, which in turn is contributing to the revived interest in the use of natural hot springs (Erfurt-Cooper 2010; Lazzerini 2013).

The importance of the concept of prevention lies in avoiding illness and maintaining good health over curing an ailment with the reasonable expectation that health can improve even without underlying medical conditions (Andrijasevic and Bartolucci 2004; Erfurt 2012; Mueller and Lanz Kaufmann 2001). The concept of prevention includes all activities and measures that may help with (a) early detection of diseases, (b) preservation of health, (c) preventing illness and (d) delaying the deterioration of health conditions (Erfurt 2012). A healthy lifestyle, together with physical exercise, relaxation and appropriate stress management, helps to gain renewed energy and strength, boosts the immune system and as a consequence positively affects a person's overall health.

To prevent illness or obtain better health, many conditions can either be avoided or eased through suitable therapies such as thermal bathing under the care of trained medical staff at hot spring clinics and spas. Therapies based on natural hot springs have the additional benefit that the amount of medication can potentially be reduced, which is not only cost effective but also lowers the possibility of adverse reactions for patients (Hecht 2007). In countries such as Germany the public health care system aims to provide primary prevention services to improve the general health and contribute towards reducing social inequality by increasing health opportunities. But while the German health care system favours preventive measures, it is expected that individuals take personal responsibility through a health-conscious lifestyle, early participation in health care measures, and through active involvement in treatment and rehabilitation to contribute to the sustainability of the system (Erfurt 2012; Pforr and Locher 2012; Sozialgesetzbuch 2019).

8.5.2 Maintenance of Good Health

Visiting a health resort or spa is a popular option to maintain good health in many countries, not just in Europe. To promote and maintain good health in Japan, balneo-therapeutic treatments are recommended to find a balance between mind and body by creating a feeling of wellbeing (Agishi and Ohtsuka 1998). Unfortunately, not every country subsidises health improvements obtained at spa destinations for the general population through national health cover. While balneotherapy or spa medicine is appreciated as a traditional form of treatment (Bender et al. 2014), the European custom of supporting spa medicine based on thermalism seems to have been considerably reduced since the collapse of eastern socialism and a generous welfare system in the west (Tzedopoulos et al. 2018). Germany is one of the countries where the public health system has continued to provide entitlements for a health spa visit of several weeks at intervals of four years. However, following extensive reforms (starting in 1997) to address cost containment of the health system as it relates to the German health and wellness industry with special consideration of the medical cure, beneficiaries are now required to contribute financially with a set daily co-payment (Busse and Riesberg 2004; Pforr and Locher 2012). Interestingly, to benefit from a spa cure it is not necessary to suffer from a physical illness because many people, who are 'taking the cure' are in need of relaxation and regeneration to be able to cope with work, family and life in general. In legal terms the German word *Kur* has been widely replaced with outpatient care and rehabilitative measures at a recognised health resort (Beer and Jobst 2016; Falkenbach et al. 2004).

With the aim of making a positive contribution to a person's health condition and general wellbeing, most treatments include natural remedies that are derived from the local environment such as healing springs, mud baths and fresh air, which constitute ancient, but successful non-invasive healing methods (Erfurt 2012). As a result of spending weeks away from their everyday environment, people are expected to and usually do regain a feeling of wellbeing and are therefore able to maintain good health for a period of time.

8.6 Concluding Discussion

Popular places for health and recreation have consistently been those with access to natural hot springs. An analysis of the practice of natural hot spring use shows that many countries have historically considered them as places of healing and for recreation (Chaps. 5 and 6). Based on studies and ongoing research, today the beneficial properties of hot springs rich in dissolved mineral components are widely recognised as a natural source of health and wellbeing. For this reason balneotherapy is included in holistic and alternative health treatments, but more importantly to successfully treat a wide range of medical conditions. This also

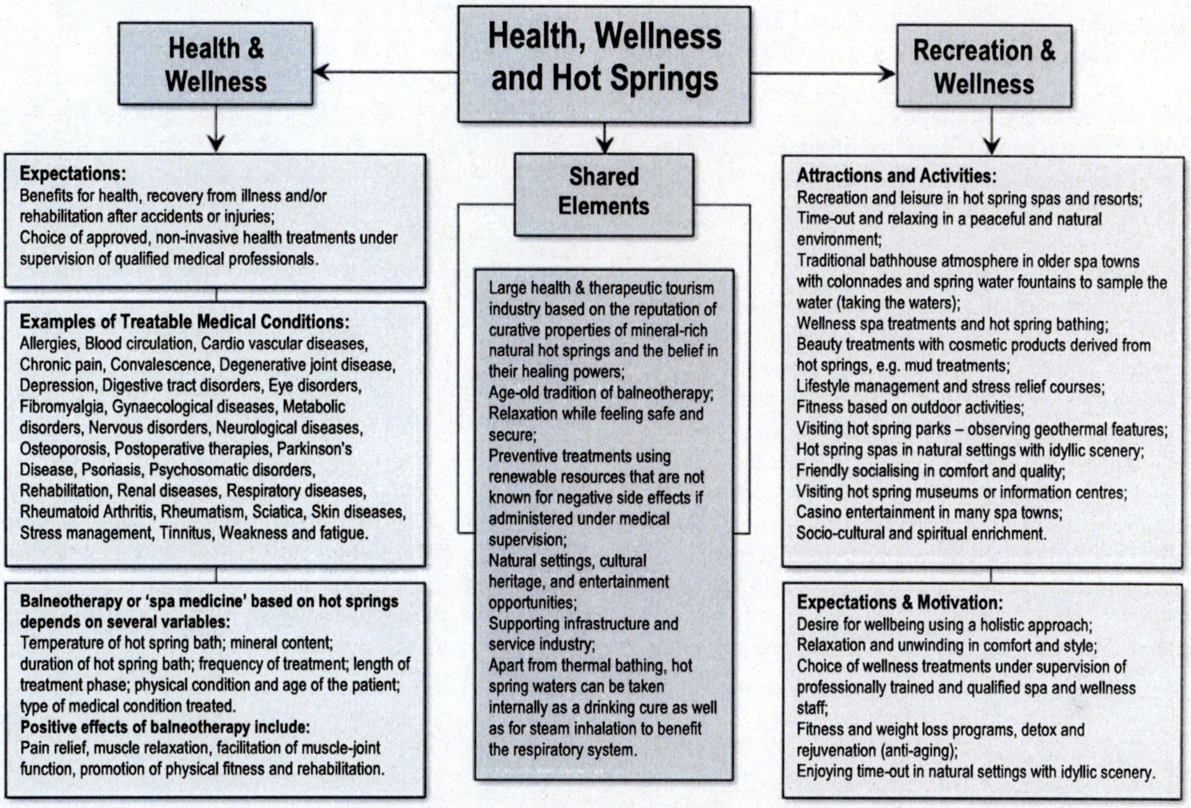

Fig. 8.16 This model represents some of the structural elements connecting natural hot springs, health, wellness and recreation. *Source* Author

includes rehabilitation after accidents and injuries or as post-operative care (Altman 2000; Bullard 2004; Cohen and Bodeker 2008; Erfurt 2012; Leavy and Bergel 2003).

A high awareness of the mineral content of hot springs and its therapeutical value is not only commonly found in the literature but is also present throughout the general population, which means that hot spring water is characterised by a firmly established belief in its curative powers (Erfurt 2012; Ross 2001). This becomes more obvious when assessing visitor preferences and expectations, and although these can vary, the main reasons for visiting hot springs usually include certain expectations for health and wellbeing including health improvements, rehabilitation after illness and maintaining good health as well as a relaxing change of scenery (Chap. 9). A conceptualised framework (Fig. 8.16) summarises some of the essential components that link natural hot springs to health and recreation.

In countries with many hot spring destinations government standards usually require certain criteria to be covered to warrant the formal recognition and approval as a natural resource for health and recreational purposes. These criteria include the type and amount of minerals in solution, the minimum temperature as well as the flow rate of the hot spring at the source (Erfurt-Cooper and Cooper 2009; Erfurt 2012; Chaps. 2 and 3).

Countries such as Austria, Bulgaria, France, Germany, Hungary and Russia (Example 8.11) have a well-established tradition of spa medicine centred around a more clinical approach, while other countries including the Americas, Australia or New Zealand seem to focus more on the wellness and recreational side of hot spring use. In the United States, spa medicine based on thermalism is labelled as 'unscientific' with complementary and alternative medicine (CAM) mainly found outside professional medical practice, more aligned with the field of wellness and beauty therapy (Erfurt 2012). In contrast to this, in Europe the use of natural hot springs in spa medicine is generally integrated into the national health care system and supported by the medical profession based on known success rates, especially in rehabilitative care. Therefore, in certain European countries spa medicine is covered either completely or partially by statutory health insurance up to a certain percentage of the overall treatment cost. This generally requires a medically documented need from a physician, who must verify that a patient will benefit from this specific type of treatment at a suitable health clinic or medical spa. However, patients are encouraged to take responsibility for their personal and health and wellbeing, which they can augment by choosing a healthier lifestyle to enhance the therapeutic benefits of thermal treatments (Erfurt 2012; Pratzel 2001a).

Example 8.11 Spa Medicine in Russia
During the Soviet era, spa medicine and thermal bathing were an integral part of the national health system, because Russia traditionally supported the use of natural hot and mineral spring water for health treatments with state run health facilities found throughout the country. At the end of the Second World War, existing health spas were redeveloped and expanded as part of the public health system. Under the guidance of the Ministry of Health, local authorities and research institutes, Russian health spas provided balneotherapy, focusing mainly on rehabilitation, which is one of the few traditions surviving today. As in many European countries, studies on the direct health benefits of thermalism were carried out in Russia in the past, supporting the use of thermalism. Many areas of the former Soviet Union are known for their natural hot springs, which have been used for thermal bathing, cooking and washing for centuries. By the end of the Soviet era circa 2500 state-run health spas had the capacity to cater for up to 500,000 visitors at a time. Unfortunately, following the collapse of the USSR in 1991 many of these health facilities went into decline.

Source Erfurt (2012), Oddy (1999), Petroune and Yachina (2009), Simkins (1986).

While hot spring therapy is beneficial in the treatment of many health conditions, there may be some exemptions, depending on a person's individual biochemical makeup (Clark 1999; Erfurt-Cooper 2010). Still, on a global scale this has never had a bearing on the popularity of hot springs and their successful application in the health and wellness industry or in the recreational sector.

Characteristic for most traditional spa towns are recreational centres of spa activities such as the Kurhaus (Spa Centre) in Baden Baden. *Source* Savin (2015a) (Public Domain)

8.7 Appendices

Appendix 8.1 Graduation Tower

1—Graduation towers for brine vapour inhalation have always been part of the 'Kurpark' in the spa town Lüneburg, Germany; 2—Due to the size of these towers the area directly surrounding them is transformed into a special microclimate that is beneficial for respiratory conditions; 3—Insert: The salt deposits on the blackthorn twigs are increasing over time and the whole structure is refilled with fresh brushwood

Appendix 8.2 Hot Spring Water and Steam for Wellbeing

1—A line of 'wellness' stations at one of the hot spring parks in Beppu (Japan) invites visitors to sample the water, the vapour and the warmth of heated rocks; 2—The hot spring steam emitting at 100 °C is said to have cosmetic effects, moisturises the skin and helps in the prevention of the flu; 3—Visitors can put their feet into a small compartment and enjoy the hot steam; 4—This station is the equivalent to European drinking fountains—mineral-rich water is dispensed for internal use with the suggestion that half a ladle is enough; 5—This 'foot spa' is made from geothermally heated natural stone and encourages blood circulation and helps to warm cold feet

References

Abels DJ, Even-Paz Z, Efron D (1996) Bioclimatology at the dead sea in Israel. Clin Dermatol 14(6):653–658. https://doi.org/10.1016/S0738-081X(96)00100-9

Agishi Y (2001) Clinical usefulness of long-term thermohydrotherapy (balneotherapy). In: Kosaka M, Sugahara T, Schmidt KL, Simon E (eds) Thermotherapy for neoplasia, inflammation, and pain. Springer, Tokyo. https://doi.org/10.1007/978-4-431-67035-3_56

Agishi Y, Ohtsuka Y (1998) Present features of balneotherapy in Japan. In: Global environmental research 2, English Edition, pp 177–185

Alçiçek H, Bülbül A, Alçiçek MC, Yavuzer I (2016) Hydrogeochemistry of the thermal waters from the Pamukkale and Karahayıt geothermal fields (Denizli Basin, Southwestern Anatolia, Turkey). European Geothermal Congress Strasbourg, France, 19–24 Sept 2016

Altman N (2000) Healing springs: the ultimate guide to taking the waters—from hidden springs to the world's greatest spas. Healing Arts Press, Rochester, Vermont

Andreassi L, Flori L (1996) Mineral water and spas in Italy. Clin Dermatol 14(6):627–632. https://doi.org/10.1016/S0738-081X(96)00095-8

Andrijasevic M, Bartolucci M (2004) The role of wellness in contemporary tourism. Acta Turistica 16(2):125–142

Arvigo R, Epstein N (2003) Spiritual bathing: healing rituals and traditions from around the world. Celestial Arts, Berkeley, CA

Bäderland Deutschland (2018) Wie lange dauert eine Kur? Retrieved from https://tinyurl.com/y6unflo9

Barabanov LN, Disler VN (1968) Principal regularities of the formation of nitrogenous thermal waters in the U.S.S.R. and some other countries. In: Kačura G (ed) Report of the twenty-third session Czechoslovakia 1968. International geological congress. Proceedings of symposium II genesis of mineral and thermal water. Academia, Prague

Bar-On R (1989) Cost-benefit considerations for spa treatments, illustrated by the Dead Sea and Arad, Israel. Revue de Tourisme (Tour Rev) 4:12–15

Baysal R, Leblebicioğlu H, Khorshid L, Sarı D (2018) Why individuals choose balneotherapy and benefit from this kind of treatment. Complement Ther Clin Pract 32(157–162):1. https://doi.org/10.1016/j.ctcp.2018.06.010

Beer A-M, Jobst D (2016) Primäre und sekundäre Präventionsmaßnahmen - Kur und Rehabilitation - das sollten Sie wissen! MMW Fortschritte der Medizin 158(19)

Bender T, Bálint G, Prohászka Z, Géher P, Tefner IK (2014) Evidence-based hydro- and balneotherapy in Hungary—a systematic review and meta-analysis. Int J Biometeorol 58:311–323. https://doi.org/10.1007/s00484-013-0667-6

Bender T, Karagülle Z, Bálint GP, Gutenbrunner C, Bálint PV, Sukenik S (2005) Hydrotherapy, balneotherapy, and spa treatment in pain management. Rheumatol Int 25(3):220–224. https://doi.org/10.1007/s00296-004-0487-4

Bernstein JE (1996) Dermatologic aspects of mineral water. Clin Dermatol 14(6):567–569. https://doi.org/10.1016/S0738-081X(96)00086-7

Bishoff M (2001) Touring New Mexico hot springs. a Falcon guide. Globe Pequot Press, Guilford, CT

Boer J (1996) The influence of mineral water solutions in phototherapy. Clin Dermatol 14(6):665–673. https://doi.org/10.1016/S0738-081X(96)00102-2

Bullard L (2004) Healing waters—Missouri's historical mineral springs and spas. University of Missouri Press, Columbia, MO

Busse R, Riesberg A (2004) Health care systems in transition: Germany. Copenhagen, WHO Regional Office for Europe on behalf of the European Observatory on Health Systems and Policies

Cantista P (2012) Balneology in Europe: results of a new survey. Balnea 6:43. ISBN: 978-84-669-3482-4

Cataldi R, Hodgson SF, Lund JW (eds) (1999) Stories from a heated Earth—our geothermal heritage. geothermal resources council. International Geothermal Association, Sacramento, CA

Clark S (1999) Japan, a view from the bath. University of Hawaii Press, Honolulu, HI

Clark-Kennedy J, Cohen M (2017) Indulgence or therapy? Exploring the characteristics, motivations and experiences of hot springs bathers in Victoria, Australia. Asia Pacific J Tour Res 22:501–511. https://doi.org/10.1080/10941665.2016.1276946

Cohen M, Bodeker G (2008) Understanding the global spa industry: spa management. Butterworth Heinemann, London

Deutscher Heilbäderverband (DHV) (2019a) Die Kur. Retrieved from https://deutscher-heilbaederverband.de/die-kur/

Deutscher Heilbäderverband (DHV) (2019b) Heilbäder und Kurorte - gesund werden und gesund bleiben. Retrieved from https://tinyurl.com/yahqmlxw

Deutscher Heilbäderverband (DHV) (2019c) Zahlen, Daten, Fakten. Retrieved from https://deutscher-heilbaederverband.de/themen/zahlen-daten-fakten/

Erfurt P (2012) An assessment of the role of natural hot and mineral springs in health, wellness and recreational tourism. Unpublished Doctoral Dissertation, School of Business, James Cook University, Cairns, Queensland, Australia

Erfurt-Cooper P, Cooper M (2009) Health and wellness tourism: spas and hot springs. Channel View Publications, Bristol, UK

Erfurt-Cooper P (2010) The importance of natural geothermal resources in tourism. In: Proceedings world geothermal congress, Bali, Indonesia, 25–29 Apr 2010

Eysteinsdóttir JH, Ólafsson JH, Agnarsson BA, Lúðvíksson BR, Sigurgeirsson B (2014) Psoriasis treatment: faster and long-standing results after bathing in geothermal seawater. A randomized trial of three UVB phototherapy regimens. Photodermatol Photoimmunol Photomed 30:25–34. https://doi.org/10.1111/phpp.12090

Fabiani D, Partsch G, Casale R, Cerinic MM (1996) Rheumatologic aspects of mineral water. Clin Dermatol 14(6):571–575. https://doi.org/10.1016/S0738-081X(96)00087-9

Falkenbach A, Karagülle MZ, Bender T, Agishi Y, Hartmann B (2004) Kurmedizin, Balneologie und Klimatologie. In: Falkenbach A (ed) Morbus Bechterev: Beratung, Betreuung, Behandlung, chap 36. Springer, Berlin, pp 613–629

Fioravanti A, Karagülle M, Bender T, Karagülle MZ (2017) Balneotherapy in osteoarthritis: facts, fiction and gaps in knowledge. Eur J Integr Med 9:148–150. https://doi.org/10.1016/j.eujim.2017.01.001

Frazier D (2000) Colorado's hot springs, 2nd edn. Pruett Publishing Company, Boulder, CO

Fytikas M, Leonidopoulou and Cataldi R (1999) Geothermal energy in ancient Greece: from mythology to late antiquity (3rd Century AD). In: Cataldi R, Hodgson SF, Lund JW (eds) Stories from a heated Earth. Geothermal Resources Council, International Geothermal Association, Sacramento, CA, pp 68–101

Ghersetich I, Lotti TM (1996) Immunologic aspects: immunology of mineral water spas. Clin Dermatol 14(6):563–566. https://doi.org/10.1016/S0738-081X(96)00085-5

Grassberger M, Hoch W (2006) Ichthyotherapy as alternative treatment for patients with psoriasis: a pilot study. Evid Based Complement Altern Med 3(4):483–488. https://doi.org/10.1093/ecam/nel033

Grether-Beck S, Mühlberg K, Brenden H, Felsner I, Brynjólfsdóttir Á, Einarsson S, Krutmann J (2008) Bioactive molecules from the Blue Lagoon: in vitro and in vivo assessment of silica mud and microalgae extracts for their effects on skin barrier function and prevention of skin ageing. Exp Dermatol 17(9). https://doi.org/10.1111/j.1600-0625.2007.00693.x

Größchen H-W (1998) Heilbäder und Kurorte in Deutschland. Conradi Bäder-Lexikon. Conradi Verlag, Altenbeken, Germany

Gutenbrunner C (2014) Balneology and physical and rehabilitation medicine—differences, similarities and synergies. Ann Phys Rehabilit Med 57(1):e159. https://doi.org/10.1016/j.rehab.2014.03.574

Gutenbrunner C, Hildebrandt G (eds) (1998) Handbuch der Balneologie und medizinischen Klimatologie. Springer, Berlin

Hann S-K (1996) Mineral water and spas in Korea. Clin Dermatol 14 (6):633–635. https://doi.org/10.1016/S0738-081X(96)00096-X

Hatton J (c. 1910) The hot springs of Bath—medical handbook. Published by the Hot Mineral Baths Committee of the Bath Corporation with the Approval of the Bath Medical Committee of the British Medical Association. Printed by Fyson and Company

Havins PJN (1976) The spas of England. Robert Hale and Company, London

Hecht K (2007) Kardiologenkongress: Kuren helfen. Deutsches Ärzteblatt 104(50)

Heilbad Baden Baden (2020) Die Geschichte der Badekultur in Baden-Baden. Retrieved from http://www.bad-bad.de/gesch/badkultur.htm

Huang A, Seité S, Adar T (2018) The use of balneotherapy in dermatology. Clin Dermatol 36(3):363–368. https://doi.org/10.1016/j.clindermatol.2018.03.010

Japanese Society of Balneology, Climatology and Physical Medicine (2019) The ABCs for the safe and secure use of onsen. What should we be aware of when bathing in onsen (hot springs) and drinking onsen mineral water? Retrieved from https://www.env.go.jp/nature/onsen/pdf/top_en.pdf

Jorden E (1631) A discovrse of natvrall bathes and minerall waters. Printed by Thomas Harper. Reproduced and published by Theatrum Orbis Terrarum Ltd. Amsterdam, London, UK

Kangal Balikli Kaplicasi (2019) Doctor fishes—treatment. Retrieved from http://kangaldoctorfishes.com/treatment/

Kangal Fish Balikli Kaplica Treatment Center (2019) A most unique place in the world, famous for its healing waters and effective psoriasis fish treatment pools. Retrieved from https://www.kangalfishbalikli.com/

Karagülle MZ (2012) What's new in balneology? Balnea 6:41–42. ISBN: 978-84-669-3482-4

Karagülle M, Karagülle MZ, Karagülle O, Dönmez A, Turan M (2007) A 10-day course of SPA therapy is beneficial for people with severe knee osteoarthritis—a 24-week randomised, controlled pilot study. Clin Rheumatol 26(12):2063–2071. https://doi.org/10.1007/s10067-007-0618-x

Karam P (1996) Mineral water and spas in France. Clin Dermatol 14 (6):607–610. https://doi.org/10.1016/S0738-081X(96)00091-0

Katsambas A, Antoniou C (1996) Mineral water and spas in Greece. Clin Dermatol 14(6):615–618. https://doi.org/10.1016/S0738-081X(96)00093-4

Kelly A (2011) Why it's wrong to use Garra rufa fish in pedicures. The Ecol. Retrieved from https://tinyurl.com/y82xn8nw

King T (1966) Water: miracle of nature. Collier Books, New York

Küffer J (1625) Beschreibung des Marggrävischen Warmen Bades. Reprint (1976) of the original edition. Antiqua Verlag, Lindau, Germany

Kubota K, Kurabayashi H, Tamura K, Kawada E, Tamura J, Shirakura T (1992) A transient rise in plasma β-endorphin after a traditional 47 C hot-spring bath in Kusatsu-spa, Japan. Life Sci 51 (24):1877–1880

Kusatsu Onsen (2016) Bathing style: keeping the Kusatsu tradition alive. Retrieved from https://www.kusatsu-onsen.ne.jp/guide/en/style/

Lazzerini F (2013). Thermalism in Brazil. The new frontiers of the European healthcare system. The implementation of the EU Directive on the application of patients' rights in cross-border traditional and thermal healthcare. Convegno [Convention] Montegrotto Terme. Retrieved from https://tinyurl.com/yhuahbdb

Leavy HR, Bergel RR (2003) The spa encyclopedia: a guide to treatments and their benefits for health and healing. Thomson Delmar Learning, New York

Ledo E (1996) Mineral water and spas in Spain. Clin Dermatol 14 (6):641–646. https://doi.org/10.1016/S0738-081X(96)00098-3

Lund JW (2005) Basic principles of balneology and examples in the United States. In: Proceedings world geothermal congress 2005. Antalya, Turkey, 24–29 Apr 2005

Mackaman DP (2007) Leisure settings: Bourgois culture, medicine and the spa in modern France. University of Chicago Press, Chicago

McGeary D, Plummer C, Carlson D (2001) Physical geology—earth revealed, 4th edn. McGraw Hill, USA

Mellilo L (1995) Thermalism in the ancient world. Medicina nei Secoli 7(3):461–483

Ministry of the Environment—Japan (2016) The criteria for the notification, etc. on contraindications and cautions for bathing and drinking as specified in Article 18, Paragraph 1 of the hot springs law. Retrieved from http://www.env.go.jp/en/nature/nps/appendix_criteria.pdf

Mitsanobu F, Hosaki Y, Ashida K, Tsugeno H, Okamoto M, Nishida N et al (2003) Long-term spa therapy prevents the progressive pathological changes of the lung in patients with pulmonary emphysema. J Jpn Soc Balneol Climatol Phys Med 66:91–98

Monti S (2003) Thermalism between past and future. In: Proceedings of the conference 'the cultural turn in geography', 18–20th Sept 2003 —Gorizia Campus. Part VI: Tourism, Sustainable Development and Culture Turn. Retrieved from https://tinyurl.com/y8azp3bl

Mueller H, Lanz Kaufmann E (2001) Wellness tourism: market analysis of a specific health tourism segment and implications for the hotel industry. J Vac Market 7(1):5–17. https://doi.org/10.1177/135676670100700101

Nagasawa Y, Komori S, Sato M, Tsuboi Y, Umetani K, Watanabe Y, Tamura K (2001) Effects of hot bath immersion on automatic activity and hemodynamics: comparison of the elderly patient and the health young. Jpn Circ J 65(7):587–592. https://doi.org/10.1253/jcj.65.587

Nahrstedt W (2004) Wellness: a new perspective for leisure centers, health tourism and spas in Europe on the global health market. In: Weiermair K, Mathies C (eds) The tourism and leisure industry: shaping the future. Haworth Hospitality Press, New York, pp 181–198

National Center for Complementary and Alternative Medicine (NCCAM) (2010) What is complementary and alternative medicine? U.S. Department of Health and Human Services

Oddy J (1999) Sanatoriums: collective memory. Retrieved from https://tinyurl.com/y9mgajtn

Office National du Tourisme (1921) The spas of France. Ministère de Traveaux Publics, Paris

Ohtsuka Y (2012) The state of the art, health resort medicine in Japan. Balnea 6:50–52. ISBN: 978-84-669-3482-4

Ohtsuka Y, Nakaya J, Nishikawa K, Agishi Y (2003) Effects of drinking hot spring water from Kawayu Onsen on blood glucose levels. J Jpn Soc Balneol, Climatol Phys Med 66(4):227–230. https://doi.org/10.11390/onki1962.66.227

Ökçesiz M (2014) Karahayit thermal spa center. Retrieved from http://www.tourmakerturkey.com/karahayit.html

Ólafsson JH (1996) The Blue Lagoon in Iceland and Psoriasis. Clin Dermatol 14(6):647–651. https://doi.org/10.1016/S0738-081X(96)00099-5

Omulecki A, Nowak A, Zalewska A (1996) Spa therapy in Poland. Clin Dermatol 14(6):679–683. https://doi.org/10.1016/S0738-081X(96)00104-6

Oumeish OY (1996) Climatotherapy at the dead sea in Jordan. Clin Dermatol 14(6):659–664. https://doi.org/10.1016/S0738-081X(96)00101-0

Özçelik S, Polat HH, Akyol M, Yalçin AN, Özçelik D, Marufihah M (2015) Kangal hot spring with fish and psoriasis treatment. J Dermatol 27(6):386–390. https://doi.org/10.1111/j.1346-8138.2000.tb02188.x. Originally published in 2000

Özkul M, Kele S, Gökgöz A, Shen C-C, Jones B, Baykara MO, et al (2013) Comparison of the quaternary travertine sites in the Denizli extensional basin based on their depositional and geochemical data. Sediment Geol 294:179–204. https://doi.org/10.1016/j.sedgeo.2013.05.018

Parish LC, Lotti TM (1996) Commentary. Clin Dermatol 14(6):547–548. https://doi.org/10.1016/S0738-081X(96)00081-8

Parish LC, Witkowski JA (1994) Dermatologic balneology: the American view of waters, spas, and hot springs. J Eur Acad Dermatol Venereol 3(4):465–467. https://doi.org/10.1111/j.1468-3083.1994.tb00401.x

Pattanaik S, Mishra SP, Swain S, Pattnayak S (2019) Hydro-geo-bio-chemical investigation of geothermal springs for balneotherapy and thermophile studies: Odisha, India. Int J Sci Res (IJSR) 8(4):1419–1435

Pesce A (2002) Thermal spas: an economic development alternative along both sides of the Uruguay River. In: Geo-heat center quarterly bulletin. Geo–Heat Center, Oregon, pp 22–28

Peterson C (2001) Exercise in 94 F water for a patient with multiple sclerosis. Health module. Phys Ther 81(4):1049–1058. https://doi.org/10.1093/ptj/81.4.1049

Petroune I, Yachina E (2009) Health tourism in Russia: medical tourism. In: Smith M, Puczkó L (eds) Health and wellness tourism. Butterworth Heinemann, Oxford, UK

Pforr C, Locher C (2012) The German spa and health resort industry in the light of health care system reforms. J Travel Tour Market 29:298–312. https://doi.org/10.1080/10548408.2012.666175

Picoto A (1996) Mineral water and spas in Portugal. Clin Dermatol 14 (6):637–639. https://doi.org/10.1016/S0738-081X(96)00097-1

Pratzel HG, Schnizer W (1992) Handbuch der Medizinischen Bäder - Indikationen – Anwendungen – Wirkungen. Karl F. Haug Verlag, Stuttgart, Germany

Pratzel HG (2001a) Outcome-research in thermalism—the new challenge of medicine. In: International Congress on Natural Spa Therapy of the I.S.M.H. Dead Sea Israel, 21–24 Nov

Pratzel HG (2001b) Health tourism and medical treatment—two parts in health resorts. In: International society of medical hydrology and climatology (I.S.M.H.) Conference, Argentina

QE Health (2019) Un établissement dédié au bien-être. Retrieved from https://www.qehealth.co.nz/cure-thermale

Roques CF (2012) Balneology research in France. Balnea 6:47–48. ISBN: 978-84-669-3482-4

Roques C, Mancret R-C, Tabone W (2012) Beneficial effect of balneotherapy: achievements of the French association of balneotherapy research and methodological difficulties. Ann Phys Rehabil Med 55(1):e351. https://doi.org/10.1016/j.rehab.2012.07.891

Ross K (2001) Health tourism: an overview. (HSMAI Marketing review). Retrieved from https://tinyurl.com/yx9gjf77

Routh HB, Bhowmik KR (1996) Basic tenets of mineral water: a glossary of concepts relating to balneology, mineral water, and the spa. Clin Dermatol 14(6):549–550. https://doi.org/10.1016/S0738-081X(96)00092-2

Ryrie C (1999) The healing energies of water. Journey Editions, Boston–Tokyo

Sanner B (2000) Baden-Baden—a famous thermal spa with a long history. GHC Bull, PDF. Retrieved from https://tinyurl.com/y2nnxejo

Savin A (2015) The Kurhaus (Spa house) in Baden-Baden (BW, Germany). Retrieved from https://tinyurl.com/ydaj85oh

Savin A (2015) The pump house in Baden-Baden (BW, Germany). Retrieved from https://tinyurl.com/ya7aa2ts

Sayili M, Akca H, Duman T, Esengun K (2007) Psoriasis treatment via doctor fishes as part of health tourism: a case study of Kangal Fish Spring, Turkey. Tour Manag 28(2):625–629. https://doi.org/10.1016/j.tourman.2006.08.010

Schnizer W (2002) Grundlagen körperlichen trainings und der Entspannung im Wasser. In: Hartmann B (ed) Aquale immersion—training durch stimulation. 1. Beurener Symposium zur Evidenz thermaler Heilwässer, Panorama Therme Beuren

Schweitzer PA (1892) A report on the mineral waters of Missouri, vol 13. Missouri Geological Survey and Water Resources, Jefferson City

Selke AC (1936) Geographic aspects of the German tourist trade. Econ Geogr 12(2):205–216. https://doi.org/10.2307/140711

Sigurgeirsson B, Olafsson JH (2003) The Blue Lagoon and Psoriasis. Blue Lagoon, Iceland

Simkins KL (1986) Physical therapy and spa treatment. Med Anthropol Q 17(5):146–147. https://doi.org/10.1111/j.1937-6219.1986.tb01068.x

Shirakura T, Sugai Y (1986) Effect of balneotherapy in high water temperature using hot spring water (Jikan-yu in Kusatsu, Japan) on platelet function. Abstracts of the twenty-third annual meeting of the japanese society of biometeorology. Int J Biometeorol 30 (4):359–385

Solanki (2019) Brine graduation towers. Retrieved from http://www.solanki.pl/en/brine-graduation-towers

Sozialgesetzbuch (2019) Solidarität und Eigenverantwortung. Fünftes Buch - Gesetzliche Krankenversicherung § 1 SGB V. Retrieved from https://www.sozialgesetzbuch-sgb.de/sgbv/1.html

Stevens F, Azara I, Michopoulou E (2018) Investigating community's perceptions and attitudes towards Thermalism as a tool to facilitate wellness and wellbeing. In: Tourism naturally conference (TNC 2018), Austria, 16–18 May 2018

Tabone W, Mancret R-C, Roques C-F (2012) Balneotherapy's actual medical benefit assessment: epidemiological and organisational basis. Ann Phys Rehabil Med 55(1):e351. https://doi.org/10.1016/j.rehab.2012.07.890

Tarita Salt Spa and Wellness (2019) Salt inhalation therapy in the graduation tower. Retrieved from https://tinyurl.com/ycll388n

Therme-Thermen.de (2019) Gradierwerk/saline and Thermalbad. Retrieved from https://tinyurl.com/yclzta24

Tilton EM (1981) Mineral and thermal spas in France. French Rev 54 (4):566–572. https://www.jstor.org/stable/391139

Titzmann T, Balda BR (1996) Mineral waters and spas in Germany. Clin Dermatol 14(6):611–613. https://doi.org/10.1016/S0738-081X(96)00092-2

Tombleson W (c.1840) Town of Wiesbaden, Old Kurhaus. Steel engraving from 'Views of the Rhine'. Retrieved from https://tinyurl.com/wnxcezr

Tsankov NK, Kamarashev JA (1996) Spa therapy in Bulgaria. Clin Dermatol 14(6):675–678. https://doi.org/10.1016/S0738-081X(96)00103-4

Tzedopoulos Y, Kamara A, Lampada D, Ferla K (2018) Thermalism in Greece: an old cultural habitus in crisis. Int J Prof Bus Rev 3(2):205–219. http://dx.doi.org/10.26668/businessreview/2018.v3i2.83

Undar L, Akpinar M, Yanikoğlu A (1990) "Doctor fish" and psoriasis. Lancet 335(8687):470–471. https://doi.org/10.1016/0140-6736(90)90699-6

Valenza JM (2000) Taking the waters in Texas: springs, spas and fountains of youth. University of Texas Press, Austin, TX

Vassileva S (1996) Mineral water and spas in Bulgaria. Clin Dermatol 14(6):601–605. https://doi.org/10.1016/S0738-081X(96)00090-9

Vereinigung für Bäder- und Klimakunde (VBK) (2016) Arbeitsgemeinschaft Kurortmedizin/Physiomedizin. Retrieved from https://tinyurl.com/y8jruyet

Vitale M, Vaccarezza M (2012) Research in Italy: research products and research funding. Balnea 6:46. ISBN: 978-84-669-3482-4

Vu K, Mitsanobu F (2004) Spa therapy for bronchial asthma: studies at the Misasa Medical Center. Altern Complement Ther 10(3):144–150. https://doi.org/10.1089/1076280041138243

Weisz G (2001) Spas, mineral waters, and hydrological science in twentieth-century France. Isis (J Hist Sci Soc) 92(3):451–483. https://doi.org/10.1086/385278

Weisz G (2002) Thermalism and mineral waters in twentieth-century France. Med Sci 18:101–108

Wiesbaden (2019) The Kurhaus Wiesbaden. Image. Retrieved from https://www.wiesbaden.de/microsite/kurhaus-en/

Williams PW, Andestad G, Pollock A, Dossa KB (1996) Health spa travel markets: Mexican long-haul pleasure travellers. J Vac Market 3(1):10–31. https://doi.org/10.1177/135676679600300102

Witt SF (1990) Reviving spa tourism—the AIEST congress. Tour Manag

Wolf R (1996) Mineral waters and spas in Israel. Clin Dermatol 14 (6):619–626. https://doi.org/10.1016/S0738-081X(96)00094

Hydrothermal features in the Krafla caldera at Leirhnjúkur (Iceland) include many hot springs and mud ponds

Visitor Expectations and Risk Management at Hot Spring Destinations

9

Contents

P. Erfurt, *The Geoheritage of Hot Springs*, Geoheritage, Geoparks and Geotourism,
https://doi.org/10.1007/978-3-030-60463-9_9

This ‘healing pool’ in the Chalice Well Garden in Glastonbury (UK) is used as a wishing well by people who throw coins in the water in the hope of good health and better fortunes

9.1 Introduction

Hot springs are visited by millions of people every year, which makes hot spring tourism a commercial sector based entirely on a unique natural resource. Visitors of hot springs include people from all walks of life, who appreciate interesting destinations that at the same time are beneficial for their health and provide recreational opportunities. With thousands of hot spring locations to choose from, there is no easy formula to categorise different aspects of hot spring tourism. Some key variables include (1) shifting consumer trends, (2) the geographical location, and (3) the geophysical attractions of natural hot spring sites. To be more specific, the wellness trend that started in the 1990s, has also contributed significantly to the popularity of existing hot spring destinations as well as instigating new developments together with upgrades and redevelopments of older hot spring infrastructure.

9.1.1 Visitor Statistics

To put hot spring tourism into a global perspective, we need to look at visitor statistics, although obtaining reliable data, especially consecutive records for recent years, is always a challenge. Some up-to-date visitor numbers are reflected in Figs. 9.1, 9.2 and 9.3 to highlight the popularity of hot springs for health, recreational as well as adventurous purposes. For example, **Japan’s** hot spring visitor numbers of (on average) over 150 million (overnight stays) annually are more of an exception, made possible by an abundance of volcanic hot springs and their associated onsen facilities throughout the country, which is unmatched elsewhere (Erfurt 2012).

In **Germany,** visitor numbers at hot spring spas and health resorts *(Mineral- und Thermalheilbäder)* have steadily increased and amount to approximately a third of the overall number of visitors to all health and recreational destinations.

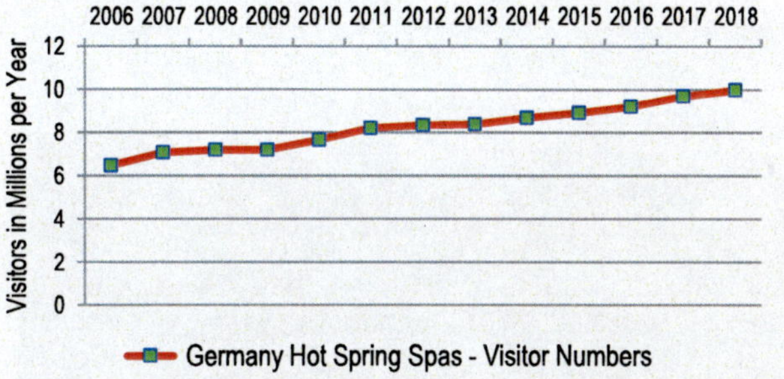

Fig. 9.1 Graphic representation of visitor numbers of Germany's hot spring spas and thermal mud spas combined from 2006 to 2018. *Source* Deutscher Heilbäderverband (2020)

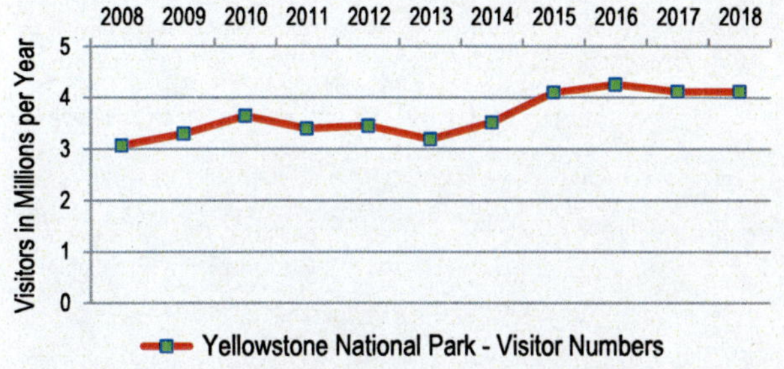

Fig. 9.2 Graphic representation of visitor numbers of Yellowstone National Park (Wyoming, USA) from 2008 to 2018. *Source* NPS Stats (2020a, b)

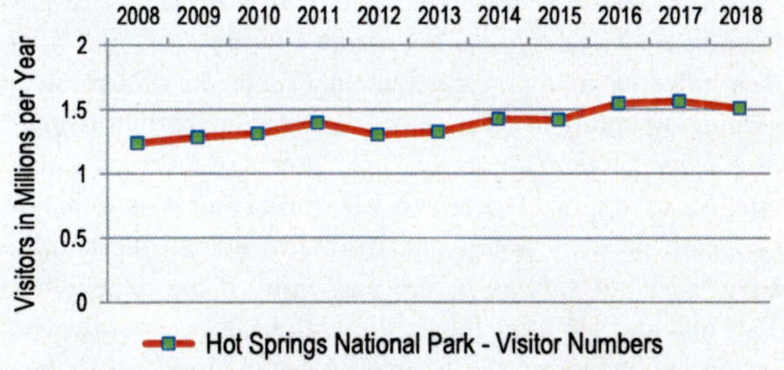

Fig. 9.3 Graphic representation of visitor numbers of Hot Springs National Park (Arkansas, USA) from 2008 to 2018. *Source* NPS Stats (2020a, b)

For example, from January to December 2018 Germany's hot spring and thermal mud spas (*Moorheilbäder*) received 9,978,803 visitors, who stayed on average 4.5 nights (Fig. 9.1). For the same time, the total number of visitors of all spa destinations, including hot spring spas, seaside spas, Kneipp spas, and spa towns known for their healthy climate, reached 28,191,087 guests with 127,179,073 overnight stays, an average of 4.5 nights (Deutscher Heilbäderverband 2020).

In contrast to Germany's health and wellness oriented hot spring usage, more than 10,000 hydrothermal phenomena in the **Yellowstone National Park** (Wyoming, USA) together with other natural features attracted 4,115,000 visitors in 2018 (NPS Stats 2020a, b; Fig. 9.2). The **Hot Springs National Park** (Arkansas, USA) (Fig. 9.3), while not showing an increase for the year 2018, nevertheless remains a popular hot spring destination.

9.1.2 Hot Spring Tourists—Who Are They and What Are Their Expectations?

Different types of hot springs have been tourist attractions for centuries, a fact that is well documented in the literature about individual hot spring regions. This enduring fascination with natural hot springs and other related manifestations has been discussed in several chapters of this book (e.g. Chaps. 4 and 5). Used for health and recreation, hot spring areas also include active hydrothermal environments, which are often found near active and dormant volcanoes (Altman 2000; Cataldi et al. 1999; Erfurt-Cooper 2009; Hróarsson and Jónsson 1992). With adventurous travellers always on the lookout for new experiences, hot springs seem to fit this bill perfectly, although some of the more dynamic environments can harbour potential dangers.

From an age perspective, visitors of hot spring areas range from very young to very old, which is largely determined by their individual destination preference. In Japan, approximately 50% of visitors of hot spring facilities (onsen resorts, traditional hot spring *ryokan* or inns, onsen hospitals, health spas) are generally in their 50s and 60s (Sekine et al. 2006). This is mirrored in many European countries, where traditional hot spring-based health resorts and spas are largely frequented by *baby boomers* and the next age group, commonly defined as *Generation X*. However, the earlier mentioned wellness trend is drawing *Generation Y* (Millennials) to traditional and contemporary hot spring destinations as well. While mature aged hot spring connoisseurs often prefer the comfort of hot spring resorts and spas that offer health and wellness treatments, relaxation and abundant leisure opportunities, younger generations might enjoy a strenuous cross-country hike to discover wild hot springs in remote natural settings. But in general, and mainly based on their diversity, hot springs have something to offer for all ages and abilities. For more adventure minded tourists there are many opportunities to join special interest groups, who undertake field trips and excursions to explore active hydrothermal areas to observe more remarkable hot springs such as geysers, fumaroles and mud ponds (Fig. 9.4).

Fig. 9.4 A group of tourists is walking through a hydrothermal area near the Krafla volcano (Iceland)

Natural hot springs without any appreciable infrastructure may be more appealing to people who prefer to be outdoors over a stay at a hotel or resort with access to hot spring pools. As laid out in Fig. 1.3 in the first chapter of this book, a considerable divide exists between the various stages of development at hot spring destinations. But one category of visitors is easy to determine—the hot spring enthusiasts who travel to discover new hot springs and add them to their collection of favourite destinations. This is especially true in Japan, where hot springs or *onsen* are so popular, that they are part of most, if not all domestic travel itineraries.

To determine visitor expectations, it is necessary to differentiate between typical hot spring environments and what they offer as a tourist destination (Table 9.1). Visiting a natural hot spring is often considered an added bonus feature in a range of tourism sectors that offer day tours of a destination or region. Access to hot spring pools is a major drawcard used by hotels and resorts to attract guests who have used hot springs before and enjoyed the experience. This includes facilities for people with health concerns, who regularly spend time at a hot spring health clinic where they receive hydrothermal treatments such as balneotherapy (Chap. 8).

Table 9.1 Main target groups and their expectations at different types of hot spring destination. The risk factor refers to potential harm from hot springs and/or other hydrothermal features, not from incidents that can occur in natural environments in general

Hot spring Resorts and Spas	
Target groups	All age groups who can afford visits to their country/destination of choice depending on disposable income and who like destinations that are independent from seasons and climate Everybody who enjoys hot springs together with natural surroundings and complementary forms of entertainment offered at multipurpose facilities The majority of visitors are non-residents and can include repeat visitors who prefer locations they are familiar with and where they feel comfortable
Expectations	Authentic hot springs—'healing waters'—safe resort setting surrounded by comfort and quality Relaxation, stress reduction, rejuvenation, improved wellbeing and prevention of illness, boosting the immune system Recreational entertainment including fitness training, sports activities, beauty salons, golf courses, tennis courts, water sports, bush walking, sightseeing tours
Destinations	Common in most European countries, Middle Eastern countries, North Africa, South American countries, Canada, China, Japan, Korea, New Zealand, USA as well as many others
Risk factor	Very low
Hot spring health resorts and rehabilitation clinics	
Target groups	People of all age groups recovering from illness, accidents or injuries who need to spend several weeks under medical supervision while receiving balneotherapy based on applications with thermal spring water rich in beneficial mineral elements The majority of visitors are non-residents—repeat visitors of popular hot spring health resorts are very common in European countries
Expectations	Authentic hot springs and treatment facilities with qualified medical staff in a safe environment Successful rehabilitation and convalescence after illness, accidents or post-surgery Recovery including the regaining of movement after injuries together with an overall improvement of wellbeing and the ability to return to work Therapeutical or healing benefits based on the curative value of mineral-rich hot spring water Recreational entertainment including social activities, restaurants, sightseeing tours, historical and cultural hot spring heritage, e.g. traditional bathhouse architecture, hot spring museums and visitor centres
Destinations	Common in most European countries, some Middle East countries, some African countries, Mongolia, China and New Zealand as well as numerous National Hot Spring (Onsen) Hospitals in Japan
Risk factor	Low
Hot spring-based aquatic theme parks/waterparks	
Target groups	Depending on the hot spring facilities, families with children of all ages are the main target group More specialised hot spring parks such as for example the Japanese *Ooedo Onsen Monogatari* appear to cater mainly for 'young professional' consumers, although amusement arcades are available for younger generations as well Visitors consist of local residents as well as non-residents—repeat visitors
Expectations	Authentic hot spring pools for recreational use (see Chap. 7) Apart from aquatic entertainment visitors also seek relaxation, getting away for a day and take time out, finding stress release or just meeting up with friends Many hot spring parks also have separate smaller hot pools for quiet relaxation
Destinations	Some examples: Aquapark Tatralandia (Slovakia), Spa Resort Hawaiians (Japan), Waterpark Resort Glenwood (USA), Water Worlds and Spas in Tyrol (Austria), Thermal Riviera (Slovenia), Thermal Waterpark Bešeňová (Slovakia), Waiwera Hot Pools (New Zealand), Therme Erding (Germany), Termálfürdő and Aquapark (Hungary), Thermal Parks and Gardens Ischia (Italy)
Risk factor	Low to medium (e.g. persons with medical conditions)
Hot springs in undeveloped natural areas	
Target groups	Originally mainly local residents who know the area and visit secluded hot springs regularly Adventure minded people interested in natural hot springs who enjoy being outdoors and have the required level of fitness and experience to explore remote and potentially hazardous areas Organised adventure tours and study groups, scientists carrying out field research, artists and photographers, recreational hot spring enthusiasts
Expectations	Finding authentic and chemical-free natural hots springs for a relaxing soak after a long walk or other exercise Unique landscapes and peaceful natural environment Natural features other than hot springs to observe, e.g. unusual flora and/or fauna
Destinations	Many different countries and regions, e.g. New Zealand, Japan, Iceland, São Miguel (Azores), western states of USA, Australia Undeveloped hot springs are often located in national parks or other protected sites, e.g. Kamchatka (Russia)
Risk factor	High
Active hydrothermal areas	
Target groups	Tour groups and individual domestic and international visitors (e.g. cruise ship passengers) interested in exciting natural destinations—frequently combined with other interesting features on a sightseeing agenda Every person able to join either organised tours (recommended in extreme areas) or prefers their own method of transport Highly active and potentially dangerous areas should not be explored without experienced guides

(continued)

Table 9.1 (continued)

Expectations	Recreational adventure, excitement, and an authentic experience Observation of active hydrothermal phenomena in a natural environment Digital documentation of unusual scenery and activity Views of magnificent geyser eruptions, boiling mud ponds and hissing fumaroles
Destinations	Hydrothermal national parks, world heritage sites and geoparks in many countries (e.g. Africa, Azores–Portugal, Chile, China, Iceland, Japan, New Zealand, Greece, Latin America, Russia, Taiwan, USA) Usually located in high temperature areas close to active volcanoes or related to large magma chambers that provide heat for the hydrothermal system Frequently located in hostile environments and/or difficult to access
Risk factor	High to very high

Source Author

Fig. 9.5 A large crowd of spectators is lined up, waiting for the famous geyser Old Faithful to erupt. Yellowstone National Park, Wyoming. *Source* Nhu (2019) (Public Domain)

A common denominator at hot spring destinations, whether remote wilderness or highly developed, is the opportunity to enjoy a break to relax, either surrounded by nature or with access to other forms of entertainment, while socialising with likeminded people. Whereas Japanese hot spring visitors combine the viewing of hydrothermal features with the pleasures of hot spring inns, German hot spring visitors mainly focus on the health aspect of natural thermal springs that are widely used for medical treatments and rehabilitative applications. Visitors of national parks like Yellowstone on the other hand are more interested in the natural environment with the opportunity to observe remarkable hydrothermal features such as the Old Faithful (Fig. 9.5) and Steamboat geysers or the travertine terraces of the Mammoth Hot Springs.

Reasons for visiting active hydrothermal environments can include sightseeing as part of a tour agenda with unique photo opportunities, but it can also be just curiosity to see unusual natural features. Other motivations may include scientific research such as studying certain hot spring sites or

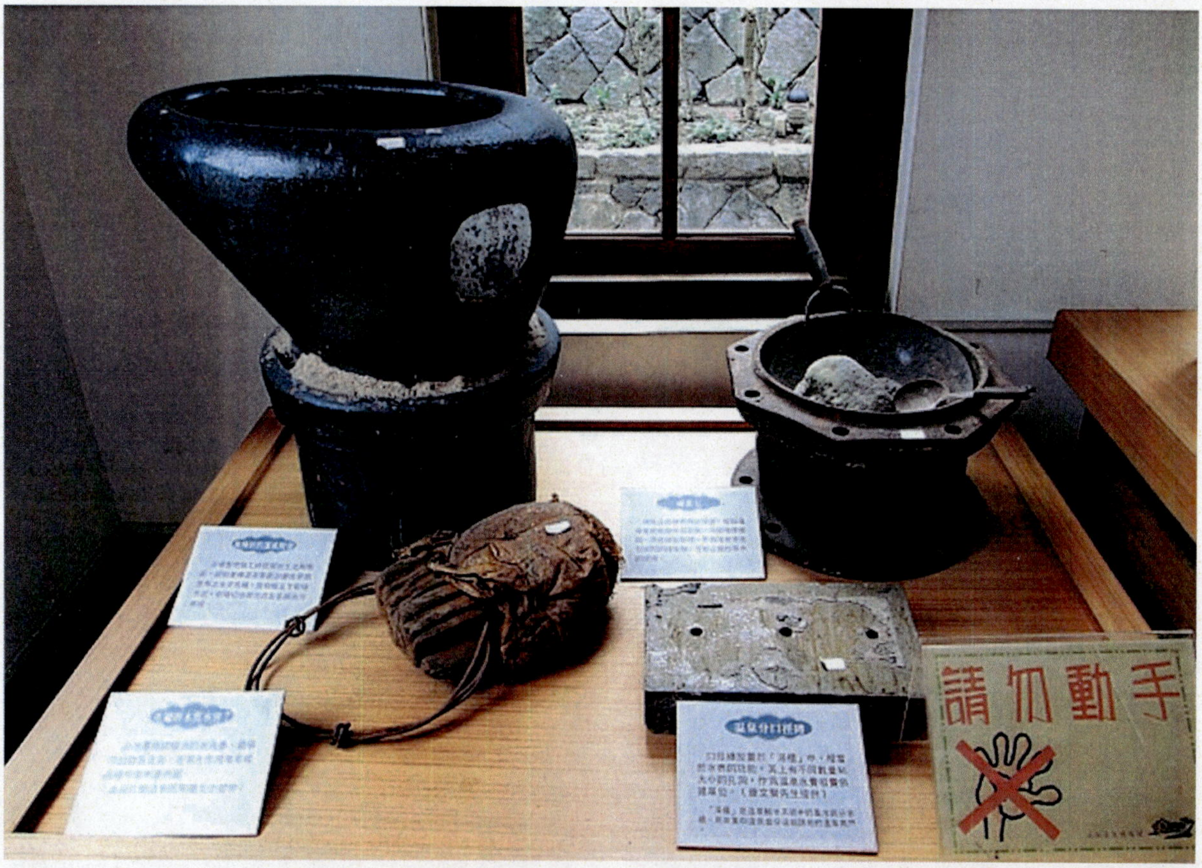

Fig. 9.6 Curiosities on display in the Beitou Hot Spring Museum (Taiwan). From a wooden bath plug (front left) to strangely shaped ceramic products and other paraphernalia these items were discovered during the renovation of the original bathhouse

Fig. 9.7 At the turn of the last century the upstairs verandah of the Beitou Bathhouse was an inviting place where people would gather. Now a museum, the same space is devoid of any furniture

Fig. 9.8 The Thermal Hotel Gellért in the Hungarian capital Budapest is a well-known hot spring resort, with early records about the healing hot springs dating back to the fifteenth century AD. The current Gellért Baths were established in 1918 with indoor and outdoor pools

the quest for education related to natural environments in general. To assist with visitor expectations at hot spring tourist destinations, visitor information centres or interpretive centres (where available) usually provide interesting facts and figures about the local hydrothermal phenomena as well as other unusual landscape features and their geological background. The Hot Spring Museum in Beitou (Taiwan) is a great place to learn about the complete hot spring history of the Beitou Thermal Valley (Figs. 5.7, 5.8; 9.6, 9.7). On display are not only exhibits relating to the cultural and geophysical aspects, but also many references to the health applications and recreational purposes for which the hot springs in this area are still used.

Expectations and customer demands are the foundations on which the hot spring tourism industry has continued to build a supply of facilities for either rest and recreation, or 'time-out' settings to cope with mounting stress levels (Erfurt-Cooper 2009). The main appeal of established hot spring destinations is the promise of relaxation, together with a feeling of rejuvenation and an improved sense of health and wellbeing after repeated immersion in mineral-enriched hot springs (Fig. 9.8). Another important factor influencing destination preferences of individuals or groups is the general independence of hot springs from seasons and climate, which means that they can be visited anytime throughout the year.

Over time the trend of relaxing in a natural hot spring environment has changed from the original concept of using hot springs for health purposes with a main focus on rehabilitation and cure (Strauss-Blasche et al. 2002) to include the prevention of disease and the maintenance of good health, with high expectations for health improvements, even without any specific medical conditions (Andrijasevic and Bartolucci 2004; Mueller and Lanz Kaufmann 2001). This sustained popularity and attractiveness of hot springs for health and recreation supports a tourism industry of worldwide significance.

To shed some light on the consumer demand for hot spring destinations worldwide requires a look at the global health and wellness spa tourism industry as a whole. In the year 2017 the hot spring tourism sector alone contributed USD 56.2 billion to the overall spa industry (Table 9.2). As

Table 9.2 The total revenue from 2015 to 2017 increased from US$51.04 billion to US$56.16 billion. The estimated figures include revenues earned from bathing and swimming, spa/wellness services and treatments, as well as recreational activities and other services related to hot spring tourism

Country/Region	Thermal and mineral spring establishments		Economic revenue
	2015	2017	in US$
Asia Pacific	20,146	25,916	$31.6 billion revenue
Europe	5,613	5,967	$21.7 billion revenue
Latin America–Caribbean	1,148	1,406	$1.6 billion revenue
Middle East–North Africa	324	416	$0.4 billion revenue
North America	237	302	$0.7 billion revenue
Sub-Saharan Africa	39	50	$0.08 billion revenue
Total			$56.16 billion (2017)

Source Global Wellness Institute (2018)

Table 9.3 The top twenty 'hot and mineral spring' destination countries represented 92% of the global hot spring tourism market in 2017, based on figures compiled by the Global Wellness Institute

Country	Hot spring facilities	Revenue in US$ millions
Japan	20,972	$12,795.5
China	3,900	$17,501.1
Germany	1,265	$7,151.9
Russia	838	$3,667.2
Italy	768	$1,718.3
Hungary	546	$798.2
Turkey	295	$667.3
Spain	293	$683.1
United States	280	$631.0
Taiwan	251	$434.5
Brazil	207	$718.0
Poland	203	$674.5
Austria	181	$961.1
France	175	$627.7
Iceland	154	$432.6
Czech Republic	131	$627.8
Slovakia	116	$439.8
South Korea	108	$336.1
Slovenia	86	$451.0
Switzerland	73	$479.5

Source Global Wellness Institute (2018)

per definition by the Global Wellness Institute (GWI) the hot spring tourism industry consists of *revenue-earning business establishments associated with the wellness, recreational, and therapeutic uses of waters with special properties*. Included in this definition are not just thermal springs but also cold springs containing mineral elements. Therefore, the figures may not exclusively refer to hot springs per se but take into account some facilities that also use artificially heated cold spring water.

Countries with a longstanding tradition in the use of hot springs for health, rehabilitation and recreational purposes are responding to the growing demand for natural hot spring destinations and are either investing in upgrading existing hot spring resorts and clinics or are establishing new facilities, subsidised and/or promoted by their governments, e.g. Brazil, European countries, China, Japan (Erfurt 2012). According to the GWI, several countries with undeveloped hot spring resources have started to establish thermal bathing

Fig. 9.9 The Caldeira do Esguicho is one of the many boiling springs of the Caldeira das Furnas and is used for cooking corncobs, which are sold to visitors at a nearby kiosk in Furnas, São Miguel (Azores)

facilities to advance their existing tourism industry sectors (e.g. Africa, India and Cambodia) and to attract more international tourists (Global Wellness Institute 2018). Table 9.3 presents a global overview of the distribution of the top twenty 'hot and mineral spring' destination countries.

9.1.3 Viewing of Extreme Hot Spring Manifestations

Reasons to visit extreme hot spring environments are usually determined by the individual traveller, their specific personal interests but also by their level of fitness. While hydrothermal attractions are frequently included in general sightseeing, due to an increasing interest in the diversity of hot springs, tourism operators have specialised in tours that focus on visiting active hydrothermal regions to cater for these special interest groups, e.g. visits to countries including Iceland, Italy, Japan, New Zealand, China, the Americas or the Azores (Fig. 9.9). One of the reasons for visiting active environments includes the desire to experience (almost) untouched natural areas with tourists prepared to overcome various inconveniences when participating in excursions for example to the Valley of Geysers on the Kamchatka Peninsula (far East of Russia) (Erfurt 2010a).

The visual excitement of erupting geysers and steaming fumaroles has remained a major highlight for tourists travelling in active hydrothermal areas for centuries. Prior to the development of digital cameras and the current smart phone technology, visitors of natural areas had only one option to memorise their observations—they had to draw a picture (Fig. 9.10—left image). Natural phenomena such as geysers are often located in an area that tourists can visit on their own, but they are also frequently a part of organised tours. For many people this is a convenient way of participating in activities they do not feel confident to undertake on their own.

A number of examples of famous geysers and their approximate maximum spouting height are listed in Fig. 9.11. While geysers are distributed all over the world the majority

Fig. 9.10 Hydrothermal tourism in Iceland. The left image is a drawing of the geyser Strokkur erupting c.185 years ago, while the picture on the right shows Strokkur in winter 2019. *Left image source* M. A. Mayer, 1835–36; *Right photo credit* Rita Erfurt (2019)

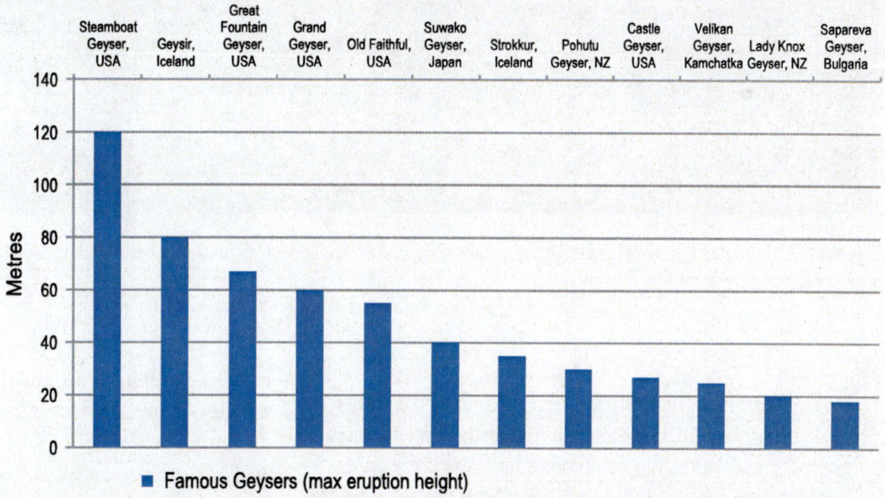

Fig. 9.11 The eruption height of geysers can vary between a few centimetres and up to over one hundred metres. The higher the eruption column and the more frequently the eruptions occur, the more popular is the geyser. *Source* Various

of the larger geysers is located in the Yellowstone National Park.

Some of the larger well-known active hydrothermal tourist destinations are listed in Table 9.4, although there are many smaller areas in other countries. Also, numerous geyser fields in southeast Asia were discovered during explorations for oil and gas but were not considered as significant, and many of them were destroyed or ceased their activity when geothermal power stations started operating in the area.

Table 9.4 The list below shows hydrothermal areas that are frequently referred to as 'hells' because of their infernal activity, which is often combined with the smell of sulphur

Hydrothermal tourist areas	
Azores—Caldeiras das Furnas, São Miguel **Chile**—El Tatio Geyser Fields **Iceland**—Geyser Fields (Haukadalur, others) **Italy**—Phlegraean Fields (Campi Flegrei) **Japan**—Beppu Jigoku (Hells) Group, Kyushu **Japan**—Jigokudani (Hell Valley) Noboribetsu, Hokkaido	**New Zealand**—Thermal Parks (Whakarewarewa, others) **Russia**—Valley of Geysers—Kamchatka **Taiwan**—Beitou Thermal Valley (Hell Valley) **USA**—Lassen Volcanic National Park–California **USA**—Yellowstone National Park–Wyoming

Fig. 9.12 The city of Beppu (Kyushu, Japan) is not only known for several thousand hot springs but also for constant steam emissions, which are more visible on overcast and humid days

Note Religious institutions are said to have transported the element *sulphur* all over Europe in centuries gone by and burnt it to demonstrate what they imagined as the putrid stench of hell (Sæmundsson and Gunnlaugsson 2002).

In the city of Beppu in Japan (Fig. 9.12) ten of these hydrothermal hells (Table 9.5) are major tourist attractions, where people can observe geysers, boiling crater lakes and bubbling mudpots in small geoparks, which are located in the city centre, shrouded in steam emissions (Appendix 9.1).

Beppu is a sister city of Rotorua in New Zealand, where 'hellish' tourist parks are equally popular with some of the major tourist attractions located on New Zealand's north island. The area around Rotorua hosts several hydrothermal sites, which have been developed into attractive parks (Table 9.6), although they are much larger than the Jigoku in Beppu. Also, apart from the Whakarewarewa Thermal

Table 9.5 Jigoku No. 1–7 are located in the same area off Beppu's Route 500 within walking distance of each other. No. 8 is located further down along Route 500 and No 9 and 10 are in a different area along Route 218. For some reason No 6 is rarely mentioned and No 8 is always left out of the general tourist information

Beppu City—Jigoku Meguri and the Ten Hells	
1. Oniishi Bouzu Jigoku Shaved Monk's Head Hell	Several mud pots bubbling quietly creating ever changing forms A foot spa is available for visitors (Fig. 7.4)
2. Umi Jigoku Sea Hell	Boiling crater lake (Appendix 9.1) with a permanent geyser (Fig. 2.7), surrounded by steam vents and a red pond with a high-temperature fumarole (Fig. 3.7). A foot spa is available for visitors and a geothermally heated greenhouse
3. Yama Jigoku Mountain Hell	Steaming fumaroles (Figs. 3.1 and 10.5; Appendix 9.1, Image 4) and hot springs combined with a small zoo where different animal groups are housed in separate enclosures with their own hot spring fed pools
4. Kamado Jigoku Oven or Cauldron Hell	Hot ponds, fumaroles, mud pots, steam features for inhalation (Fig. 8.11), foot spas for relaxation (Appendix 8.2), hot steam cooked eggs and dumplings (Figs. 3.8 and 6.19)
5. Oniyama Jigoku Devil's Mountain Hell	An impressive boiling pool (Appendix 9.1, Image 3) and warm water basins full of alligators
6. Kinryu Jigoku Golden Dragon Hell	Boiling pond (Appendix 9.1) surrounded by fumaroles; steam piped into a dragon's head to create the impression of a fire breathing hellish creature
7. Shiraike Jigoku White Pond Hell	A small lake fed by hot springs surrounded by typical Japanese gardens—the lake changes colour during the year (Appendix 9.1, Image 5)
8. Bouzo Jigoku Monks Hell	Boiling hot spring pond (Appendix 9.1, Image 2) and many small erupting mud pots (Fig. 2.17), located at the junction of Route 500/Route 11—not mentioned in tourism advertising or online although this is one of the nicer jigoku
9. Chinoike Jigoku Blood Pond Hell	Also known as blood pond, which refers to the red colour of the water (Appendix 9.1, Image 7) as well as to a legend related to this hot spring. A foot spa is available for visitors
10. Tatsumaki Jigoku Spouting Geyser Hell	The main attraction is a geyser erupting at regular intervals spouting water at 150 °C (Appendix 9.1, Image 8)

Table 9.6 Hydrothermal parks between Rotorua and Taupo on New Zealand's north island and their unique selling points

Rotorua and surrounds—Hydrothermal parks	
1. Orakei Korako Geothermal Park and Cave The Hidden Valley	Around 23 active geysers, boiling hot springs, bubbling mud pools, silica terraces covered in microbial mats—thermal Ruatapu Cave Located between Rotorua and Taupo
2. Whakarewarewa Thermal Village (Te Puia) Geothermal Valley	Geysers, boiling hot pools, steam vents, mud pools and Māori cultural experiences in 'The Thermal Village', the home of the Tūhourangi Ngāti Wāhiao people
3. Tokaanu Thermal Pools Thermal Walk	Hot mineral pools, bubbling mud and steam vents—thermal pools with healing waters
4. Waimangu Volcanic Rift Valley Thermal Valley	The world's youngest geothermal system—spectacular geothermal wonders—Inferno Crater—Frying Pan Lake
5. Wairakei Terraces and Thermal Health Spa Craters of the Moon	Geothermal walkway—silica terraces, geysers and hot springs—Wairakei Tourist Park
6. Kuirau Park	Hydrothermal activity close to the city centre of Rotorua—walking tracks—fumaroles, mud pools, boiling ponds—foot spas
7. Hells Gate and Wai Ora Spa Thermal Park Geothermal Reserve	Māori culture, hot springs, fumaroles, mud pools, steaming cliffs, cooking pools—guided tours—spa and wellness centre
8. Wai-O-Tapu Thermal Wonderland	Visitor centre—silica terraces—boiling mud pools—Lady Knox geyser—described as 'out of this world' and a 'Wonderland of stunning geothermal activity'

Village and Kuirau Park, they are located in the surrounding areas along the Thermal Explorer Highway between Rotorua and Taupo, not within the city limits of Rotorua.

Once inside the hydrothermal parks, safe walkways and barriers are generally provided for protection with park management committed to ensuring the safety of all visitors. However, it is equally expected that visitors take responsibility for their actions by following the marked trails and heeding the warning signs.

9.2 What Could Possibly Go Wrong?

The answer to this question is easy: nothing or everything. Again, depending on the hot spring area there may be no risk involved at all, while on the other hand the risk can be considerable. Apart from geophysical hazards, tourists can significantly increase their personal risk factor by trying to get as close to hydrothermal activity as they can; often without being aware of the possible consequences for their own health and safety. Such environments are known to harbour potential hazards based on their dynamics and/or their proximity to an active volcano. This means that exploring these areas comes with a certain risk, which all visitors must take into consideration. Some of the more spectacular features should only be observed from a safe distance and should not be approached due to their unpredictable nature.

Many hazardous situations are not anticipated and can be at best an unpleasant, and at worst a dangerous surprise with a fatal outcome. When considering potential hazards in active hydrothermal environments it becomes clear that it is essential to plan ahead and obtain as much information as possible beforehand. Unexpected hydrothermal eruptions, while rare, can happen in every active geothermal field. In extreme circumstances, where hydrothermal features are located near volcanoes or inside a volcanic crater, there is a real potential for additional risk factors. Table 9.7 provides a brief overview of the most common hazards in active hydrothermal environments.

Discovering a natural spring after a long walk through the wilderness can be very inviting but can also harbour certain dangers. Without knowing exactly the source of the hot water and whether the water temperature remains constant, the risk must be calculated very carefully. For example, hot springs submerged in lakes or rivers can change their location or their temperature without warning, or they can erupt like a geyser and cause thermal burns, which can be life threatening (Example 9.1; Fig. 9.13).

Example 9.1 Hot Creek Gorge, California

The beautiful blue pools and impressive boiling fountains along Hot Creek in east-central California have provided enjoyment to generations of visitors, but they have also been the cause of injury or death to some who disregarded warnings and fences. The springs and geysers in the stream bed and along its banks change location, temperature, and flow rates frequently and unpredictably (Fig. 9.13). The hot springs and geysers of Hot Creek are visible signs of dynamic geophysical processes in this volcanic region, where underground heat drives thermal spring activity.

Source Directly quoted from USGS Fact Sheet 2007–3045 2007.

Apart from the risk of thermal burns from boiling hot springs, active fumaroles can release extremely hot gas or steam, which can also lead to severe burns as well as respiratory problems. Even though geysers, steam vents and mud ponds offer great photo opportunities, they remain a hydrothermal hazard with the potential to cause serious injuries. In addition to hazards associated with hydrothermal areas, especially in isolated regions, there are a number of other factors such as unforeseen changes in weather conditions resulting in hostile temperatures caused by rain or snow. Lack of shelter is another problem in many areas, as is the lack of vehicle access in case of a medical emergency (Erfurt-Cooper 2018).

Conditioned by nearly unlimited digital access in their daily lives, people hiking in remote areas increasingly rely on their 'smart' phones to enable them to communicate in case of an emergency. While in many countries phone

Table 9.7 Key hazards related to hydrothermal activity

Hazards in hydrothermal environments
Increased hydrothermal steam discharge (fumaroles and geysers)
Toxic fumes and gas emissions (e.g. H_2S, CO_2, SO_2)
Unexpected hydrothermal or volcanic eruptions in high-temperature fields
Unexpected seismic activity (earthquakes, tremors)
Unexpected change of water temperature
Thermal burns as submerged hot springs change location or direction
Sudden change of flow rate of hot spring or strength of underwater currents
Thermal burns from extreme hot springs (geysers)
Health hazards from thermophilic micro-bacteria (N. fowleri)
Unstable ground—thin crusts covering underlying hydrothermal systems
Extremely radioactive hot springs (Paralana Hot Springs, Australia)

Fig. 9.13 Hot Creek is located in the Long Valley Caldera in California, USA. *Source* USGS Fact Sheet 2007–3045 (Public Domain)

coverage now includes isolated regions, this is not the case in every wilderness area. But these are exactly the conditions where communication barriers can make the difference between life and death. In fact, lack of communication is a problem that can affect tourists in every remote and undeveloped area, not just around hot springs.

9.3 Hazards and Risks

The definition of a hydrothermal hazard is the risk to life and property from hydrothermal eruptions, gas emissions and toxic fumes, extreme hot springs, and other hydrothermal features such as geysers and steam vents (fumaroles). In more general terms a hazard is the probability of a natural event happening and developing into a potential source of exposure to danger or vulnerability (Example 9.2, 9.3). Being vulnerable means that humans and/or physical infrastructure are susceptible to impacts caused by natural hazards with the inability to withstand them. Exposure to danger in active hydrothermal areas therefore takes into consideration the prospect of injury, structural damage and potentially the loss of life (Aspinall and Blong 2015; Barclay et al. 2015; Jolly and De La Cruz, 2015; McGuire et al. 2009; Newhall 2014; UNISIDR 2016). Individual risk perception and risk assessment are contributing factors in both risk creation and risk prevention. Every person's decision-making process is influenced in different ways how they consider their individual responsibility, their options for protection and their preparedness for visiting hazardous environments (Erfurt-Cooper 2018; Paton et al. 2001).

Example 9.2 Rotorua, New Zealand

A humble mud pool can unexpectedly generate a sizeable hydrothermal eruption, which happened in 2001 in *Kuirau Park* in Rotorua. The steam vent feeding into the mud pool erupted into clouds of vapour, hot mud and large ballistic rocks. This hydrothermal event caused the closure of adjacent roads by covering everything in mud and debris and destroyed the surrounding vegetation. Other hydrothermal eruptions have happened in Kuirau Park in 2003 and 2006, and more recently in 2016, fortunately without any accidents or injuries.

Source Kenny and Bathgate (2016)

Hydrothermal areas are often part of volcanic regions where seismic activity such as ground tremors or earthquakes should be expected and considered in personal risk assessments before embarking on a visit to a potentially dangerous region. Mud eruptions, flashing steam from geysers and toxic gas emissions can occur without clear warning signs (Erfurt-Cooper 2010b). Slowly degassing fumarolic vents can suddenly increase their steam discharge and scalding steam plumes can change direction with the potential to cause thermal burns as well as other health problems for observers standing too close (Erfurt-Cooper 2018; Example 9.3).

Example 9.3 Whakaari—White Island, New Zealand

The tragic events that unfolded on 9 December 2019 on the volcanic island Whakaari (White Island) clearly demonstrated the devastating effects an unexpected hydrothermal eruption can have on tour groups visiting at the time. The volcanic alert level (VAL 2) did not deter daily visits by boat or helicopter, weather permitting. But as a hydrothermal hazard this volcanic island ticked all the boxes: a location in the middle of the ocean 50 kms from the mainland with no time to seek shelter (in fact, there are no shelters near the crater), while dozens of tourists were suddenly confronted with the discharge of volcanic gas and steam mixed with hot mud and rocks. The result was catastrophic and 22 people lost their lives, either during the eruption or later from their injuries. The crisis was further compounded because access for rescue teams was delayed due to the perceived risk of further eruptive activity. While this example demonstrates a worst-case scenario, similar risks and hazards can be encountered at other locations as well.

9.3.1 Toxic Hazard—Hydrogen Sulphide

Frequently detected in volcanic and hydrothermal environments, hydrogen sulphide (H_2S) is one of the naturally occurring gases, which can fatally poison a human within moments at high concentrations (Chap. 3; Erfurt-Cooper. 2018; Hansell et al. 2006). H_2S is known by a number of different names mainly in reference to its unpleasant smell frequently compared to rotten eggs, which is a key indicator for the presence of the gas at low level concentrations. Common terms to describe H_2S include sour gas, swamp gas, sewer gas and stink gas. Being 20% heavier than air this gas collects in caves, craters and low lying areas, where it pools at the lowest possible point. If the gas is not dispersed by wind it can quickly reach lethal concentrations (Erfurt-Cooper 2018; USGS 2014; Whittlesey 2014; Williams-Jones and Rymer 2015). As an invisible danger H_2S can also be present at hot spring facilities that are used for health and recreational purposes, although in controlled environments gas emissions are most likely monitored and elevated concentrations are detected. As one of the principal components of the natural sulphur cycle H_2S is dissolved in many different types of hot springs at varying concentrations (Giampaoli et al. 2013) and can find its way into enclosed spaces with poor ventilation such as some hot spring bathhouses. One tragic example of H_2S poisoning is a case from Japan (Example 9.4), although other hot spring areas also have a history of exposure to this particular gas.

Example 9.4 Akita Prefecture, Japan

In December 2005, a family of four lost their lives at a Japanese hot spring resort in Yuzawa after inhaling hydrogen sulphide. The toxic gas had accumulated in a depression covered by thick layers of snow, which the family's children entered to retrieve their frisbee. Based on the level of toxicity in the hollow the two boys collapsed and would have died instantly, which caused their mother to follow her children, also perishing in the lethal cloud of hydrogen sulphide. The father, who was searching for his family, on discovering their bodies in the snow, attempted to rescue them but was overcome by the gas as well and had to be taken to hospital, where he later passed away. Five years earlier, at the same hot spring resort 21 people suffered eye irritation from hydrogen sulphide caused by the use of piped hot spring water to melt snow near the building.

Source Japan Times (2005).

Shortly after this unfortunate incident the Japanese government requested about 80 hot spring (onsen) facilities to take preventive measures to avoid further accidents caused by toxic gas emissions. To prevent hydrogen sulphide poisoning at hot spring bathhouses, strict guidelines in accordance with the Japanese Hot Spring Act must be followed by all onsen operators. These guidelines determine that the total sulphur contained in the air inside public bathing facilities is not supposed to be more than 20 parts per million (ppm) at a position 10 cm above the hot spring pool surface and 10 ppm at a position 70 cm above the floor of indoor pools. This requires the ongoing monitoring of levels of toxicity with special gas detecting equipment. Other safety measures include a set number of ventilation holes and aerating equipment to be installed to disperse any gas accumulations. Furthermore, the position of the hot water taps and the height of the pool edge above the floor are required to comply with specified measurements (Ministry of the Environment Japan 2019). Upholding safety standards in controlled environments therefore does not present unsurmountable obstacles compared to the wide open spaces in hydrothermal areas, where gas emissions can virtually happen anywhere, anytime and at any volume and concentration.

The human detection threshold for hydrogen sulphide (H_2S) depends on the level of build-up in the air. At concentrations below 1ppm H_2S has a strong 'rotten egg' smell, which increases up to a level of approximately 30 ppm. Between 30 and 100 ppm the smell changes to a sickly sweet odour, while above 100 ppm humans generally lose their sense of smell and the gas is no longer detectable. However, prolonged exposure to low concentrations of H_2S can also cause the loss of the sense of smell. It is therefore extremely dangerous and impossible to rely on olfaction to detect the presence of H_2S.

Typical symptoms of H_2S poisoning vary according to exposure time and the level of concentration. At levels between 0 and 10 ppm people may experience irritation of eyes, nose and throat, while moderate concentrations between 10 and 50 ppm can cause dizziness, headaches, nausea and vomiting as well as coughing and difficulty in breathing. At higher concentrations between 50 and 100 ppm people most likely experience severe respiratory tract irritation, eye irritation, shock and convulsions followed by coma and ultimately death. At extreme concentrations of >300 ppm, asphyxiation is certain after a few inhalations. H_2S is not just a toxic gas, it is also dangerous for other less than favourable characteristics, which include causing corrosion, igniting at 270 °C as well as being an explosive gas. This requires continuous monitoring of environments where this gas is likely to exceed permissible levels.

9.3.1.1 Other Dangerous Gases

Other gases common in hydrothermal and volcanic environments include carbon dioxide (CO_2) and sulphur dioxide (SO_2), which all are a threat to human health (Example 9.5). Hazards such as steam and gas emissions therefore pose a danger that needs to be taken seriously and not underestimated.

Example 9.5 Lipari Island, Southern Italy

Lipari, a UNESCO World Heritage listed site and a popular tourist destination, is the largest of the Aeolian islands just north of Sicily. Visitors enjoy the mud baths and the thermal waters along the coast, courtesy to the proximity of two active volcanoes, Stromboli and Vulcano. Another contribution of volcanic processes is the risk of a sudden build-up of carbon dioxide (CO_2), which can accumulate to reach dangerous levels in the absence of sufficient wind to disperse the gas. In the year 2015 a nine-year old child was hospitalised in Lipari, suffering from carbon dioxide poisoning. Research on the neighbouring island Panarea has revealed that emissions of residual gas containing H_2S and CO_2 are not uncommon in the marine sector around the Aeolian Islands and result from the interaction of discharged volcanic gas with seawater. Emissions of gas with a high CO_2 content and submarine hydrothermal phenomena in this area were already known at Roman times.

Source Chiodini et al. (2006), Philipson (2015)

Other than unexpected eruptions of steam vents potentially causing thermal burns from super-heated water, rock fragments and mud, risk factors also include inherent hazards such as ground deformation and instability as well as seismicity, all posing additional hazards which should never be ignored.

9.3.2 Health Risks at Hot Springs—Naegleria Fowleri

While natural hot springs are susceptible to disturbances and contamination caused by human activities, there are other factors that can have a negative effect on humans. With respect to health risks, hot springs can occasionally harbour free-living Amoebae (FLA), which may cause serious health issues, including death. One of these health hazards is the possibility to contract primary amoebic meningoencephalitis (PAM), a rare but usually fatal infection of the brain that is frequently misdiagnosed (Booth et al. 2015; Lopez et al. 2012; Tung et al. 2013; Yoder et al. 2010). PAM, also known as *Naegleriasis,* is caused by the thermophilic amoeba *Naegleria fowleri*, which can exist in warm freshwater bodies including hot springs, swimming pools, lakes, ponds, and streams (Lekkla et al. 2005; Lopez et al. 2012; Yoder et al. 2010). FLA such as *N. fowleri* have been detected at developed and undeveloped hot spring tourist destinations throughout the world (Booth et al. 2015; Erfurt-Cooper 2009; Heggie 2010; Lekkla et al. 2005), although with relatively few reported cases. Out of more than 40 species of Naegleria, only *N. fowleri* is causing a rapidly developing infection of the brain that does not respond to conventional treatment methods and has a mortality rate of around 95% (Abrahams-Sandi et al. 2015; Yoder et al. 2010). Infections occur mainly while swimming and diving in hot springs and streams or pools fed by hot spring water. Warning signs (Fig. 9.14; Appendix 9.3, Images 1, 2) are placed at some locations where it is known that FLA such as *N. fowleri* inhabit the water.

N. fowleri is thermotolerant up to 45 °C, which means it can exist anywhere in the world in untreated warm freshwater. However, natural hot springs are not always used for activities such as swimming and diving, which may reduce the risk of exposure (Yoder et al. 2010) because avoiding head immersion is the main protection against PAM (Examples 9.6 and 9.7). While casual exposure to *N. fowleri* via the skin does not cause infections, the inhalation of water contaminated with the pathogenic amoebae can cause serious problems (Barnett et al. 1996). Infections only occur when water containing *N. fowleri* enters the nasal passages and via ascending the olfactory nerve migrates to the central nervous system in the brain (Abrahams-Sandi et al. 2015; Barnett et al. 1996; Yoder et al. 2010). As a rare disease PAM is difficult to diagnose because the symptoms are similar to bacterial meningitis (Heggie 2010; Lopez et al. 2012). This can lead to delays in the diagnosis and correct

Fig. 9.14 A warning sign at Kerosene Creek (New Zealand) observed several years ago is alerting people to the danger of contracting amoebic meningitis and to keep the head above water

treatment, which is the reason for the high fatality rate. Incubation times for PAM range between five and seven days resulting in the rapid onset of symptoms, which usually start with headaches, fever and stiffening of the neck. As the infection develops further, nausea and vomiting are signs of the deteriorating condition together with fatigue, seizures and coma. Death follows usually within a week (Abrahams-Sandi et al. 2015; Heggie 2010; Yoder et al. 2010).

Example 9.6 French West Indies

On the island of Guadeloupe, a fatal case of PAM occurred, involving a child swimming and diving in a hot spring fed pool a week before symptoms of the disease appeared. As it is common in Guadeloupe to visit the many hot springs around the Soufrière volcano, people regularly come into contact with untreated water. Subsequent sampling of numerous hot springs on the island revealed the presence of thermophilic amoeba such as *N. fowleri*, but only at low concentrations.

Source Moussa et al. 2013.

The majority of reported cases of PAM are associated with aquatic activity prior to infection, frequently involving immersion in hot springs or other warm freshwater bodies (Abrahams-Sandi et al. 2015).

Example 9.7 Costa Rica

In 2014 an 11-year old boy from Florida was hospitalised with a suspected PAM infection after he returned from Costa Rica where he was travelling with his family. His condition was severe and quickly deteriorated, resulting in his death 72 h after he was admitted to hospital and diagnosed with PAM. Water samples taken at the hot spring resort where the family had stayed in Costa Rica revealed the presence of *N. fowleri*.

Source Abrahams-Sandi et al. (2015), Booth et al. (2015)

Cases of infection with *N. fowleri* also include some of the southern **US**, mainly Florida and Texas (Lopez et al. 2012), as well as **Taiwan**, where the first confirmed case of PAM was reported in 2011 after a visit to a hot spring one week before hospitalisation was required. To ascertain the presence of *N. fowleri,* water samples were taken from the hot spring facility. As a consequence, all recreational hot spring pools were drained, cleaned and disinfected to monitor the situation (Tung et al. 2013).

Other reports from countries where *N. fowleri* was detected include **Japan**, where water analyses from hot springs revealed the presence of FLA (Izumiyama et al. 2003), and **Thailand**, where sampling of 69 hot spring sites in 13 provinces found 35.3% of the water samples testing positive for *N. fowleri* (Lekkla et al. 2005). In **New Zealand** a common hot spring safety warning carries the following advice:

> *When swimming in natural hot pools, where the water comes out of the ground, keep your head above water because there is a small risk of contracting an illness called amoebic meningitis. While very rare, this illness is serious.*

Apart from *Naegleria,* various other types of FLA, e.g. *Acanthamoeba* and *Balamuthia,* are a serious cause for health concerns due to their ability to invade the central nervous system. In the Ardebil Province in **Iran**, an area famous for hot spring spas used for health and recreation, such FLA were found in natural hot springs. Water samples taken from springs used predominantly for recreational purposes, were analysed for FLA (e.g. *Acanthamoeba*) with 42.9% of the samples returning positive results (Badirzadeh et al. 2011). Other than causing an infection of the central nervous system, Acanthamoeba can cause medical conditions including skin problems (dermatitis), eye irritations (keratitis), inflammation of lung tissue (pneumonitis) and inflammation of the sinuses (sinusitis) (Lekkla et al. 2005).

But there is more. At a **French** hot spring spa several cases of *legionellosis*, more commonly known as Legionnaires' Disease (LD), occurred over the decade between 1986 and 1997. Although the source of the contamination could not be clearly identified, incidences are frequently linked to hot springs, showers, fountains, cooling towers and/or whirlpool baths (Chaabna et al. 2013; Lin et al. 2017; Molmeret et al. 2001). Legionellosis is a severe form of pneumonia caused by the human pathogen *Legionella pneumophila,* a member of the *Legionellaceae* family that consists of at least 43 different species (Molmeret et al. 2001). Legionellosis outbreaks related to natural hot springs have also been reported from Japan in recent years, which caused Taiwan to carry out a survey at 19 hot spring resorts to investigate a potential presence of *L. pneumophila,* although no reported cases of legionellosis were known at the time in Taiwan (Lin et al. 2017).

9.4 Visitor Safety at Hot Spring Tourist Sites

Hot springs and their surrounding areas are so diverse that it is difficult to include every hazard that can be encountered. To prevent injury or death in active hydrothermal areas all visitors must be made aware of potential problems and adequate warnings must be met with personal responsibility instead of relying on the presence of rescue services. Therefore, with the main focus on active hydrothermal tourist sites, hazard education and developing a healthy risk perception is one of the first steps. The challenge of providing visitor safety depends on many variables, often including the underestimation of how quickly a situation can change, especially when visitors are exposed to additional risks from nearby volcanic activity. It can certainly be an advantage if hot springs are located in protected areas such as national parks with a ranger presence and other ways of monitoring, although there is no guarantee for safety if access is difficult due to remoteness and other adverse factors.

9.4.1 Visitor Behaviour and Risk Perception

A book titled *Death in Yellowstone: Accidents and Foolhardiness in the First National Park* highlights the importance of the right balance between sufficient warnings and individual responsible behaviour by visitors of such environments. The author argues that accidents and injuries are almost every time due to visitors' irresponsible behaviour, thereby endangering themselves and others (Whittlesey 2014). In highly active areas such as Yellowstone the main causes for injuries and fatal accidents include reckless and foolish conduct by park visitors (Table 9.8).

When dealing with stupidity and careless actions there is very little that park management can do to prevent incidents. If people choose to ignore warning signs and announcements to show off daring behaviour they are taking a risk and frequently pay the price. Most high temperature areas have a history of incidents, usually due to the lack of concern by individual tourists for their personal health and safety. In such cases the best risk management concepts cannot prevent human casualties. To demonstrate the point, visitors of hydrothermal areas are quite often curious how hot the thermal springs are and reach down to touch the water. People, who sustain minor burns to their hands this way, are probably too embarrassed to

Table 9.8 Some examples of careless and ill-considered behaviour in hazardous areas

Irresponsible actions in hydrothermal environments	
Being distracted	Leaving designated boardwalks
Careless running	Walking in darkness
Entering off-limit areas	Intoxication
Ignoring warnings	Over-confidence
Losing balance	Showing off

Source Whittlesey (2014)

report this as an injury they caused themselves by giving in to the temptation of checking the temperature of a hot spring. Back in the year 1928 this temptation caused the death of a man from Texas (Example 9.8).

Example 9.8 Midway Geyser Basin, Yellowstone, 1928

On a summer afternoon a group of people approached one of the hot springs 'immediately east of Excelsior Geyser' at Midway Geyser Basin. One of the group members bent over to stick his hand into the water, when he lost his balance and fell into the hot spring and was entirely submerged in the boiling water. His two sons managed to remove him from the water but severely burnt their arms in the process. The unfortunate man was rushed to the nearest hospital, which took over four hours. The victim died before he arrived there, while reportedly remaining conscious for the whole time until he died from his severe thermal burns.

Source Whittlesey 2014

For more authentic reports about deaths and injuries in Yellowstone's hydrothermal areas, Whittlesey's book provides a deep and disturbing insight into visitor behaviour with dire consequences (Example 9.9). In recent years other fatal accidents and injuries caused by thermal burns were reported from Yellowstone, which could have been avoided, if the people involved would have followed the safety instructions and obeyed the warning signs.

Example 9.9 Norris Geyser Basin, Yellowstone

In June 2016 a park visitor slipped and fell to his death in a hot spring near Porkchop Geyser. He and his sister had illegally left the boardwalk and walked a few hundred metres into the Norris Geyser Basin when the accident happened. Park officials later confirmed his death, but were unable to retrieve any remains of the dead man from the hot acidic spring, because nothing could safely be recovered. Rangers only found some of his belongings. The sister, who witnessed the accident and reported it to authorities, was not injured.

Source Arrandale 2018; Helsel 2016; My Yellowstone 2016

9.5 Risk Management

Because the focus of this book is on the geoheritage of hot springs, the topic of risk management is discussed in relation to natural environments and covers only to a lesser degree the relative safety of controlled environments such as hot spring resorts, clinics and spas.

9.5.1 Challenges and Strategies

Reducing the risk of accidents and injuries of visitors is the main focus of risk assessment in active hydrothermal areas and includes identifying the potentially damaging effects of a hazard (Aspinall and Blong 2015; UNISDR 2014); however, due to the potential hazards in active environments, risk management in hydrothermal areas is a demanding task (Fig. 9.15). To reduce the exposure to hazards, visitors need to be aware of the risk and respond to warnings and safety instructions at all times to prevent injury or death, which is only possible with effective risk communication and hazard education as the basis for risk management (Leonard et al. 2008; Paton et al. 2001).

Note The risk of exposure increases with the length of time visitors are spending in hydrothermal areas, which are frequently combined with volcanic activity (Bratton et al. 2013).

Personal risk assessment is a process that depends entirely on an individual's perception of the potential risk for health and safety in hazardous environments. Having identified the risk is a step in the right direction, but does not necessarily stop anybody from disregarding safety advice and common sense, which leads back to visitor behaviour. Even if visitors of hydrothermal areas may be aware of probable dangers, they may find themselves in an inaccessible location, which could be difficult to reach in an emergency. In situations like this it is essential that people can rely on their mobile phones to be able to communicate with rescue services, which requires adequate phone reception in these areas (Erfurt-Cooper 2018).

Many hydrothermal tourist attractions are located close to active volcanoes, where alert level systems developed by volcanologists are in place (Fearnley et al. 2012; Gregg et al. 2015; Jolly and De La Cruz 2015; McNutt 2015; Williams-Jones and Rymer 2015), and which could be adapted to hydrothermal areas. The task of risk management in hydrothermal environments, especially at high temperature fields is challenging to say the least (Table 9.9). In addition to hydrothermal hazards, sudden weather changes in remote regions and perhaps together with difficult terrain can generate adverse conditions, further complicating an emergency situation. While these are different types of hazards, they nevertheless must be expected in a natural setting together with the potential lack of reliable communication services and should therefore be part of any personal risk assessment before entering isolated areas (Erfurt-Cooper 2009, 2010a, 2014, 2018).

To provide satisfactory safety measures, the development of risk management strategies largely depends on the area in question and on any existing infrastructure. In less active

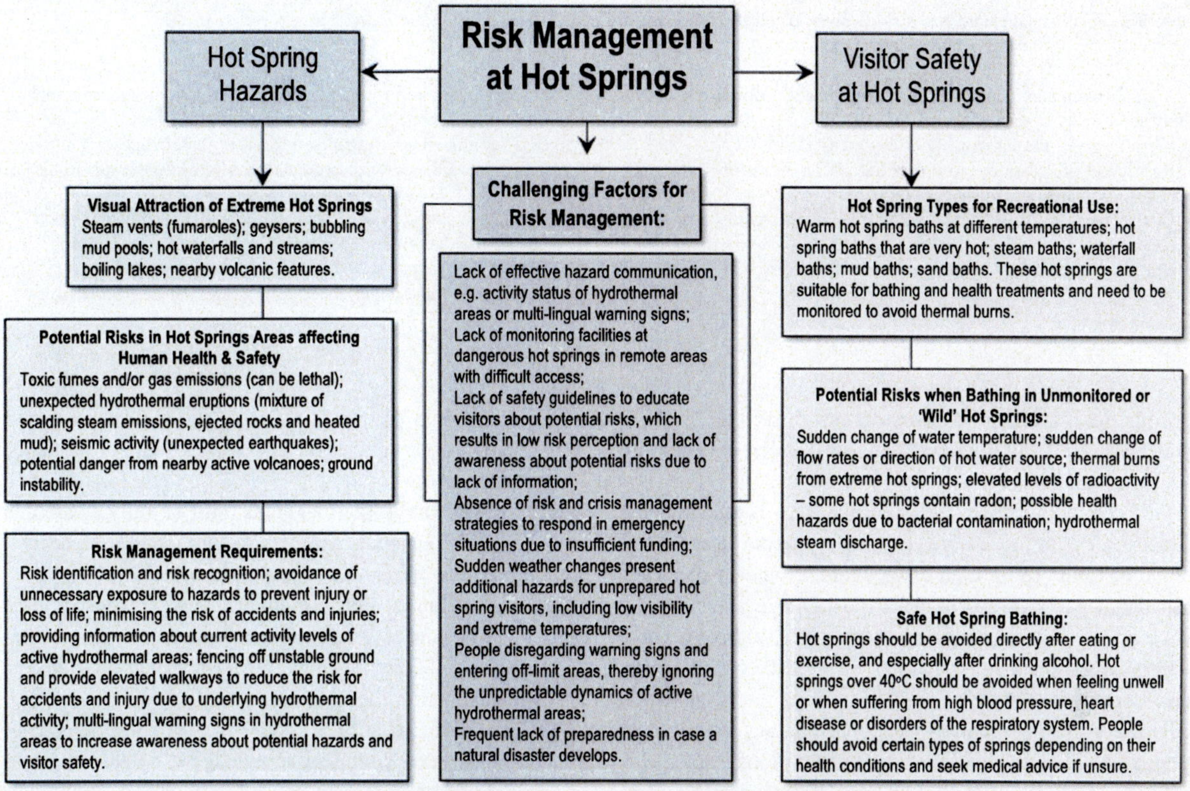

Fig. 9.15 Risk management is essential at natural hot springs, especially in active hydrothermal areas. *Source* Author

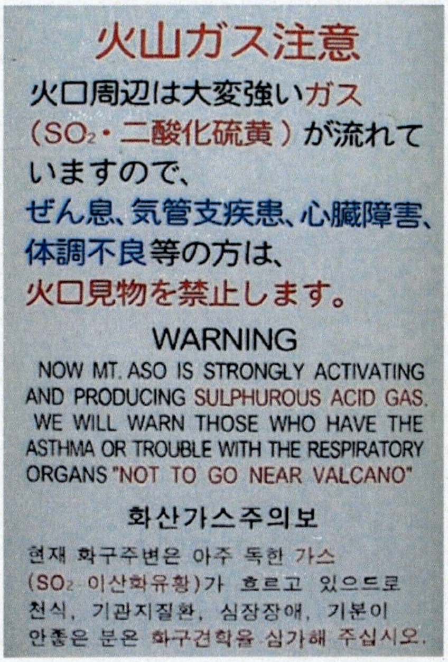

Fig. 9.16 One of the multilingual warning signs at Mount Aso, a popular tourist destination with the ever-present potential of toxic gas emissions. If visitors are at the summit at the time of elevated gas levels, evacuations are swiftly executed, and access roads are closed

areas with lower injury potential, risk management does not require the same actions than in high temperature hydrothermal environments that are also closely connected to active volcanism. In contrast, areas with a high visitor capacity require different additional risk management plans to deal with larger numbers of tourists. While most countries have their own risk management strategies to address complex emergency situations and monitor hazardous areas in case a crisis develops, not all areas are in a position to fund permanent surveillance equipment (Erfurt 2018; Williams-Jones and Rymer 2015). For this reason, it is paramount that visitors follow all safety instructions and are well informed before entering areas they are not familiar with.

9.5.2 The Importance of Safety Recommendations and Guidelines

Being unsuspecting or ignorant of potential risks in remote and/or unstable hydrothermal areas does expose visitors to considerable vulnerability that can be avoided or at least minimised. Taking a risk therefore increases the probability of getting injured in the process, especially in the absence of protective shelters. To effectively reduce the risk factor in

Table 9.9 Risk management can include some or all the following measures depending on the individual hot spring location

Risk management measures	
1. Visitor Information Centres to provide updated information and educational talks at start of visit 2. Warning signs throughout hazardous areas 3. Restriction of access or closure of hazardous areas for visitors in case gas emissions reach toxic levels (Fig. 9.16) 4. Gas sensors—GPS stations—thermal monitoring equipment 5. Emergency services on standby 6. Presence of park rangers and staff trained in First Aid	7. Concrete shelters to protect from fallout in case of hydrothermal eruptions 8. Access for emergency rescue services (vehicle, helicopter) 9. Webcam surveillance of dangerous areas—permanent recording stations 10. Electronic warning systems and public announcements systems 11. Reliable mobile phone reception to enable communication in emergencies 12. Only qualified and experienced tour guides for dangerous areas

hydrothermal areas it is essential to raise the level of awareness about any imminent dangers that can develop into a crisis situation.

Safe conduct in active hydrothermal areas is essential and tourists need to be aware that safety recommendations should be followed for their own benefit. Protected sites such as national parks, geoparks or world heritage areas usually provide their own site specific safety rules, although some visitors may choose not to follow this advice for their personal protection. While up-to-date information is generally provided at visitor centres (where available) or by park rangers, this is not the case at many undeveloped hot spring destinations. Some people may also ignore helpful information, while at the same time overestimating their personal ability to cope with unexpected emergencies. Especially in large national parks like Yellowstone this can lead to dangerous situations for all involved, including the rescue teams. The following revised list (based on earlier publications by the Author) contains safety recommendations specifically for hot spring tourists worldwide.

1. All visitors of active hydrothermal areas should be encouraged to obtain essential information in advance. Updated information for much frequented hot spring locations is usually provided online and/or on-site either by park management, tour operators, geological surveys or local governing bodies.
2. In case of potentially dangerous hydrothermal environments precise information is essential and visitors need to be informed about potential hazards related to the area they are visiting.
3. In the case of hot spring areas phone apps can include real time status updates for tourists. Special hot spring apps can be developed (some may already be available) to provide all essential information for a specific region including emergency advice. Merely stating an alert level or risk level is not sufficient.
4. Social media sites are increasingly used for information updates and can be applied to every active hydrothermal area with updated relevant information. Other social media platforms (SMS, twitter, Instagram) can also be used to update hot spring tourists in active environments.
5. Visitors of hydrothermal areas must be fully aware that it is their personal choice to decide what is an acceptable and what is an unacceptable risk.
6. Tour operators and tour guides, travel agents, tourist organisations and all information centres must communicate clearly any possible risk factors involved at the destination they are promoting.
7. Tour guides must be specially trained for emergencies and must have sufficient knowledge to be able to assess situations of imminent danger in active environments. During an emergency situation tour guides must be able to calculate the timeframe needed to reach the next shelter(s) or a safe escape route out of the danger zone. Tour guides must also be aware that in an emergency they may be equally incapacitated and therefore not able to lead their tour group to safety.
8. Guidelines and safety instructions must be available in ALL major languages consistent with known visitor patterns at any destination.
9. In many regions additional signage is still needed in more than one language as well as images depicting danger for instant visual recognition.
10. Another option are fact sheets with symbolic signage or pictograms, which could be handed out to visitors when entering active hydrothermal areas. They should contain essential information such as emergency phone numbers, a hazard map with colour coded danger zones, clearly marked escape routes and shelter locations.
11. Signposts or guideposts should be colour coded to indicate whether the tourist is in a safe (green) zone or in a dangerous (yellow, orange, red) zone and point in the direction where the nearest shelter is located. They should also be numbered and reflective to be useful as markers after dark, which would assist rescue efforts. Tourists must to be aware of the zone they are in at all times.

12. Emergency shelters must be constructed of reinforced concrete and are required in highly active hydrothermal areas. Shelters must contain signboards/posters with visual interpretation of indicators of imminent danger that can lead to a crisis situation.
13. In particularly dangerous areas webcams/CCTV surveillance should be installed at strategic points as well as at emergency shelters (if any) to monitor the area.
14. Emergency phones, directly linked to rescue services, could be installed at certain distance intervals and in the shelters. Visual explanation of how to use them in a crisis situation is important. Arrows on colour coded guideposts should indicate in which direction the closest phone is located.
15. First aid kits with instructions for their use (especially for treating thermal burns) and clean drinking water should be available at emergency shelters.
16. Communication services (e.g. local telecom companies) need to guarantee mobile phone reception in all remote areas frequented by hot spring tourists. Visitors worldwide rely on their mobile phones and must be able to contact emergency services without delay.
17. Before visiting risk prone areas such as active hydrothermal environments, which are often part of general sightseeing itineraries, people should check with their insurance companies whether they are covered in the event of injury or death. Policy holders should be encouraged to seek detailed information about any potential risk beforehand and whether they require extra cover in the event of an accident or disaster, especially if they incur costly rescue missions.
18. While there is an obligation for authorities to provide up-to-date information, it is up to the individual traveller to follow advice. To encourage this, any rescue service expenses should be recovered from those individuals (and tour operators/guides) whose irresponsible and reckless action causes rescue personnel to risk their lives to save them. This may encourage more responsible behaviour.
19. To minimise the overall risk, only vehicles suitable for access to hydrothermal environments should be permitted for use. They need to contain first aid kits and necessary communication equipment in case of emergency.
20. In extremely dangerous hydrothermal environments (e.g. inside craters of active volcanoes) visitors need to wear protective clothing and safe footwear (closed sturdy shoes), hard hats (helmets) and carry a gas mask that is suitable as protection from toxic fumes.
21. In such environments concrete shelters should be placed in strategic locations to provide the best option to reach them quickly in an emergency (e.g. unexpected hydrothermal eruption).
22. To keep count of all participants of tours in active hydrothermal environments, every group member should be supplied with a GPS tracking bracelet with their details, which allows for them to be located should they go missing, especially in a crisis situation.

It is a prime concern to make hot spring tourism as safe as possible with the proposed recommendations intended as a constructive and encouraging contribution. Ongoing research related to visitor safety in active hydrothermal areas is central to the development of appropriate management strategies in the event of an unexpected crisis unfolding. The above listed recommendations for the improvement of safety in active hydrothermal environments could be incorporated either into current strategic tourism management frameworks or used to establish them in areas without existing or insufficient health and safety management strategies for tourism. One question however remains whether dangerous environments such as active volcanic and hydrothermal areas should be promoted as tourist destinations for the sake of economic benefits and do these benefits outweigh any potential risk (Erfurt-Cooper 2009, 2010b, 2018)? This question can only be answered with safety measures that are so stringent that they do not leave room for any unnecessary risks to be taken in extremely hazardous environments, neither by tour operators nor by individual travellers.

9.5.3 Signage—The Good and the Bad

In hazardous areas adequate signs are necessary to provide visitors with relevant instructions for their safety combined with essential warnings not to leave the designated walkways and raised boardwalks or climb fences while they are observing hydrothermal activity. Part of effective safety communication in hydrothermal environments therefore comes in the form of informative signage. Examination of advisory signboards at hot spring sites over several decades and at countless locations has shown that, while there have been improvements, many signs are still only in one or two languages, despite an ever increasing number of international travellers (Appendices 9.2 and 9.3). It is understood from communication with relevant stakeholders that there is a concern with having too many signs; however, when it comes to necessary safety instructions there cannot be too much advice, which needs to be enforced as well. The question here is, why are people so laissez-faire in active hydrothermal and volcanic environments where they are frequently observed side-stepping protective fencing and completely ignoring advice for their personal safety (Fig. 9.17) and that of their potential rescuers? (Compare Sect. 9.4.1—Visitor Behaviour).

Relevant information should be available in multiple languages to ensure the safety of all tourists in active areas. The colour of warning signs should also be a giveaway—red always signifies alarm and caution. Often red is combined with yellow, definitely not an encouraging colour combination and one, which every person driving a car, is familiar with. In Japan for example, well-known hot spring tourist destinations generally offer safety advice in Japanese, Korean and English (Fig. 9.18), which is mainly determined by research into the origin of foreign visitors. In Hokkaido for example, the proximity to Russia means that information boards can include Russian as well, and areas that receive many visitors from China, use Chinese also. In Latin American countries signage is usually provided in Spanish and English, while Indonesia and China commonly use English translations for important information boards. The Azores Geopark (Portugal) also provides several translations of the most important warning signs (Fig. 9.19).

To increase the effectiveness of warnings, signs are essential in raising hazard awareness of the public (Dengler 2005; Erfurt-Cooper 2018; Leonard et al. 2008). Signs that indicate extreme caution should never be disregarded because they have been placed there for a reason. Ignoring warnings can lead to personal injuries and in extreme cases has caused the loss of lives in the past (Erfurt-Cooper 2009, 2018; Whittlesey 2014).

Quite often signs have been damaged, are defaced with graffiti or are in a neglected condition. Sometimes signs are covered by vegetation or are positioned too low or on an angle, which obstructs their visibility. Therefore, all signage, especially warning signs, must always be positioned where they can be clearly seen. Signs also need to be large enough, even if park management may have concerns that too many larger signs could have a negative impact on the scenery. In areas with natural hazards such as hydrothermal activity this should never be considered as too intrusive or as a visitor deterrent, but as an important way of communicating the need for safe conduct by displaying multilingual warnings together with effective symbols. But the fact remains that if warning signs, safety recommendations and hazards maps that are made available for hazardous areas are ignored, it can result in serious injuries or worse, the unnecessary loss of lives. According to Yellowstone's park officials there have been 22 deaths related to hydrothermal features since 1890 in this national park alone with the majority of fatalities due to disregarding warning signs.

9.5.3.1 Bathing Rules—Don't Drink and Soak

Even in developed hot spring environments signage is essential, although for different purposes as Fig. 9.20 shows. Signs like this are not uncommon at Japanese onsen facilities, where they are frequently placed at the entrance to the male section of public bathhouses to deter visitors who may be sick or have consumed alcohol.

Safe conduct at natural hot springs includes common sense and following safety advice. Hot spring pools are often

Fig. 9.17 The tourists standing on the rim of a large degassing mud pond inside the crater of an active volcano have ignored the barriers and warning signs to get a closer look and to take photos. Apart from the potentially unstable ground there is the additional danger of toxic gas emissions or a sudden eruption of boiling mud

Fig. 9.18 Warning Sign with safety advice in Japanese, Korean and English

Fig. 9.19 Warning sign for visitors of the hot springs in the Caldeira Velha Natural Monument, São Miguel (Azores)

accessible in hydrothermal areas and unfortunately bathing accidents are not uncommon in these environments. Consuming alcohol before or while using hot spring pools is an unsafe practice that can have serious consequences (Erfurt-Cooper 2018). Generally, hot spring spas, resorts or public pools advise their guests on the risk of excessive soaking in hot water, especially when suffering from medical conditions (Chap. 8).

Note When using hot springs avoid alcohol, drugs, and do not overeat prior to a soak.

Fig. 9.20 Infectious diseases are banned, so are drunk people who are bothering other bathers. While the third picture is also self-explanatory, the direct translation refers to people who 'pollute the bath and disturb the public health and morals'

9.6 Conclusion

Hot spring destinations all over the world are visited by millions of people every year, which is evident from visitor statistics provided at the beginning of this chapter. Because of the popularity of natural hot springs in their various stages of development and their diverse natural phenomena, safety is an important issue.

As a significant sector of the health, wellness and recreational tourism industry (total revenue in 2017 of US$56.16 billion) the economic value of hot spring tourism is considerable. Therefore, education plays a major role in raising awareness about safe conduct in hazardous environments to prevent accidents and to assist with effective risk management.

The prevention of accidents and/or injuries can be further supported through a personal perception of the risk, the appropriate assessment and evaluation of all potential risks that people may encounter in hazardous areas, and if required, through the avoidance of the risk altogether.

9.7 Appendices

Appendix 9.1 Tour of the Hells' (Jigoku Meguri), Japan (see Table 9.5)

1, 2—Umi Jigoku or Sea Hell and Bouzo Jigoku (Monks Hell), not to be confused with the Oniishi Bouzu Jigoku (not pictured); 3, 4—Oniyama Jigoku (Devil's Mountain Hell) and Yama Jigoku (Mountain Hell); 5, 6—Shiraike Jigoku (White Pond Hell) and Kinryu Jigoku (Golden Dragon Hell); 7, 8—Chinoike Jigoku (Blood Pond Hell) and Tatsumaki Jigoku (Spouting Geyser Hell)

Appendix 9.2 Warning Signs at Hot Springs

A collection of warning signs in hydrothermal areas with high temperature hot springs and other manifestations (see Table 9.7). Only one sign displays five languages.

Appendix 9.3 Warning Signs at Hot Springs

Hazards in hydrothermal environments can include the risk of contracting infections such as amoebic meningitis, a disease that is often misdiagnosed and has a high fatality rate. Apart from extreme temperatures other hazards may also include elevated emissions of different types of gas common near hot springs, especially in volcanic areas.

References

Abrahams-Sandi E, Retana-Moreira L, Castro-Castillo A, Reyes-Batlle M, Lorenzo-Morales J (2015) Fatal meningoencephalitis in child and isolation of Naegleria fowleri from hot springs in Costa Rica. Emerg Infect Dis 21(2):382–384. https://doi.org/10.3201/eid2102.141576

Altman N (2000) Healing springs: the ultimate guide to taking the waters—From hidden springs to the world's greatest spas. Healing Arts Press, Rochester, Vermont

Andrijasevic M, Bartolucci M (2004) The role of wellness in contemporary tourism. Acta Turistica 16(2):125–142

Arrandale T (2018) Deaths and injuries at Yellowstone's geysers and hot springs. Retrieved from https://www.yellowstonepark.com/things-to-do/cautionary-tale

Aspinall W, Blong R (2015) Volcanic risk assessment. In: Sigurdsson H et al (eds) Encyclopedia of volcanoes, 2nd edn. Elsevier, Academic Press, London, pp 1215–1231

Badirzadeh A, Niyyati M, Babaei Z, Amini H, Badirzadeh H, Rezaeian M (2011) Isolation of free-living amoebae from Sarein Hot Springs in Ardebil Province, Iran. Iran J Parasitol 6(2):1–8

Barclay J, Haynes K, Houghton B, Johnston D (2015) Social processes and volcanic risk reduction. In Sigurdsson H et al (eds) Encyclopedia of volcanoes, 2nd edn. Elsevier, Academic Press, London, pp 1203–1214

Barnett ND, Kaplan AM, Hopkin RJ, Saubolle MA, Rudinsky MF (1996) Primary amoebic meningoencephalitis with Naegleria fowleri: clinical review. Pediatr Neurol 15(3):230–234. https://doi.org/10.1016/S0887-8994(96)00173-7

Booth PJ, Bodager D, Slade TA, Jett S (2015) Primary Amebic meningoencephalitis associated with hot spring exposure during international travel—Seminole County, Florida, July 2014. Morb Mortal Wkly Rep 64(43):1226

Bratton A, Smith B, McKinley J, Lilley K (2013) Expanding the geoconservation toolbox: integrated hazard management at dynamic geoheritage sites. Geoheritage 5(3):173–183. https://doi.org/10.1007/s12371-013-0082-8

Cataldi R, Hodgson SF, Lund JW (1999) Stories from a Heated Earth—our geothermal heritage. Geothermal Resources Council, International Geothermal Association, Sacramento, California

Chaabna Z, Forey F, Reyrolle M, Jarraud S, Atlan D, Fontvieille D, Gilbert C (2013) Molecular diversity and high virulence of Legionella pneumophila strains isolated from biofilms developed within a warm spring of a thermal spa. BMC Microbiol 13:17. https://doi.org/10.1186/1471-2180-13-17

Chiodini G, Caliro S, Caramanna G, Granieri D, Minopoli C, Moretti R, Perotta L, Ventura G (2006) Geochemistry of the submarine gaseous emissions of Panarea (Aeolian Islands, Southern Italy): magmatic vs. hydrothermal origin and implications for volcanic surveillance. Pure Appl Geophys 163(4):759–780. https://doi.org/10.1007/s00024-006-0037-y

Dengler L (2005) The role of education in the national Tsunami hazard mitigation program. Nat Hazards 35(1):141–153. https://doi.org/10.1007/s11069-004-2409-x

Deutscher Heilbäderverband (2020) Jahresbericht 2019 – Oktober 2018 bis December 2019. Deutscher Heilbäderverband e.V. (DHV), Berlin. Retrieved from https://tinyurl.com/y4ztpwtr

Erfurt P (2012). An assessment of the role of natural hot and mineral springs in health, wellness and recreational tourism. Unpublished Doctoral dissertation, School of Business, James Cook University, Cairns, Queensland, Australia

Erfurt R (2019). Figure 9.10—Image of Strokkur erupting. With permission

Erfurt-Cooper P (2009) Health and wellness tourism: spas and hot springs. Channel View Publications, Bristol, UK

Erfurt-Cooper P (2010a). The importance of natural geothermal resources in tourism. In: Proceedings world geothermal congress, Bali, Indonesia, 25–29 Apr 2010

Erfurt-Cooper P (2010b) Volcano and geothermal tourism: sustainable geo-resources for leisure and recreation. Earthscan, London

Erfurt-Cooper P (ed) (2014) Volcanic tourist destinations. Geoheritage, Geoparks and Geotourism Series. Springer, Berlin, Heidelberg

Erfurt-Cooper P (2018) Active hydrothermal features as tourist attractions. In: Fearnley CJ, Bird DK, Haynes K, McGuire WJ, Jolly G (eds) Observing the volcano world: volcano crisis communication. Advances in volcanology series. Springer, Berlin, Heidelberg, pp 85–105. https://doi.org/10.1007/11157_2016_33

Fearnley CJ, McGuire WJ, Davies G, Twigg J (2012) Standardisation of the USGS Volcano Alert Level System (VALS): analysis and ramifications. Bull Volcanol 74(9):2023–2036. https://doi.org/10.1007/s00445-012-0645-6

Giampaoli S, Valeriani F, Gianfranceschi G, Vitali M, Delfini M, Festa MR, Bottari E, Romano Spica V (2013) Hydrogen sulfide in thermal spring waters and its action on bacteria of human origin. Microchem J 108:210–214. https://doi.org/10.1016/j.microc.2012.10.022

Global Wellness Institute (2018) Global wellness economy monitor—October 2018. Global Wellness Institute. Retrieved from https://globalwellnessinstitute.org/press-room/statistics-and-facts/

Gregg CE, Houghton B, Ewert JW (2015) Volcano warning systems. In: Sigurdsson H et al (eds) Encyclopedia of volcanoes, 2nd edn. Elsevier, Academic Press, London, pp 1173–1185

Hansell AL, Horwell CJ, Oppenheimer C (2006) The health hazards of volcanoes and geothermal areas. Occup Environ Med 63(2):149–156. https://doi.org/10.1136/oem.2005.022459

Heggie TW (2010) Swimming with death: Naegleria fowleri infections in recreational waters. Travel Med Infect Dis 8(4):201–206. https://doi.org/10.1016/j.tmaid.2010.06.001

Helsel P (2016). Man confirmed dead after fall in Yellowstone hot spring. Retrieved from https://tinyurl.com/y5uotaon

Hróarsson B, Jónsson SS (1992) Geysers and hot springs in Iceland. Mál og Menning, Reykjavik, Iceland

Izumiyama S, Yagita K, Furushima-Shimogawara R, Asakura T, Karasudani T, Endo T (2003) Occurrence and distribution of *Naegleria* species in thermal waters in Japan. J Eukaryot Microbiol 50(s1):514–515. https://doi.org/10.1111/j.1550-7408.2003.tb00614.x

The Japan Times (2005) Hydrogen sulfide kills trio at spa. Retrieved from https://tinyurl.com/vyjfvys

Jolly G, De La Cruz S (2015) Volcanic crisis management. In: Sigurdsson H et al (eds) Encyclopedia of volcanoes, 2nd edn. Elsevier, Academic Press London, pp 1187–1202.

Kenny K, Bathgate B (2016) Rotorua experiences another hydrothermal eruption. Retrieved from https://tinyurl.com/y9xjl9rb

Lekkla A, Sutthikornchai C, Bovornkitti S, Sukthana Y (2005) Free-living ameba contamination in natural hot springs in Thailand. Southeast Asian J Trop Med Public Health 36(4):5–9

Leonard GS, Johnston DM, Paton D, Christianson A, Becker J, Keys H (2008) Developing effective warning systems: ongoing research at Ruapehu volcano, New Zealand. J Volcanol Geoth Res 172(3–4):199–215. https://doi.org/10.1016/j.jvolgeores.2007.12.008

Lin YE, Lu W, Huang H, Huang W (2017) Environmental survey of legionella pneumophila in hot springs in Taiwan. J Toxicol Environ Health Part A 70(1):84–87. https://doi.org/10.1080/15287390600754987

Lopez C, Budge P, Chen J, Bilyeu S, Mirza A, Custodio H, Irazuzta J, Visvesvara G, Sullivan KJ (2012) Primary amebic meningoencephalitis: a case report and literature review. Pediatr Emerg Care 28(3):272–276. https://doi.org/10.1097/PEC.0b013e3182495589

Mayer MA (1835–36. Illustrations. In: Gaimard P (ed) Voyage en Islande et au Groenland. Paris

McGuire WJ, Solana MC, Kilburn CRJ, Sanderson D (2009) Improving communication during volcanic crises on small, vulnerable islands. J Volcanol Geoth Res 138(1–2):63–75. https://doi.org/10.1016/j.jvolgeores.2009.02.019

McNutt S (2015) Eruption response and mitigation. In: Sigurdsson H et al (eds) Encyclopedia of volcanoes, 2nd edn. Elsevier, Academic Press, London, pp 1069–1070

Ministry of the Environment Japan (2019) Standard modification about the structure of hot spring facilities used as public baths—Notice No. 66 of Ministry of the Environment 1 September 2017. Guidelines for the prevention of hydrogen sulphide poisoning accidents at hot spring facilities. Translated from Japanese. Retrieved from www.env.go.jp/nature/onsen/docs/soguide.pdf.

Molmeret M, Jarraud S, Pierre Morin J, Pernin P, Forey F, Reynolle M, Vandenesch F, Etienne J, Farge P (2001) Different growth rates in amoeba of genotypically related environmental and clinical Legionella pneumophila strains isolated from a thermal spa. Epidemiol Infect 126(2):231–239

Moussa M, De Jonckheere JF, Guerlotte J, Richard V, Bastaraud A, Romana M, Talarmin A (2013) Survey of Naegleria fowleri in geothermal recreational waters of Guadeloupe (French West Indies). PLoS ONE 8(1):e54414. https://doi.org/10.1371/journal.pone.0054414

Mueller H, Lanz Kaufmann E (2001) Wellness tourism: market analysis of a specific health tourism segment and implications for the hotel industry. J Vacation Market 7(1):5–17. https://doi.org/10.1177/135676670100700101

My Yellowstone (2016) Man confirmed dead after fall in Yellowstone hot spring. Retrieved from https://tinyurl.com/yap2m27s

Newhall C (2014) A dilemma for tourists and land managers alike: what risks to take? In: Erfurt-Cooper P (ed) Volcanic tourist destinations. Springer, Berlin, Heidelberg, pp 351–353

Nhu H (2019) The Old Faithful—spectators gathering to watch the geyser at Yellowstone National Park. Retrieved from https://tinyurl.com/5aljqqey

NPS Stats (2020a) Yellowstone National Park. NPS Public Use Statistics Office. Retrieved from https://tinyurl.com/y7cdvuty

NPS Stats (2020b). Hot Springs National Park. NPS Public Use Statistics Office. Retrieved from https://tinyurl.com/y7cdvuty

Paton D, Millar M, Johnston DM (2001) Community resilience to volcanic hazard consequences. Nat Hazards 24(2):157–169. https://doi.org/10.1023/A:1011882106373

Philipson A (2015) Italy warns tourists to keep away from Lipari's toxic volcanoes. Retrieved from https://tinyurl.com/y57uvyof

Sæmundsson K, Gunnlaugsson E (2002) Icelandic rocks and minerals. Mál og menning, Reykjavík, Iceland

Sekine M, Nasermoaddeli A, Wang H, Kanayama H, Kagamimori S (2006) Spa resort use and health-related quality of life, sleep, sickness absence and hospital admission: The Japanese civil servants study. Complement Therapies Med 14:133–143. https://doi.org/10.1016/j.ctim.2005.10.004

Strauss-Blasche G, Ekmekcioglu C, Vacariu G, Melchart H, Fialka-Moser V, Marktl W (2002) Contribution of individual spa therapies in the treatment of chronic pain. Clin J Pain 18(5):302–309

Tung M-C, Hsu B-M, Tao C-W, Lin W-C, Tsai H-F, Ji D-D, Shen S-M, Chen J-S, Shih F-C, Huang Y-L (2013) Identification and significance of *Naegleria fowleri* isolated from the hot spring which related to the first primary amebic meningoencephalitis (PAM) patient in Taiwan. Int J Parasitol 43(9):691–696. https://doi.org/10.1016/j.ijpara.2013.01.012

UNISIDR (2016) Hyogo Framework for Action (HFA)—Building the resilience of nations and communities to disasters. The United Nations Office for Disaster Risk Reduction. Retrieved from www.unisdr.org/we/coordinate/hfa

USGS (2014) Volcanic gases and their effects. Volcanic gases can be harmful to health, vegetation and infrastructure. Volcano hazards program. Retrieved from https://volcanoes.usgs.gov/hazards/gas/

Whittlesey LH (2014) Death in Yellowstone: accidents and foolhardiness in the first National Park, 2nd edn. Amazon Digital Services, Inc., Roberts Rinehard

Williams-Jones G, Rymer H (2015) Hazards of volcanic gases. In: Sigurdsson H et al (eds) Encyclopedia of volcanoes, 2nd edn. Elsevier, Academic Press, pp 985–992

Yoder JS, Eddy BA, Visvesvara GS, Capewell L, Beach MJ (2010) The epidemiology of primary amoebic meningoencephalitis in the USA, 1962–2008. Epidemiol Infect 138(7):968–975. https://doi.org/10.1017/S0950268809991014

Degassing fumarole with thick sulphur incrustations. White Island/Whakaari, New Zealand

10 Glossary of Terms Related to the Geoheritage of Hot Springs

Contents

Hot mud ponds contain a mixture of hot spring water, dissolved rock and hydrogen sulphide, resulting in a very fine clayey fluid of varying viscosity. Boiling mud can reach extreme temperatures and is constantly in motion with ever changing variations of steaming hot bubbles bursting

10.1 Introduction

The glossary compiled for this chapter is intended to assist the reader with a collection of definitions and technical terms related to the subject of natural hot springs and their geoheritage. Because some of the terms may not be familiar to every reader, definitions have been selected to present a quick reference as well as to clarify similar sounding terminology. The following list includes terms not only from Earth science disciplines such as geology, volcanology, mineralogy and hydrogeology, but also from other fields of study including environmental science, conservation and sustainability as well as definitions related to hot spring spa

P. Erfurt, *The Geoheritage of Hot Springs*, Geoheritage, Geoparks and Geotourism,
https://doi.org/10.1007/978-3-030-60463-9_10

medicine. The glossary contains more than 200 terms, which are central to the geoheritage of hot springs and other directly related subject areas. The content was compiled from a variety of relevant literature sources, which are listed at the end of this chapter. For further information, recommended readings together with the literature referenced throughout this book offer a choice of additional sources.

10.2 Glossary of Terms

Acidic Springs
Springs that exist in extreme, often volcanic environments, and can support acidophilic and thermophilic microbes as well as certain viruses and bacteria. The pH level of acidic hot springs used for bathing can be as low as 1.2 (*Tamagawa* Hot Spring, Japan) and 1.4 (*Rehai*, China).

Advection
The horizontal transfer of properties or substances by the flow of a current. In relation to hot springs, advection refers to the process of transporting dissolved and suspended solids along with flowing groundwater. The path and rate of movement are depending on the geological setting and hydraulic gradient.

Alkaline Springs
This type of spring can be found worldwide in areas with volcanic activity such as Yellowstone National Park (USA), Iceland and Japan. Like acidic hot springs, some alkaline springs are also able to host a diversity of thermotolerant microbes, regardless of the hostile conditions caused by extreme pH levels high temperatures.

Aquifer
A geological formation containing a groundwater system that consists of water bearing strata such as permeable rock formations or layers of unconsolidated materials, for example sand or gravel. Aquifers that are porous and permeable and allow the groundwater to move freely are *unconfined aquifers,* from which hot and cold springs can rise. Aquifers overlain by solid, impermeable rock or clay are classed as *confined aquifers*. To extract groundwater from confined aquifers, wells must be drilled down to the water-bearing stratum.

Aquiclude
An impermeable, solid layer of rock or clay confining an aquifer by overlying or underlying the water-bearing rock formation.

Artesian Aquifer
A confined aquifer containing groundwater that is trapped between layers of impermeable rock and/or clay, keeping the water under positive pressure.

Artesian Basin
A large-scale geological formation that contains groundwater systems confined under pressure, similar to smaller aquifers. Depending on depth and pressure, the temperature of the confined water can vary between 25 and 100 °C. Some examples of the largest artesian basins include the Great Artesian Basin (GAB) in Australia (largest worldwide) and the Guarani Aquifer System (GAS) in South America (second largest). Artesian basins are a source of cold, warm and hot springs.

Artesian Bore or Well (Fig. 2.20)
Access to water in an artesian aquifer via a drilled well, in which the water rises under pressure.

Artesian Springs
Free flowing springs, which rise from the ground due to varying degrees of pressure in the underlying artesian basin or aquifer. Not all artesian springs are free flowing, but require bores or wells sunk into the aquifer, which causes the water to rise until reaching a hydrostatic equilibrium. Artesian hot springs are generally non-volcanic springs and can be found worldwide.

Arsenic (As)
A chemical element that occurs naturally in the Earth's crust, in various minerals, e.g. sulphur, as well as in certain types of hot springs.

Bacterium, Bacteria
Bacteria are unicellular microorganisms and agents of decay, fermentation and mineralisation. While bacteria have important functions in nature and in the biosphere, they can cause serious infections in humans and potentially result in death. A bacterial agent is a live pathogenic organism that occurs in a large variety of habitats with some resilient bacteria surviving at extreme temperatures and at great depth within the Earth.

Balneae
From Greek *balaneion* - Roman bath, also known as *thermae* from the Greek word *thermos* for 'hot'.

Balneology
From the Latin word *balneum*. Balneology is the scientific study of the therapeutic use of thermal baths and the curative benefits derived from naturally occurring mineral-rich hot and cold springs.

Balneotherapy
A therapy based on the recognition that natural hot spring water, rich in naturally occurring minerals and trace elements, has medicinal benefits and is effective in the

treatment and rehabilitation of a wide range of medical conditions.

See Chap. 8.

Balneum
Latin for place of bathing, bath.

Bathhouse
A building that provides bathing facilities for public or communal use. Many countries have their own traditional bathhouse culture, e.g. Roman thermae, Middle Eastern hammams, Japanese onsen. Not all bathhouses have access to hot springs, although their location is frequently chosen based on an available source of natural hot water nearby.

Bicarbonate Springs
Carbonic water containing carbon dioxide (CO_2), nitrogen (N) and oxygen (O) at varying amounts that can be used internally (drinking cure) as well as externally (bathing cure). Carbonic springs occur worldwide, many of them in volcanic regions, while others rise from artesian groundwater basins.

Black Smoker (Figs. 2.19 and 11.16)
Black smokers are deep-sea vents with chimney-like structures and are found in geologically active regions on the sea floor, especially at mid-ocean ridges and in the abyssal (4000–6000 m BSL) and hadal (>6000 m BSL) zones. Black smokers discharge continuous jets of particle-laden superheated water (up to 400 °C) containing high levels of suspended matter of sulphur-bearing minerals including black iron and sulphur, which combine to become iron monosulphide precipitates. The vents or chimneys are created by the deposition of solidified iron monosulphide. Black and white smokers can occur in close vicinity; however, their temperatures and their pH levels are different.

See Chap. 2.

See Hydrothermal Vent; Compare White Smoker.

Blue-Green Algae (Cyanobacteria) (Fig. 10.1)
Blue-green algae are unicellular organisms and a bacteria group with the ability to use photosynthesis, a form of microbial activity, to produce oxygen in warm and hot springs. Cyanobacteria thrive in thermal waters close to hot springs and geysers, where they colonise runoff channels and develop into microbial communities, potentially influencing the deposition of minerals.

See Chap. 3.

Boiling Lakes, Ponds
Volcanic craters, lakes or ponds which are fed by submerged hot springs that are emitting large volumes of extremely hot water and/or steam.

Boiling Mud Pools, Mud Pots, Mud Volcanoes
The boiling and bubbling mud is the result of dissolved rock, water and hydrogen sulphide, which reacts with the water and forms sulphuric acid, slowly dissolving the surrounding rock into a hot mud of fine clay and silica particles.

Boiling Springs
A spring emitting water at high temperatures, causing the water to boil. For example, hot springs in Yellowstone, which discharge at extreme temperatures from the subsurface, are causing lakes and rivers to boil and pose a potential hazard to visitors.

Brine Springs
Brine springs or saline springs contain extremely high concentrations of sodium chloride (NaCl). Also known as chloride springs, they occur at varying temperatures depending on their origin. Brine springs can be either of artesian (including seawater) or volcanic origin with a wide temperature range. Cold brine springs are generally heated before being used for recreational bathing, balneotherapy and rehabilitative treatments.

See also Saline Springs.

Calcareous Sinter
See Tufa, Travertine.

Calcic Springs
A natural hot spring that contains calcium hydroxide ($Ca(OH)_2$) and which frequently results in the creation of travertine or limestone terraces, e.g. Pamukkale in Turkey or Mammoth Hot Springs in Yellowstone, USA.

Calcium (Ca)
A mineral that plays an essential role in human body functions, e.g. liver, muscles, and heart. Natural hot springs containing calcium are therefore considered to be beneficial for human health and wellbeing.

Carbon Dioxide (CO_2)
A heavy, colourless, odourless, naturally occurring gas that dissolves in water to form carbonic acid. Commonly known as 'greenhouse gas'.

Carbonated Springs
A spring that contains carbon dioxide gas and is common in active volcanic areas.

Chalybeate Springs (Fig. 10.2)
A term used for natural springs containing salts of iron. Due to the high concentration of iron compounds, chalybeate spring water or ferruginous water has a reputation for

Fig. 10.1 Algae growth is encouraged by warm water runoff in drainage channels connected to a hot spring system (Landmannalaugar Iceland)

therapeutic value and is considered an important natural resource for health and wellness spa treatments.

Chloride Springs
Occurring worldwide, these springs are rich in sodium chloride that originates from subterranean rock salt deposits. Chloride springs are used internally (drinking, inhaling of vapour) as well as externally (bathing).

See also Saline or Brine Springs.

Classification of Hot Springs
Hot springs can be categorised according to their mineral content, their temperature, their magnitude of discharge from the source (flow rate) or by their pH level (acidic, neutral, alkaline). Further categories can be based on their location and geologic setting, e.g. rock type and structure and embedded mineral deposits. Another classification can be based on their type of usage and their medicinal benefits.

See Chap. 2.

Conduction
The process by which heat is transferred or radiated from a solid source (e.g. magma chamber, cooling igneous body) directly to a fluid (or another substance), thereby transmitting heat by conduction.

Confined Aquifer
Groundwater confined within permeable rock units, which are commonly located lower than unconfined aquifers, but are overlain by impermeable rock or clay limiting groundwater movement from or to the confined aquifer.

See Aquifer.

Connate Water
Fluids entrapped in the interstices of a sedimentary or extrusive igneous rock at the time of its formation. Connate fluids often have a high mineral content or can consist of highly saline groundwater if formed in marine environments.

Fig. 10.2 Visitors of this chalybeate spring in Glastonbury (UK) are encouraged to sample the water, which is said to have healing powers

Conservation
The strategic protection, maintenance, management, sustainable use and restoration of natural resources and their surrounding environment to preserve them for future generations.

Contamination
The act of polluting or contaminating e.g. the groundwater system by introducing hazardous substances or contaminants. Degradation of natural resources is usually a result of anthropogenic activities.

Convection
The transfer of heat within a fluid by its movement, whereby the heated fluid expands, becomes less dense and rises to the surface while the denser cold fluid moves down driven by gravity. The vertical circulation of subterranean groundwater and/or hydrothermal systems is driven by differences in density and thereby facilitating the transport of heat. The flow path and rate of movement of the heated fluid depend on the geological formation and the hydraulic gradient. Convection also applies to air or molten rock within the Earth.

Crenotherapy
A term that describes healing with natural hot spring water rich in minerals and essential trace elements. Treatments can incorporate internal use of mineral water (drinking, inhalation) and external applications such as hot mineral spring baths and mud therapy.

Cyanobacteria
See Blue-green Algae.

Deep-Well Injection
Deposition of hazardous waste by pumping it deep underground, where it is supposed to be stored in the permeable subterranean rock strata. A risky strategy at best and a

dangerous option with the potential to contaminate groundwater aquifers or artesian basins.

Compare Well Injection.

Discharge Area
An area in which hot or cold subsurface water is discharged at the land surface.

Discharge Rate
The flow rate of hot or cold groundwater from a source measured as volume per unit of time.

See Flow Rate; Specific Capacity.

Dissolved Solids
The term refers to solids suspended and carried as dissolved constituents in solution. In the case of hot springs dissolved solids include minerals and trace elements, e.g. sulphate, chloride, bicarbonate, sodium, or calcium.

See Solubility; Total Dissolved Solids (TDS).

Drawdown
The decrease of the water level in an underground reservoir or a well that is supplied by groundwater through extraction. In technical terms drawdown refers to the vertical distance by which the water elevation is lowered, or a reduction of the pressure head caused by the removal of groundwater.

Drinking Cure (Hydropinic Therapy)
Also referred to in the past as *'taking the waters'*, which involves the consumption of mineral water usually directly from the source of the spring. To benefit from the curative value, the water should be taken under the supervision and guidance of a professional health care provider.

Explosion Crater
Hydrothermal (phreatic) eruption craters are caused by violent eruptions of steam, hot water, mud and rocks.

See Hydrothermal Eruption; See Phreatic Eruption.

Extremophiles
Microorganisms that survive the high temperatures of hydrothermal springs in acidic or alkaline chemical concentrations. Extremophiles are also tolerant to high pressure environments such as subaqueous hydrothermal vents as well as to extreme cold.

See Thermophiles.

Fangotherapy
Fango is a special type of clay that is rich in minerals and sulphur of volcanic origin, which in Europe is derived mainly from hot spring areas in Italy. Also known as pelotherapy, it involves either heated mud packs to cover certain body areas to relieve pain and heal skin conditions, or the complete immersion in warm mud baths.

See also Mud Therapy, Pelotherapy.

Filtration Springs (Seepage Springs)
Natural springs that are formed in depressions or at the bottom of hillsides where groundwater is flowing or seeping through sand and gravel. Springs of this type can appear at higher elevations as well as in depressions and are fed by precipitation. The water of filtration springs is percolating from numerous small openings in porous rock formations, and the discharge of hot or cold spring water can range from small to large volumes. While filtration springs and seepage springs are similar, the water discharge from a seepage spring can be slower than from a filtration spring; however, seepage springs can discharge or 'seep' groundwater from shallow aquifers over larger areas.

Fish Cure (Fig. 10.3)
A small breed of thermotolerant fish (*garra rufa*) thrive in water at temperatures up to 43 °C. Based on their unique feeding habit (they eat dead skin) and their increasing popularity more and more wellness facilities (including airport terminals) are offering basins with 'doctor fish' as a treatment option for skin conditions including psoriasis. Available now at many hot spring spas, with or without hot spring access, they are believed to originate in Kangal, Turkey.

See Chap. 8; See also Hot Spring Fish; Ichthyotherapy.

Fissure Springs
Springs that emerge from a crack in the rock or along a fault line with the water usually highly mineralised and commonly warm or hot.

See also Fracture Springs.

Flow Characteristics
The magnitude of discharge or flow rate of a spring is based on physical characteristics such as consistency, quantity and seasonal conditions.

Flow Rate
Flow velocity or the rate of discharge of water from a hot or cold spring. This rate can be measured in different units of volume and time, e.g. litres or gallons per second or per minute.

See Discharge Rate; Specific Capacity; Velocity.

Fluoride (F^-)
An organic compound also sometimes referred to as *Halide*, fluoride is a naturally occurring compound in both groundwater and surface water. It is commonly found in nature, e.g. as the mineral fluorspar (calcium fluorite CaF_2).

Fig. 10.3 Fish spas are becoming increasingly common, even at international airports like Singapore, where transit passengers pass the time while resting their tired feet. *Photo credit* Sandra Harrison (2020)

Fountains of the Great Deep

This term refers to subterranean water reservoirs mentioned in the Bible in Genesis 7:11 and Genesis 8:2, which ruptured the Earth's crust and flooded the whole planet. From a hydrogeological viewpoint 'great depth' implies that these waters were discharged from the fractured ground under great pressure at supersonic speed and at high to supercritical temperatures, similar to geysers in hydrothermal areas, although on a much larger scale to have unleashed what is known as the biblical flood.

See also Hydroplate Theory.

Fountain of Youth (Fig. 10.4)

The legend of the fountain of youth goes back to several centuries BC with similar versions from different countries across the world. The myth of the fountain of youth refers to a natural spring that is supposed to restore a person's youth and health forever after drinking from its water or bathing in it. Many explorers embarked on the quest for the legendary spring, but it was never found.

Fracking

A process that requires the drilling of wells to reach rock formations, where temperatures are sufficient for the generation of geothermal power (150–200 °C). By setting off a series of small explosive charges, a network of fractures and fissures is created to serve as passages or conduits. To extract heat from underground, cold water is injected into the production wells above a geothermal reservoir, where it circulates through the fissures and returns to the surface as hot water or steam.

Compare Hydraulic Fracturing.

Fracture Springs

Fracture springs discharge from faults or joints in the ground, where the water has followed a natural course of openings or fissures in the underlying non-porous rock formation.

Compare Fissure Springs.

Fig. 10.4 The Fountain of Youth visualised in a painting by Lucas Cranach from 1546. The aged and infirm enter the pool on the left side, and after a miraculous transformation exit unassisted on the right. It remains unclear whether the fountain of youth was envisioned as a hot or cold spring

Fumarole (Fig. 10.5)

A hydrothermal vent of either volcanic or geothermal origin discharging steam or gas, which, depending on location, can include sulphur dioxide (SO_2), carbon dioxide (CO_2) as well as hydrogen sulphide (H_2S). Fumaroles are commonly present in active volcanic environments, but can also be present around dormant and extinct volcanoes. Based on the Latin origin for Sulphur, fumaroles are also known as Solfatara in reference to their frequent emission of sulphurous gases. The discharge of volcanic fumaroles can reach > 200 °C, while the discharge of geothermal fumaroles is commonly around the boiling point of water but can potentially reach higher temperatures.

See Hydrothermal Vent; Solfatara.

Geochemistry

The scientific study of the chemical composition of the Earth as well as the abundance of elements and their distribution in the different spheres of the Earth, e.g. atmosphere, biosphere, hydrosphere and lithosphere.

Geology

The study (Greek *logia*) of the Earth (Greek *geo*).

Geothermal

A term that has its origin in the Greek language with the prefix *geo* referring to Earth. The word geothermal is applied to any system that transfers heat from the Earth's interior to the surface involving water, both as a liquid and as steam.

Geothermal Cooking (Figs. 7.8, 10.9 and 10.22)

See Hot Spring Cooking; See Chap. 11.

Geothermal Energy

Thermal energy contained in the Earth, which can be used directly to supply heat or can be converted to mechanical or electric energy. Useable energy can be derived from geothermal features reaching the surface as steam or hot water or through accessing superheated fluids in subterranean reservoirs heated by processes including compression at depth, radioactive decay of rocks or by a magma chamber.

Geothermal Gradient

A term that refers to the temperature increase with depth in the Earth's interior. While the average temperature increase is 25 °C per kilometre or 75 °F per mile of depth, some

Fig. 10.5 What resembles a campfire is a fumarole degassing in a high-temperature area (Kyushu, Japan)

regions have a significantly higher gradient, which can be utilised for geothermal energy generation.

See Chap. 2.

Geothermal Reservoir
Accessible high temperature fluids in permeable conduits are part of the essential components of geothermal reservoirs. High temperature systems are used in many countries to supply steam turbines with geothermal energy to generate electricity.

Geothermal Resource
A reservoir of naturally occurring dry heat and/or thermal water in the Earth's crust that may be exploited for energy generation.

Geothermal System
A localised geological setting or system where the Earth's internal heat flow is transferred to the surface by circulating steam and/or hot water to be harnessed for use, e.g. power generation. Hot springs, geysers, boiling streams and lakes are surface phenomena of geothermal/hydrothermal systems.

Geyser (Fig. 10.6)
A natural hot spring that discharges columns of water and steam at extreme temperatures from the ground. Most geysers erupt at intervals, while some continuously discharge steam and water and create spectacular sights.

Geyser Basin
An area that contains clusters of springs, fumaroles and geysers, which are fed by the same subterranean source (e.g. Valley of Geysers, Kamchatka; El Tatio Geyser Field, Chile; Norris Geyser Basin, Yellowstone, USA).

Geyser Cone
A mound of siliceous sinter shaped by the precipitation of silica deposits and results in a build-up of incrustations around the discharge area of a geyser.

See Geyserite.

Geyser Pool
Heated water that fills the geyser crater and covers the outlet or geyser pipe.

Fig. 10.6 Because erupting geysers are a popular feature, an artificial vent was created next to the Perlan building on Öskjuhlíð Hill in Reykjavík (Iceland). This geyser is programmed to erupt in regular intervals with water temperatures up to 125 °C

Geyserite (Fig. 10.7)

Also known as siliceous sinter, geyserite is a silica deposit occurring around geysers and hot springs, forming mounds and layers of incrustations.

Gravity Springs

Springs formed by gravity are mostly found on hillsides, where water moves horizontally after encountering confining layers until it finds an outlet from which it can emerge. Unless the water is circulating near a magma body, these springs are usually cold springs.

Groundwater

The term refers to all subsurface water as distinct from surface water. Groundwater is an essential part of the hydrological system and a natural resource that requires strict protection from pollution and contamination. The upper surface of groundwater systems is marked by the water table.

Haematite (Fe_2O_3)

A mineral form of iron oxide and one of the principal ore minerals. Haematite can be found as a common deposit around fumaroles in high-temperature geothermal areas.

Hammam

A public bathhouse of Middle Eastern origin, which is usually gender segregated. If hammams are located near natural hot springs, they are used for thermal baths and to heat the steam rooms (sauna). The major focus is on cleansing, revitalisation and socialising, often accompanied by a relaxing massage.

Hazard

The term hazard is frequently used as an alternative for danger or risk; however, correctly defined a hazard is the probability of a natural event occurring and a potential source of exposure to danger resulting in vulnerability.

Fig. 10.7 Layers of geyserite have accumulated in the area surrounding the boiling springs and fumaroles at the Caldeiras das Furnas in São Miguel (Azores, Portugal)

Hazard Map
Indication of areas with potential hydrothermal hazards such as seismic activity, unstable ground, toxic fumes and steam eruptions. Hazard maps are an important method to reduce the risk in areas with hydrothermal activity for residents and temporary visitors such as tourists. Hazard maps must include shelter locations, escape routes as well as emergency contacts.

See Chap. 9.

Hazardous Substance
Any material that poses a threat to human health or the environment, in particular to natural water resources. Typical hazardous substances are toxic, corrosive, ignitable, explosive or chemically reactive and can harm the environment beyond recovery.

Healing Springs
Natural hot springs with a mineral content that is considered to be beneficial for the human body in the treatment of various medical conditions. Hot springs rich in essential mineral and trace elements have been used for therapeutic purposes for thousands of years.

See Chap. 8.

Fig. 10.8 This large block of Hokutolite is one of the key exhibits in the Hot Spring Museum of Beitou (Taiwan)

Hokutolite ((Ba,Pb)SO_4) (Fig. 10.8)
A rare, radioactive mineral deposit that is found at only two hot spring locations to date; in the Thermal Valley of Beitou, Taiwan and at the Tamagawa Hot Springs in Senbuko, Akita Prefecture, Japan. Both hot springs are used for therapeutical purposes.

See Chap. 3.

Holocene
The current geological epoch, which followed the Pleistocene to the present, also known as the Postglacial period.

Hot Lakes
Volcanic crater lakes, ponds or lagoons, fed by one or more submerged hot springs at different temperatures and different volumes, e.g. the *Blue Lagoon* in Iceland or the thermal lake *Lake Hévíz* in Hungary. Lake Hévíz allows for therapeutic use all year round due to a source temperature of 38.5 °C and a high flow rate, which replenishes the whole lake completely every 72 h. Lakes that are fed by boiling springs such as the *Frying Pan Lake* in the Waimangu Rift Valley (Fig. 2.13), New Zealand, on the other hand have only visual appeal.

Hot Rivers and Streams
Rivers or streams fed by one or more hot springs at varying temperatures exist at several locations and are generally related to volcanic hydrothermal systems, e.g. Yellowstone National Park (Boiling River), New Zealand (Hakereteke Stream or Kerosene Creek). However, a hot river in Peru exists without a volcanic source, fed by hot springs emerging from faults, which generate temperatures over 80 °C, making this 'Boiling River' a unique body of water.

See Chap. 2.

Hot Springs
Natural springs that are fed by a constant flow of warm to hot groundwater, which is circulating underground, following faults and cracks in the surrounding rock. Many hot springs are located in volcanic areas and derive their heat from magma chambers or cooling magma intrusions. However, hot springs can also be unrelated to volcanic activity and derive their heat through convection of groundwater. Springs are generally classed as hot springs, if the water temperature is higher than that of the human body (37 °C or 98 °F), although classifications vary from one country to another. The correct terminology for hot spring is hydrothermal spring.

Hot Spring Classification
See Classification of Hot Springs.

Hot Spring Cooking (Figs. 7.8, 10.9 and 10.22)
Cooking in boiling springs and over steam vents is used in many countries and goes back in time to early civilisations with access to hot springs. In north America, native tribes have historically used the hydrothermal features at Yellowstone for food preparation. At cooking pools in New Zealand people can observe ancient Māori cooking techniques and in Iceland geyser cooking is a tourist attraction, while in Furnas (São Miguel Island) geothermal cooking is used to prepare typical Azorean cuisine.

Hot Spring Fish (Fig. 10.3)
A unique breed of fish (*garra rufa)* that live in water at temperatures of up to 43 °C. Based on their feeding habits (they thrive on dead skin) ever more health and wellness spas are offering a 'doctor fish' treatment (Ichthyotherapy) for skin conditions including psoriasis. These types of spas (usually connected to a hot spring) are by now popular in many countries (e.g. Canada, China, Indonesia, Japan, Korea, Malaysia), but are thought to were first discovered in Kangal, Turkey.

See Fish Cure.

Hot Spring Law
The remarkable features of natural hot springs can be protected by a variety of special laws that apply to the management of hot springs as a resource as well as by environmental laws to prevent unsustainable development in their proximity. The sustainable use of water resources can be controlled by regulations that concentrate on water rights, the control of ownership and environmental sustainability. Apart from governments, national and international non-government organisations and non-profit associations are also advocating for the sustainable management and pollution control of all freshwater resources. While there are water laws in most countries, government policies and legislations do not always include a provision for the protection of hot spring water from over-exploitation or pollution.

See Chap. 4 for Hot Spring Laws and Regulations in different Countries.

Hot Spring Spa/Resort
A destination that offers recreational as well as health and wellness facilities based on hot springs as a natural resource. Thermal waters sourced from hot springs are used for balneotherapy to treat medical conditions that respond well to bathing in warm, mineral rich water.

Hydraulic Fracturing
A process to create additional permeability through man-made fractures for unconventional extraction of natural gas and shale oil. The process involves the injection of a pressurised fluid to fracture the rock to increase its permeability. This water-based fluid contains a blend of chemicals to boost the flow rate of the extracted fluids and has been a source of controversy since hydraulic fracturing has become more prevalent in many countries.

Compare Fracking. See Chap. 4.

Fig. 10.9 Steam vents are used by the community in a special outdoor cooking area in the Kannawa district in Beppu (Kyushu, Japan)

Hydrochemistry
The study of the chemical composition of water. Hydrochemical data are important in analysing the mineral content of groundwater aquifers and natural hot springs. Hydrochemistry analysis has increased the understanding of groundwater processes and is used to determine discharge and recharge rates and is essential for the sustainable management of groundwater reservoirs.

Hydrodynamics
A branch of science that deals with the mechanical properties of liquids and studies the forces exerted by fluids in motion.

Hydrogen Sulphide (H_2S)
A colourless, flammable, toxic natural gas with a smell similar to rotten eggs at lower concentrations. H_2S is common in volcanic and hydrothermal environments but is also emitted during organic decomposition.

Hydrogeology
The science that deals with subsurface water and with related geologic aspects of surface water, a term frequently used interchangeably with geohydrology. In theory hydrogeology is the study of geology from the perspective of its role and influence on hydrology, while geohydrology is the study of hydrology from the perspective of the influence on geology.

Hydrologic Cycle (Fig. 10.10)
Also known as water cycle or hydrogeologic cycle, which is the perpetual natural circulation of water that involves evaporation, condensation and precipitation, thereby redistributing water in its liquid, solid and vapour phases on, above and under the Earth's surface.

Hydrology
The study of the properties, circulation and distribution of water and water vapour in the hydrological system in the context of the atmosphere, climate and geography. Hydrology includes the science and study of subterranean water resources and the treatment of surface water and its systems. While hydrology refers to the science of surface water as well as ground water, hydrogeology focuses mainly on ground water.

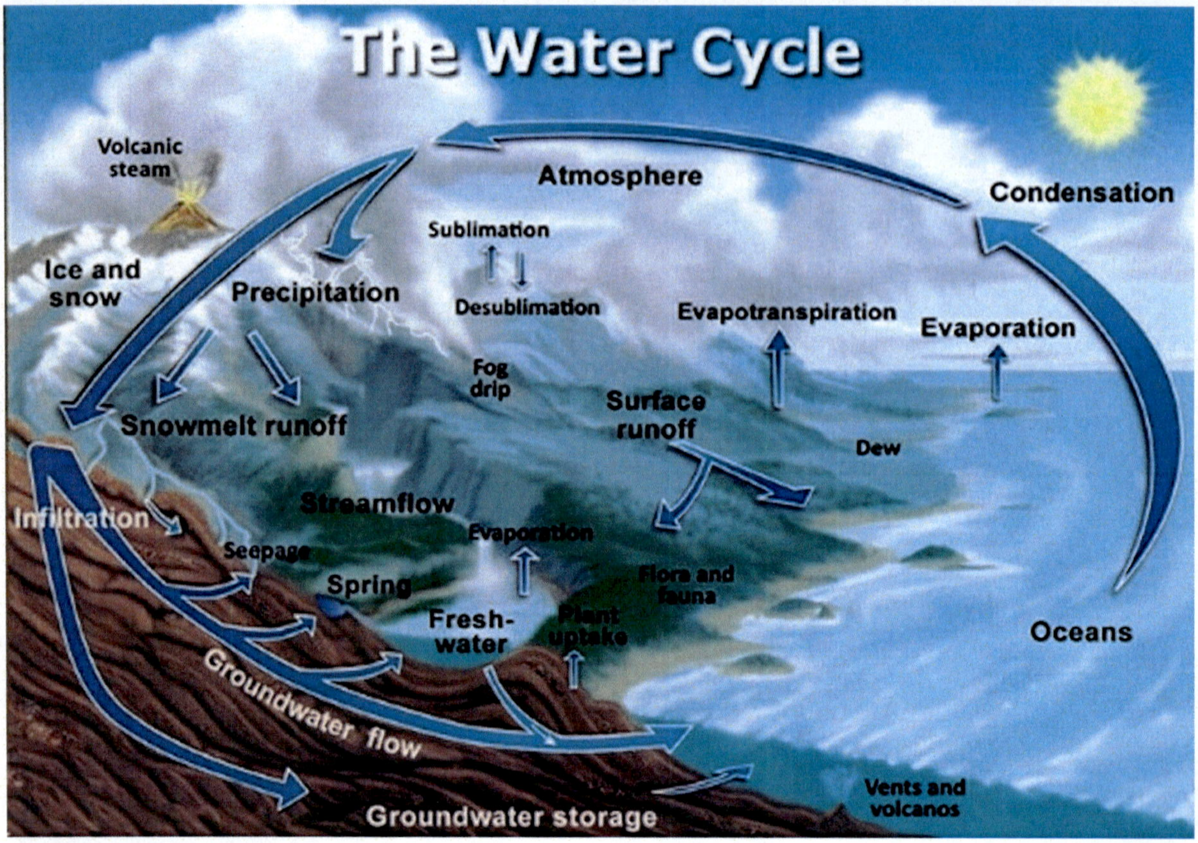

Fig. 10.10 Hydrologic Cycle. *Source* Evans and Perlman, USGS (2013)

Hydropinic Therapy
The consumption of natural hot spring water as a drink for therapeutic purposes at specific temperatures and times.
See Drinking Cure; Taking the Waters.

Hydroplate Theory
This theory is an alternative hypothesis based on the assumption that the Earth, before the biblical flood, used to have enormous reservoirs of groundwater placed between the outer crust of the Earth and the lower mantle. According to this theory the water was released under the immense pressure that had built up over time through subterranean tidal movements. An interesting theory, which would explain many different landforms and climate events, and of course the extent of Noah's flood.
See Fountains of the Great Deep.

Hydrotherapy
An essential part of balneotherapy, where water at varying temperatures is applied in the form of baths, showers and underwater massage to treat and rehabilitate different medical conditions.

Hydrothermal
Relating to the heated water and steam generated either in association with igneous activity or through convection underground, trapped in fractured or porous rock within the Earth and which contain dissolved minerals.

Hydrothermal Activity
A range of geophysical processes that involve the movement of hot subsurface water and steam, particularly the alteration and emplacement of minerals and the formation of hot springs and geysers.

Hydrothermal Alteration (Fig. 10.11)
A process in which rocks interact with hydrothermal solutions passing through and thereby causing an alteration of the original rock structure and contribute to metamorphic processes. Hydrothermal alteration also includes the

Fig. 10.11 Hydrothermal alteration of a hillside due to constant degassing in the Natural Park Furnas do Enxofre in Terceira (Azores, Portugal)

formation of hydrothermal veins in a host rock as a result of the precipitation of ore minerals.

See Metasomatism.

Hydrothermal Circulation

The movement of hot water in the Earth's crust is resulting from thermal and density gradients, either in the vicinity of heat sources such as volcanic activity or deeper in the crust close to granite intrusions. Hydrothermal circulation at the seafloor occurs when seawater enters the underlying crust through fractures where it is heated to several hundred degrees Celsius, generally along mid-ocean-ridges with high volcanic activity. While being heated, the circulating seawater chemically reacts with the surrounding rock before rising back to the seafloor where the enriched fluids enter the overlying water column through vents or chimneys known as black or white smokers.

Hydrothermal Eruption

A hydrothermal explosion occurs when superheated groundwater suddenly changes into steam, thereby causing a violent eruption of hot water and mud, steam and fragmented rock. Earthquakes or volcanic activity can cause instability of a hydrothermal system and trigger an eruptive event. A hydrothermal explosion can cover an area from a few meters up to several kilometres in diameter. They are a potential risk to tourism.

See Explosion Crater; See Phreatic Eruption.

Hydrothermal Deposit

Mineral deposits that are precipitated from hot, aqueous solutions and predominantly include copper, iron, zinc, but also silver, gold and manganese.

Compare Volcanogenic Massive Sulphide Ore.

Hydrothermal Fluid, Solution

An aqueous solution and the most important source of metallic ore deposits, generated by geophysical processes in the Earth's interior and derived from magmas. These hot, residual fluids contain large amounts of dissolved metallic elements, mineral compounds, salts and gases. The pH levels of hydrothermal fluids can range from strongly acidic to strongly alkaline and can change depending on the interaction with surrounding rock material. During the crystallisation process, ore deposits accumulate along faults and fractures in subsurface rock formations. Hydrothermal fluids discharging from subaqueous vents such as black smokers, precipitate fine-grained ore minerals once the hot solutions come into contact with cold seawater. During this process layers of dissolved metallic elements are accumulated on the seafloor, forming large hydrothermal mineral deposits, potentially of economic value.

See Chap. 3.

Hydrothermal Minerals

Minerals that are precipitated from a hydrothermal solution and are forming a mineral deposit.

See Hydrothermal Deposit; Volcanogenic Massive Sulphide Ore.

Hydrothermal Manifestations

All naturally occurring hot springs, fumaroles, geysers and related features that discharge warm to extremely hot fluids, gases and steam from an underlying hydrothermal system.

Hydrothermal Springs

See Hot Springs.

Hydrothermal System

A groundwater system that has an area of recharge, an area of discharge, and a heat source. When a magma chamber supplies the heat source and volatiles, the hydrothermal system is termed a magmatic hydrothermal system.

Hydrothermal Vent

Hydrothermal vents are located on the sea floor at mid-ocean ridges created by seafloor spreading where super-heated, mineral-rich water is released at temperatures of up to and above 400 °C. The minerals in the continuously spouting jets of hydrothermal fluid settle around the vent, where they solidify after mixing with cold seawater and create chimney like structures around individual vents. These hydrothermal vents are commonly known as black and white smokers, depending on their mineral content. The first discoveries were made in 1977 near the Galápagos Islands. Hydrothermal vents

Fig. 10.12 Hyperthermophiles are thriving in a pond, which is fed by hot spring water (Beppu, Japan)

have also been observed in the shallow water of the Aira Caldera in southern Japan and on the bottom of lakes.

See Black or White Smoker. Chaps. 2 and 11.

Hyperthermophiles (Fig. 10.12)

See Thermophiles.

Ichthyotherapy

See Fish Cure; Hot Spring Fish; Fig. 10.3.

Infiltration

The downward process of water from the atmosphere into the ground or the transition from the surface to subsurface. Only a portion of water infiltrating into the subsurface percolates down to the groundwater system as some water is retained in the unsaturated zone or is lost through evapotranspiration.

Injection Well

A well that is used for injecting fluids (e.g. fracking fluids, wastewater) into the subsurface or injection zone.

See Well Injection.

Intermittent Springs

A spring that discharges only periodically, sometimes depending on seasonal rainfall or other weather/climate conditions. Another type of intermittent spring are geysers that erupt at certain intervals.

Compare Perennial Springs.

Iron-Rich Springs or Ferruginous Springs

See Chalybeate Springs.

Juvenile Water

Magmatic water stored within a magma body or in volatile magmatic fluids with a high water content. During a volcanic eruption, this water can be released into the atmosphere and enters into the hydrological system for the first time. Juvenile water can also be released as hydrothermal fluids from subaqueous springs or vents. This type of water is original water formed through chemical reactions as a result of magmatic processes and has not entered the atmosphere before. Magmatic or juvenile water can form in very large quantities and remain underground indefinitely.

Fig. 10.13 Before the Blue Lagoon was developed into the facility it is today, the clean surplus water from the Svartsengi geothermal power station flooded the surrounding lava fields and was increasingly frequented by people, who sought the healing waters for health and recreation

Lagoon (Fig. 10.13)
The basin of a hot spring or a pool or lake formed by hot spring water. The term lagoon can also refer to cold water basins and lakes.

Leaching
The natural action of hydrothermal fluids leaching minerals and metallic trace elements out of the host rock and carry the dissolved load while circulating underground.
Compare Mineralisation.

Lithium (Li)
A chemical element that occurs dissolved in sea water and brines (saline springs), but also occurs naturally in vegetables and grain. Lithium has a low melting point and is highly reactive as well as flammable. Novel technologies are currently developed to extract lithium from geothermal brine to cater for the growing demand.
See Chap. 11.

Magmatic Water
See Juvenile Water.

Magnesium (Mg)
As one of the most common chemical elements magnesium occurs naturally in combination with other elements dissolved in natural spring water.

Medicinal Spring
A spring of reputed therapeutic value based on the substances (mineral and trace elements) contained in its water.

Metasomatism
A geophysical process whereby hydrothermal fluids induce the re-equilibration of rock material through alterations f the chemical composition. Original minerals are dissolved and replaced with new mineral components that take their place. Metasomatism can occur in rock formations close to igneous intrusions or magma bodies.
See Hydrothermal Alteration.

Meteoric Water
Water of atmospheric origin reaching the Earth from above, mainly as rainfall and meltwater of snow and ice. Most groundwater is of meteoric origin. Meteoric water circulating underground is the principal fluid source of most hydrothermal systems.

Microbes, Microorganisms
Microbes include all microscopic organisms such as bacteria, fungi, protozoa, and viruses. Microorganisms living in hot springs are classed as thermophiles and include thermotolerant bacteria and archaea, which can grow into large colonies forming colourful microbial mats.

Mineral
A naturally occurring substance formed by inorganic processes with a characteristic chemical composition expressed by a chemical formula. Minerals have a highly ordered atomic arrangement and occur either as individual crystals or disseminated in other minerals or rock. The specific physical properties of minerals include chemical elements and inorganic nutrients in the form of salts, which can be absorbed by the human body.

Mineral Content
Hot springs are rich in essential mineral elements including calcium, magnesium, sodium, potassium, chloride, phosphorus, sulphur.

Mineral Deposits
See Volcanogenic Massive Sulphide Ore. See Travertine, Tufa.

Mineralisation
The introduction of chemical components into a rock results in a mineral deposit. The mineralisation of hot and cold springs is a process whereby mineral elements are introduced into the water, usually by leaching or dissolving them out of the host rock through which the water flows underground.
Compare Leaching.

Mineral Salts
As a non-metallic mineral resource, mineral salts are essential for the metabolic function of the body and contain trace elements that are also found in the human body.

Mineral Springs
Naturally occurring springs, which contain dissolved mineral and trace elements, are considered to have therapeutic value. Mineral springs contain significant amounts of iron, sulphates or chlorides, but to be classed as true mineral spring the water must contain more than 1 g of dissolved mineral components per litre.

Mineral Water
Water from a spring that contains a range of mineral elements in solution. To be classed as genuine mineral water, the content of elements in solution should be between 0.2 and 1 g per litre. If the mineral water contains gas it is termed sparkling or effervescent. Worldwide more than 3000 different types of mineral water are available.

Mound Springs (Fig. 10.14)
The mound springs of Australia's outback are natural conduits for water rising from the Great Artesian Basin (GAB), forced to the surface by artesian pressure. The typical mounds are formed by the precipitation of minerals and salts contained in the spring water, as well as from build-up of organic matter and sediments such as clay and sand. The

Fig. 10.14 The Bubbler, one of the mound springs in the Wabma Kadarbu Mound Springs Conservation Park of South Australia. The mound springs are fed by water from the underlying GAB. *Source* Lewis (2018)

water temperature of the springs varies between tepid and hot (approximately 30 to 50 °C and above).

See Appendix 4.2.

Mud Bath (Fig. 10.15)

Mud baths for health treatments consist of hot spring water combined with volcanic ash or clay and are used all over the world for mud therapy. Hot mud baths have a long history and were already known for thousands of years to the Celts and the Romans.

Mud Ponds, Mud Pots

Mud ponds can reach temperatures up to boiling point depending on the source. The boiling mud is a result of hot spring water, dissolved rock and hydrogen sulphide, which reacts with the water and forms sulphuric acid, slowly dissolving the surrounding rock into a hot mud of fine clay and silica particles.

Mud Therapy

Also known as Fangotherapy and/or Pelotherapy, this type of treatment is known to relieve joint and muscle pain. Mud packs and mud baths are also used as a detox therapy and are relaxing and rejuvenating.

Compare Mud Bath.

Naegleria Fowleri

A single-celled freshwater amoeba found in some natural hot springs, which can cause serious health problems such as Primary Amoebic Meningoencephalitis (PAM), a rare but life-threatening infection. N. fowleri are also found in warm and shallow fresh water bodies or insufficiently maintained swimming pools. This microorganism can cause an infection of the brain (encephalitis) by entering through the nose, often with fatal results.

See Primary Amoebic Meningoencephalitis (PAM); see Chap. 9.

Fig. 10.15 The thermal mud baths at Hells Gate are promoted as a traditional New Zealand experience with the healing properties of the mud and sulphur-rich spring water used by local Māori for many centuries

Onsen 温泉

The Japanese term onsen refers to natural hot springs, public and private hot spring bathhouses, hot spring spas and resorts as well as hot spring towns.

Paleo Water (Palaeo)

From the Greek *palaios*, which means 'ancient' and is a prefix for 'very old' or 'ancient'. Also known as fossil water that accumulated in undisturbed groundwater reservoirs or ancient glaciers under different environmental conditions in the past. Geological changes sealed the paleo water reservoirs, thereby preventing any recharge or discharge.

Peloid

Peloids are clay-based muds that are rich in minerals and organic compounds, and which are used as a natural remedy in balneotherapy. Peloids or muds are a by-product of natural hot mineral springs and valued for their curative properties.

Pelotherapy

Also known as Fangotherapy or Mud Therapy, a form of treatment using either thermal mud baths or mud packs.

See also Mud Therapy.

pH (Potential of Hydrogen)

A scale from 0 to 14 that specifies the level of acidity or alkalinity of a substance or solution and can vary according to chemicals dissolved in the water. A pH level below the neutral stage (pH7) indicates acidity and a level above indicates the degree of alkalinity.

Perennial Springs

Natural springs with a constant discharge that flow all year round.

Compare Intermittent Springs.

Phreatic Eruption

A phreatic eruption is also known as a steam-blast eruption that occurs when ground water or surface water comes in contact with a heat source such as magma (up to 1200°C). This leads to the sudden formation of steam and results in an explosion of hot water, steam, gas, rock and ash.

See Hydrothermal Eruption.

Phreatic Water

In hydrology phreatic is a term that refers to groundwater, and in volcanology phreatic refers to an eruption type that involves subsurface water, also known as hydrothermal eruption.

Pollution

Pollution and contamination with hazardous substances are the most serious threat to all groundwater resources and include seepage from mining operations and landfills, oil and chemical spills, leaking septic systems and mismanaged

wastewater facilities. These are all contributing factors that are causing not just environmental degradation on the Earth's surface but can affect groundwater reservoirs as well as hydrothermal systems permanently.
See Chap. 4.

Primary Amoebic Meningoencephalitis (Fig. 10.16)
A rare, but life-threatening infection of the brain caused by *Naegleria fowleri*, an amoeba inhabiting some hot springs.
Compare Naegleria Fowleri.

Primary Hot Springs
See Volcanic Hot Springs.

Primordial Water
Water contained in the Earth's interior in the deep mantle.
Compare Palaeo Water.

Radioactive Springs
Springs that contain traces of radioactive substances such as radon gas, radium or uranium. Spring water that contains only low levels of radioactivity is used in hot spring spas at various destinations (e.g. Montecatini Terme, Italy; Radium Hot Springs, Canada; Rudas Bath, Budapest, Hungary; Hot Springs Arkansas, United States) without any adverse effects.
See Chap. 3.

Radium or Radon Springs
See Radioactive Springs. See Chap. 3.

Recharge (Aquifer Recharge)
The addition of water to recharge areas (aquifers, reservoirs, artesian basins) by precipitation, surface runoff, infiltration or well injection. Recharge is essential for the replenishment of saturated zones and aquifers to maintain their sustainability, and must be protected.

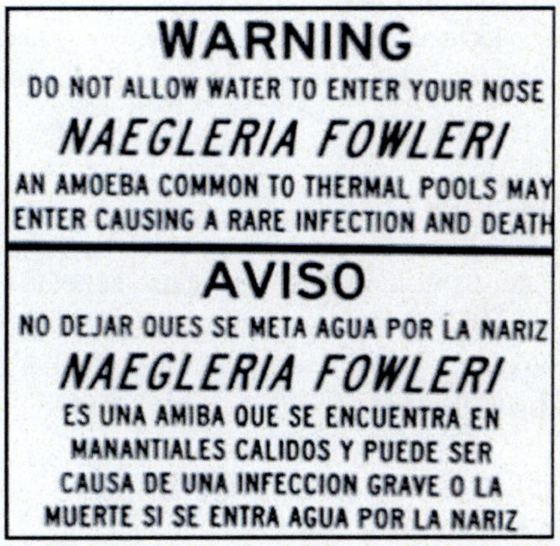

Fig. 10.16 Bilingual signs at thermal pools in Latin America are warning visitors of the potential risk of contracting amoebic meningitis. *Source* Public Domain

Recharge Capacity
The capacity of soils and subterranean rock formations to allow precipitation and runoff to infiltrate and reach the zone of saturation to replenish the groundwater aquifer.

Risk Management (in Hydrothermal Environments)
Identification and assessment of potential hazards combined with the implementation of suitable strategies to minimise the possibility for accidents and injuries. It includes raising the awareness of visitors about any risks in hydrothermally active areas and how to stay safe.
See Chap. 9.

Rock Formation
Rocks are a solid substance or aggregate of mineral matter, consolidated or unconsolidated, and form the lithosphere of the Earth. Rock formations are natural channels and conduits through which groundwater is transported and from which mineral elements are leached into the circulating solution.

Sacred Springs
See Water Worship.

Saline Springs
See Brine Springs.

Saline Water
Water that is generally considered as unsuitable for human consumption or for crop irrigation because of its high content of dissolved solids. However, saline hot springs are used for health purposes due to their beneficial effects and for industrial purposes (e.g. extraction of mineral elements).
See Chap. 5.

Seepage Springs
See Filtration Springs

Seismic Activity
Earthquake activity caused by the sudden release of accumulated stress along a fault underground. The released energy moves as seismic waves, which results in shaking of the ground. Seismic activity can affect the flow of hot springs either by blocking their conduits to the surface or by creating new flow channels from which springs can emerge.
See Chap. 5.

Siliceous Sinter
A term that refers to mineral deposits from active hot springs or geysers, which consist of a mainly white, porous silica variety. Siliceous precipitations are composed of silica and associated mineral elements as well as biological and lithological components. Siliceous deposits of carbonate minerals (travertine) occur in hydrothermal areas where carbon dioxide-rich thermal fluids discharge at the surface, either as hot springs or geysers.

Fig. 10.17 Sulphur deposits have formed thick encrustations around a steam vent (Whakaari/White Island, New Zealand)

The constant flow of hot spring water enriched with mineral elements creates travertine features such as silicified hillsides, terraced pools, petrified waterfalls, travertine mounds and microterracettes. Where the water cools and evaporates, mineralised crusts are formed.

See Chap. 3; Geyserite; Travertine; Appendix 10.1.

Solfatara (Fumarole, Steam Vent) (Fig. 10.17)
A volcanic hydrothermal manifestation that discharges steam rich in CO_2, H_2S, or SO_2 and can result in the deposition of sulphur, causing acid-sulphate alteration.

See Fumarole.

Spring[s]
Springs are points, or areas were hot or cold groundwater is flowing naturally from the ground onto the land surface or into a body of water, either as a result of gravity or pressure. Springs can be perennial or ephemeral and can yield either small or large volumes of water.

Subaerial Hot Springs
Hot springs located and/or occurring, existing and formed on or close to the Earth's surface.

Subaqueous (Subaquatic) Hot Springs
Hot springs located and/or occurring, existing and formed under water—refers to the ocean as well as to lakes.

Sublacustrine Hot Springs
Hot springs located and/or occurring, existing and formed below the surface of a lake and/or at the bottom of a lake.

Submarine Hot Springs
Hot springs located and/or occurring, existing and formed below the surface of the ocean and/or at the bottom of the seafloor.

Subsurface Water
All water occurring and circulating beneath the Earth's surface, including soil moisture, the vadose zone and artesian groundwater aquifers. Subsurface water is an integral part of the hydrological system and also includes all water contained in hydrothermal and geothermal systems.

Sulphur (S) (Figs. 3.2, 3.3 and 10.17)
Sulphur is a non-metallic solid element that ranges in colour from pale to bright yellow and occurs in many sulphide and

sulphate minerals. It is one of the most common and important elements on Earth. Sulphur (also sulfur, from Sanscrit *sulvere*) forms naturally at fumaroles in high-temperature geothermal areas and is mainly known for its use in the production of gunpowder, explosives, fireworks and matches, as well as in the vulcanisation of rubber. Sulphur is a chemical element that is of industrial importance as well as being essential in the treatment of some dermatological conditions.

Sulphur Dioxide (SO_2)
Sulphur dioxide is a colourless poisonous gas. When released into the atmosphere, sulphur dioxide is an air pollutant that can cause acid rain. Sulphur dioxide is also used as a preservative and antioxidant (E220) for foods and drinks, known to frequently causing allergic reactions in humans with asthma.

Sulphur Springs
Natural hot springs primarily of volcanic origin, which contain sulphur or sulphur compounds.

Taking the Waters
An old-fashioned expression for visiting a health resort or spa to participate in drinking cures based on the internal use of mineral-rich natural spring waters.

See also Drinking Cure; Hydropinic Therapy.

Thermalism
A term common in Europe and used for therapies or treatments based on thermal water, often derived from natural hot springs. French *thermalisme* and Spanish *termalismo*.

Thermal Springs
To class a thermal spring as either a warm or a hot spring, its temperature must be considerably above the local mean annual atmospheric temperature. The normal atmospheric temperature is 20 °C (68°F), and the standard atmospheric temperature is 0 °C (32°F). Therefore, the classification of thermal springs can be subject to interpretation based on local climate aspects and environmental considerations.

See Classification of Hot Springs.

Thermal Water
Water that discharges from a spring at temperatures above the local mean annual air temperature.

See Thermal Springs.

Thermophiles (Figs. 10.12 and 10.18)
Microorganisms with specialised enzymes that enable them to withstand high temperatures with an optimum growth temperature between 55–65 °C. For *hyperthermophiles* the optimum growth temperature is even higher at above 80 °C and up to over 110 °C. These microorganisms are also classed as extremophiles due to their adaptation to extreme environments such as the close vicinity of subaqueous hydrothermal vents or around extreme hot springs.

See Extremophiles.

Thermophilic Bacteria
See Thermophiles.

***Total Dissolved Solids* (TDS)**
The total concentration of dissolved solids in solution is determined by chemical analyses and usually expressed in milligrams per litre (mg/l). The amount of TDS can be determined by measuring the electrical conductivity of a fluid.

Total Suspended Solids (TSS)
Total suspended solids is a parameter to assess the water quality of any type of water or wastewater. TSS can be measured by analysing the filtered dry-weight residue of suspended particles in a water sample. TSS measurements refer to particles in a water column that are greater than 2 microns and can include organic and inorganic compounds. Because suspended solids can absorb light, this can cause an increase in the water temperature and a decrease in oxygen levels.

Toxic Pollutants
Substances or materials that can cause death, disease or birth defects in organisms that ingest or absorb them.

See Contamination.

Trace Elements
Elements that occur in minute (much less than 1%) but detectable quantities in minerals and rocks. All elements, except the most common rock-forming elements (e.g. O, Si, Al, Fe, Ca, Na, K, Mg, and Ti), generally occur as trace elements.

Travertine (Figs. 10.19, 10.20a, b; Appendix 10.1)
Concretionary limestone that forms calcareous sinter deposits either at the mouths of hot springs or in caves due to the process of precipitation of calcium carbonate. Travertine

Fig. 10.18 Thermophiles are colonising the mineral deposits of a hot spring in Karahayıt (Turkey)

is also found in cave deposits precipitated from carbonate-saturated waters that can be hot or cold. Travertine deposits or siliceous sinter are usually precipitates of hydrothermal origin and composed of silica and associated minerals. Examples include Mammoth Hot springs in Yellowstone, Pamukkale in Turkey and the famous Pink and White Terraces in New Zealand that vanished during a volcanic eruption.

Compare Tufa. See Chap. 3.

Tufa

A sedimentary rock also known as calcareous tufa. A porous limestone formed by the deposition or precipitation of calcium-bicarbonate from evaporating hot and cold mineral springs or in caves. Despite a comparable chemical composition with Travertine, tufa carbonates or calcareous sinter are precipitates from calcium-bicarbonate waters and are terms usually reserved for mineral deposits from cold springs or water emerging at ambient temperatures, while travertine or siliceous sinter are usually precipitates of hydrothermal origin. This differentiation is based on a much higher content of silica or silicon dioxide (SiO_2) in solutions up to and above boiling point than in colder mineral springs.

See Chap. 3; Siliceous Sinter; Travertine.

Tuff

Consolidated and lithified equivalent of a volcanic ash deposit composed of fragmented material (<4 mm) of crystals and rocks produced directly by volcanic eruptions.

Fig. 10.19 Pamukkale travertine – a sample of 'Hierapolis Stone', a dense banded fissure travertine that was used as building material (sample on left c. 50 cm wide), also known as Phrygian Marble from Turkey

Welded tuff is the result of internal heat and pressure of ash flows, also known as ignimbrite. While Tuff is unrelated to hot springs the term is included here to clarify the difference between Tuff and Tufa.

Unconfined Aquifer
See Aquifer.

Undeveloped Hot Springs (Fig. 10.21)
Also known as 'wild' hot springs as they generally lack basic facilities and should be used with extreme caution due to potential risks as well as hygiene concerns. However, the scenic value and the uniqueness of hot springs in the wild make them popular destinations for more adventurous people.

See Chap. 9.

Velocity
Velocity is an indication of the average rate at which groundwater and groundwater solutes move. It can be estimated using Darcy's law and an equation ($V = Kl/n$) that is based on the variables (a) horizontal hydraulic conductivity, (b) horizontal hydraulic gradient and (c) effective porosity to calculate groundwater velocity.

Volcanic Hot Springs
Primary hot springs that are closely related to volcanic activity and are containing minerals, gases and trace elements generated by volcanic processes. The majority of volcanic hot springs are located in the vicinity of a magma reservoir deep underground.

Volcanogenic Massive Sulphide Ore Deposit (VMS)
Metal sulphide ore deposits are associated with volcanic and hydrothermal environments and are rich in metallic sulphide minerals. VMS deposits have been discovered on every continent, and are a highly sought-after source of copper, lead and zinc as well as precious metals such as gold and silver. Resulting from hydrothermal fluids circulating in previous submarine environments or near the seafloor, where hydrothermal vents precipitate ore minerals, VMS developed into large layers of mineral deposits.

Warm Springs
See Thermal Springs.

Water (H_2O)
A colourless, tasteless and odourless liquid if uncontaminated. Water has the unique property that allows it to occur either as a liquid, a solid (ice) or as a gas (steam, vapour) depending on the ambient temperature. The density of water is also affected by temperature, for example cold water contracts and hot water expands.

Fig. 10.20 a, b Pamukkale travertine – a ridge of old mineral deposits (left) and a sample of botryoidal mineralisation from further down the hillside (scale 20 cm)

Fig. 10.21 Katherine Springs, an undeveloped hot spring in the small town of Katherine in Australia's Northern Territory

Fig. 10.22 Geothermal cooking in Furnas is a slow process—large pots with meat and vegetables are buried in the ground and heated by steam over the course of many hours until cooked. The result is delicious typical Azorean food. (São Miguel, Azores, Portugal)

Water Pollution

Pollution of a freshwater resource through anthropogenic contamination, which can cause eutrophication or acidification resulting in the pollution of a groundwater reservoir and thereby making the water unsuitable for human use.

See Chap. 4; Pollution.

Water Resource

All water resources must be protected, conserved and managed appropriately to preserve them for future generations.

Water Worship

Ancient cultures all over the world regarded water as a natural resource of divine origin and therefore sacred, and any gods and spirits related to water were worshipped as guardians of the springs. With regional and cultural variations, different forms of water worship developed, and temples and shrines were built to acknowledge the significance of water as a connection to the spiritual world.

See Chap. 6.

Well Injection

The subsurface emplacement of fluids through a drilled well, either to replenish an overexploited recharge area or for sequestration purposes. Some countries use the strategy of well injection to replenish exploited water resources to keep their hot springs flowing.

See Deep Well Injection.

White Smoker

White smokers are deep-sea vents or chimney-like structures found at the crest of ocean ridges that discharge plumes of highly alkaline hydrothermal fluids at temperatures between 260–300 °C. The hydrothermal fluid from white smokers is rich in light coloured precipitates that contain particles such as barytes and silica. The chimneys are created by the deposition of minerals in the hydrothermal plumes.

See Hydrothermal Vent; Compare Black Smoker. See Chap. 3.

Yield

The volume of water discharged and harvested from an aquifer, borehole or natural spring over a certain time period.

See Discharge Rate; Flow Rate; Specific Capacity.

10.3 Final Remarks

At this point of the book the original plan was to include a list of as many hot springs worldwide as possible, an arduous task that was left until last. However, the idea was abandoned due to the fact that an extensive list of natural hot springs around the world is already available online (Wikipedia), which is constantly updated with relevant links and references of source literature. Compared to the information publicly available less than a decade ago, this *List of Hot Springs* (https://tinyurl.com/yjtn73xs) covers most countries, and their major hot springs together with some geophysical data as well as relevant demographic and/or historical information. Therefore the intention of including a comprehensive list of hot springs in this book was abandoned with some relief.

The thermal pool of Bitter Springs in Australia's Northern Territory

10.4 Appendix

Appendix 10.1 Travertine Deposits as Building Material

1, 2—Some of the more than 1200 burial chambers and tombs at the Necropolis of the ancient city Hierapolis; 3—They were built during the Hellenic Period (323–31 BC) from the local limestone deposited by the same hydrothermal processes that created the white travertine terraces of Pamukkale

Sourced Literature

Allaby M (ed) (2013) A dictionary of geology and earth sciences, 4th edn. Oxford University Press, Oxford, UK

Altman N (2000) Healing springs—From hidden springs to the world's greatest spas. Healing Arts Press, Vermont, USA

Australian Water Information Dictionary (2020) Total suspended solids. Commonwealth of Australia, Bureau of Meteorology. Retrieved from https://www.bom.gov.au/water/awid/id-541.shtml

Boudreau AE, Lynne BY (2012) The growth of siliceous sinter deposits around high-temperature eruptive hot springs. J Volcanol Geoth Res 247–248:1–8. https://doi.org/10.1016/j.jvolgeores.2012.07.008

Bryan K (1919) Classification of springs. USGS Staff—Published research. J Geol 27(7):522–561. Retrieved from https://digitalcommons.unl.edu/usgsstaffpub/493

Capezzuoli E, Gandin A, Pedley M (2014) Decoding tufa and travertine (freshwater carbonates) in the sedimentary record: the state of the art. Sedimentology 61:1–21. https://doi.org/10.1111/sed.12075

Clugston MJ (ed) (2004) The new Penguin Dictionary of Science, 2nd edn. Penguin Books, London

Cowan D, Tuffin M, Mulako I, Cass J (2012) Terrestrial hydrothermal environments. In Bell EM (ed) Life at extremes: environments, organisms and strategies for survival, Chapter 12. CAB International, pp 219–241

Daintith J (ed) (2000) A Dictionary of Chemistry, 4th edn. Oxford University Press, New York

Department of Water Affairs (2017) The Groundwater dictionary—A comprehensive reference of groundwater related terminology, 2nd edn. Republic of South Africa. Retrieved from https://tinyurl.com/ya7evmf4

Dilsiz C (2006) Conceptual hydrodynamic model of the Pamukkale hydrothermal field, southwestern Turkey, based on hydrochemical and isotopic data. Hydrogeol J 14(4):562–572. https://doi.org/10.1007/s10040-005-0001-4

Britannica (2018) Hot spring geology. Retrieved from www.britannica.com/science/hot-spring.

Erfurt-Cooper P (2010) The importance of natural geothermal resources in tourism. In: Proceedings world geothermal congress 2010, Bali, Indonesia, 25–29 April

Erfurt-Cooper P (2014) Volcanic tourist destinations. geoheritage, geoparks and geotourism series. Springer, Berlin, Heidelberg

Erfurt-Cooper P (2018) Active hydrothermal features as tourist attractions. In: Fearnley CJ, Bird DK, Haynes K, McGuire WJ, Jolly G (eds) Observing the volcano world: volcano crisis communication. Advances in volcanology series. Chapter 27. Springer, Berlin, Heidelberg, pp 85–105. https://doi.org/10.1007/11157_2016_33

Erfurt-Cooper P, Cooper M (2009) Health and wellness tourism—spas and hot springs. Aspects of tourism series. Channel View, Bristol, UK

Evans J, Perlman H (2013) The water cycle (Natural water cycle). Credits: Howard Perlman, USGS and John Evans, USGS. Public domain. Retrieved from https://tinyurl.com/vwyvg3s

Fouquet Y (2011) Black smoker. In: Gargaud M et al (eds) Encyclopedia of astrobiology. Springer, Berlin, Heidelberg

Friends of Mound Springs (FOMS) (2020) About South Australia's Far North Mound springs. Retrieved from https://www.friendsofmoundsprings.org.au/about-mound-springs/

Fridleifsson IB, Bertani R, Huenges E, Lund JW, Rybach L (2008) The possible role and contribution of geothermal energy to the mitigation of climate Change. In: Hohmeyer O, Trittin T (eds) IPCC scoping meeting on renewable energy sources—Proceedings. IPCC Special Report

Guo Q, Planer-Friedrich B, Liu M, Li J, Zhou C, Wang Y (2017) Arsenic and thioarsenic species in the hot springs of the Rehai magmatic geothermal system, Tengchong volcanic region, China. Chem Geol 453:12–20. https://doi.org/10.1016/j.chemgeo.2017.02.010

Hancock PL, Chalmers RML, Altunel E, Çakir Z, Becher-Hancock H (2000) Creation and destruction of travertine monumental stone by earthquake faulting at Hierapolis, Turkey. Geol Soc Lond Spec Publ 171(1):1–14. https://doi.org/10.1144/GSL.SP.2000.171.01.02

Harrison, S (2020) Fish Spa, Singapore International Airport. Photo credit

Jones B, Renaut RW (2003) Hot spring and geyser sinters: the integrated product of precipitation, replacement, and deposition. Can J Earth Sci 40(11):1549–1569. https://doi.org/10.1139/e03-078

Kalender L, Öztekin Ökan O, İnceöz M, Çetindağ B, Yildirim V (2015) Geochemistry of travertine deposits in the Eastern Anatolia District: an example of the Karakoçan-Yoğunağaç (Elazığ) and Mazgirt-Dedebağ (Tunceli) travertines, Turkey. Turkish J Earth Sci 24:607–626. https://doi.org/10.3906/yer-1504-27

Klein C (2002) The manual of mineral science, 22nd edn. Wiley, USA

Konhauser KO, Jones B, Reysenbach A-L, Renaut RW (2003) Hot spring sinters: keys to understanding Earth's earliest life forms. Can J Earth Sci 40(11):1713–1724. https://doi.org/10.1139/E03-059

Kruse FA (1997) Characterization of active hot-springs environments using multispectral and hyperspectral remote sensing. Analytical imaging and geophysics. In: Presented at the twelfth international conference and workshops on applied geologic remote sensing, Denver, Colorado, 17–19 Nov

Kwader T (2001) Is it hydrogeology or geohydrology? National driller. Retrieved from https://tinyurl.com/y9h8h7jq

LaMoreaux PE, Tanner JT (eds) (2001) Springs and bottled waters of the world—Ancient history, source, occurrence, quality and use. Springer, Berlin, Heidelberg

Lewis S (2018) North Freeling Spring on the Peake Pastoral Lease. Friends of Mound Springs (FOMS). Photo credit

Lynne BY (2007) Diagenesis of siliceous sinter deposits in the U.S.A. and New Zealand. Doctoral dissertation. Department of Geology, University of Auckland, New Zealand

McGeary D, Plummer CC, Carlson DH (2001) Physical Geology—Earth revealed, 4th edn. McGrawHill, New York

McGraw-Hill (2003) Dictionary of Geology & Mineralogy, 2nd edn. McGraw-Hill, New York

Merriam-Webster (2017) Definition of black smoker. Retrieved from https://www.merriam-webster.com/dictionary/black%20smoker

Özkul M, Kele S, Gökgöz A, Shen C-C, Jones B, Baykara, Alçiçek MC (2013) Comparison of the Quaternary travertine sites in the Denizli extensional basin based on their depositional and geochemical data. Sed Geol 294:179–204. https://doi.org/10.1016/j.sedgeo.2013.05.018

Oxford Living Dictionaries (2017) Definition of white smoker. Retrieved from https://en.oxforddictionaries.com/definition/us/white_smoker

Park C, Allaby M (2013) A Dictionary of Environment and Conservation, 2nd edn. Oxford University Press, Oxford, UK

Pearson A, Huang Z, Ingalls AE, Romanek CS, Wiegel J, Freeman KH, Smittenberg RH, Zhang CL (2004) Nonmarine crenarchaeol in Nevada hot springs. Appl Environ Microbiol 70(9):5229–5237. https://doi.org/10.1128/AEM.70.9.5229-5237.2004

Peninsula Hot Springs (2017) Balneotherapy & balneology: the science and art of mineral water therapy. Retrieved from https://www.peninsulahotsprings.com/wp-content/uploads/2017/11/Balneotherapy.pdf

Sæmundsson K, Gunnlaugsson E (2002) Icelandic rocks and minerals. Mal og mennig, Reykjavík, Iceland

Science Dictionary (2017) The world's largest online science dictionary. Retrieved from https://thesciencedictionary.org/

Sigurdsson H (ed) (2015) The Encyclopedia of Volcanoes, 2nd edn. Academic Press, London

Soos A (2011) Ancient fossil aquifers and Nasa. Environmental Science News, Sept 19. Retrieved from https://www.enn.com/sci-tech/article/43271

Thain M, Hickman M (2000) The Penguin dictionary of biology, 10th edn. Penguin Books, London

USGS (2017) Glossary of hydrologic terms. Retrieved from https://or.water.usgs.gov/projs_dir/willgw/glossary.html

Villanueva RE, Prol-Ledesma RM, Rodriguez A, Armienta MA (2013) Arsenic in hot springs of Bahía Concepción, Baja California Peninsula, México. Chem Geol 348:27–36. https://doi.org/10.1016/j.chemgeo.2012.09.008

Zhang CL, Huang Z, Li Y-L, Romanek CS, Mills GL, Gibson RA, Talbot HM, Wiegel J, Noakes J, Culp R, White DC (2007) Lipid biomarkers, carbon isotopes, and phylogenetic characterization of bacteria in California and Nevada hot springs. Geomicrobiol J 24 (6):519–534. https://doi.org/10.1080/01490450701572515

Hot spring-fed waterfall in Kyushu, Japan

Hot Springs—A Final Overview

11

Contents

P. Erfurt, *The Geoheritage of Hot Springs*, Geoheritage, Geoparks and Geotourism,
https://doi.org/10.1007/978-3-030-60463-9_11

Private hot spring pool in a secluded backyard in Japan

11.1 Introduction

The aim of the final chapter is to draw together some essential viewpoints discussed in earlier parts of this book, as well as adding some thoughts about a few topics that warrant their inclusion here. One of these topics relates to the diversity of natural hot springs and their geothermal origins as well as their equally diverse usage for domestic and commercial purposes. More importantly and not mentioned yet are current research and development projects in the field of geothermal energy, a subject closely related to natural hot spring geosystems. The potential of hydrothermal and geothermal systems as a renewable resource is explored with brief examples from several countries to provide an overview of the current efforts and/or expansions in progress to generate clean energy from renewable resources. Also, new technological advances are currently pursued with view to recovering particular minerals like lithium from hydrothermal fluids, which further proves the versatility of this natural resource. Staying on the topic of research, a brief summary of development aspects related to hot springs used for health and recreation, together with an overview of the hot spring tourism sector, is also included. Finally, the prospects of hot springs as a natural resource are considered together with an emphasis on the need for preserving the overall geoheritage of hot spring environments for generations to come.

11.1.1 Direct Use of Hot Springs—Some Examples

Hot springs are most commonly associated with recreational activities like a relaxing soak in a natural thermal pool or centred around health and wellness at a hot spring spa, clinic or resort. At the same time, there are a number of different applications of geothermal water, which play a considerable role in domestic as well as commercial and industrial areas. More specifically, space heating and district heating are some of the most common products of geothermal energy, either using hot groundwater directly or indirectly through a heat exchanging process if the source temperatures are too high or too low.

In cold climate countries, geothermally heated water is often used for snow melting in car parks, streets and footpaths to keep them snow and ice-free. Countries reportedly using geothermal snow melting methods include Argentina, Iceland, Japan, and north America. To warm the ground sufficiently, the same technology is used as in floor or ceiling heating of buildings, where warm water is pumped through a network of pipes laid out in a grid to evenly heat the space below or above. This principle is also applied in agricultural processes including open-ground heating and the warming of greenhouses (Fig. 11.1a, b), where soil and plants require a constant temperature.

To present a clearer picture of the versatility of geothermal resources, Table 11.1 and Fig. 11.2 demonstrate that the use of geothermal water, either from hot springs or hydrothermal systems, is not restricted to health and recreation, with a wide range of industrial processes relying on heat and energy of geothermal origin in different forms.

A wealth of information can be obtained from the proceedings of the quinquennial World Geothermal Congress (WGC), where updates of geothermal energy use are presented. At the 2015 WGC, direct utilisation of geothermal energy was reported by 82 countries, with a marked increase in their installed capacity since 2010. Greenhouse heating for example increased by 19%, while space heating

Fig. 11.1 **a** Greenhouses in Japan are geothermally heated and are refitted with new covers when necessary. **b** Greenhouses are not only used for agricultural purposes (left image); walking through tropical gardens is a popular tourist attraction, especially in cold climate regions or during winter

Table 11.1 The distribution of direct utilisation of geothermal energy as reported by 82 countries at the World Geothermal Congress in 2015. *Source* Lund and Boyd (2015). Due to the worldwide Covid19 pandemic, the 2020 WGC was postponed, and no country updates were available at the time of writing

Geothermal utilisation	(%)
Ground-source heat pumps	55.3
Health and recreational facilities	20.3
Space and district heating	15.0
Agriculture—open-ground and greenhouse heating	4.5
Aquaculture, fish farming—pond heating	2.0
Industrial process heating	1.8
Snow melting	0.4
Agricultural drying	0.4
Other uses	0.3

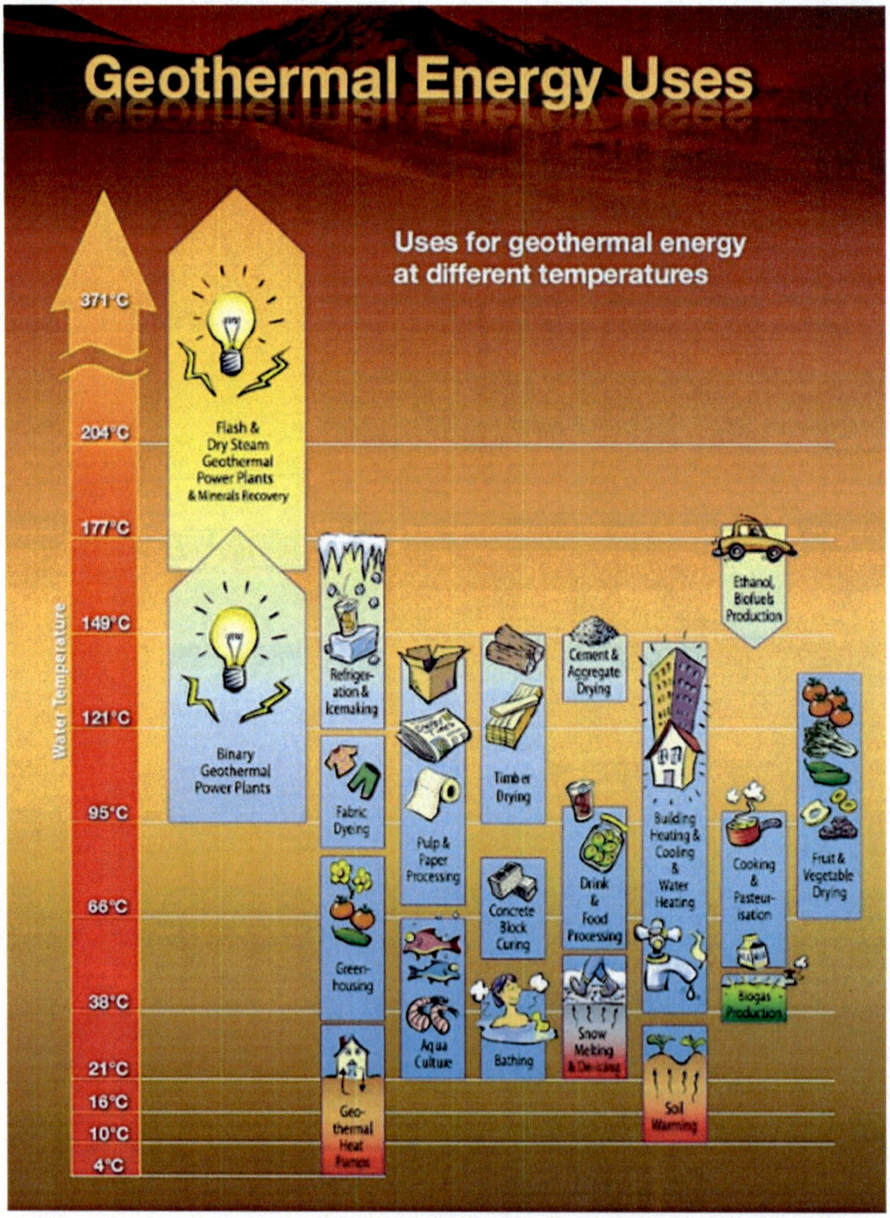

Fig. 11.2 The multiple benefits derived from natural geothermal resources. *Source* GNS Science, New Zealand, 2020 (with permission)

Fig. 11.3 This insulated steel pipeline is 27 km long and has the capacity to carry up to 1850 L per second of 100 °C hot water from the Nesjavellir geothermal power plant (Hengill high-temperature field) to storage tanks in Reykjavík (Iceland)

increased by 44%, pond heating for aquacultural purposes increased by 6.7%, and agricultural processes such as the drying of fruit, seaweed, cereal grains, Lucerne hay and vegetables increased by 28.8%. Other industrial processes based on the utilisation of geothermal energy include the curing of concrete, bottling of water and carbonated drinks, pasteurisation of milk, extraction of CO_2, processing in paper mills, extraction of salt, and the production of borate and boric acid (Ragnarsson 2015).

Aside from generating energy or using geothermal water for commercial applications, countries like Iceland also use it as a municipal hot water source. To supply Reykjavík, hot water is piped from *Nesjavellir*, the country's second-largest geothermal power plant, to the capital city, where it is stored in tanks. Effective insulation of the steel pipeline (Fig. 11.3) with rock wool and a protective aluminium sheet cover, as well as the speed of transmission of the 100 °C hot water, result in a heat loss less than 1.8 °C (Verkís 2020).

In Rotorua, New Zealand, hot spring water from geothermal activity is also available to domestic and commercial buildings, including numerous hotels who provide hot mineral pools for their guests. In some areas of Japan, private residences have the option of having hot spring water piped directly into their bathrooms where it is used for daily hot baths.

11.2 The Geophysical Diversity of Hot Springs

11.2.1 Hot Springs and Sustainable Clean Energy—A Promising Concept

Geothermal resources are generally considered as a renewable energy source if used in a sustainable manner. While not all geothermal energy is derived from wet geothermal systems, there is unlimited heat stored underground and available for use. One of the advantages of extracting heat from the Earth is a relative independence from weather and climate (Fridleifsson et al. 2008). But not everything is downright positive, and concerns for the environment remain even for geothermal energy production, although it

Table 11.2 The list below includes countries with either established geothermal energy producing facilities, currently in the process of expanding their existing power plants or exploring the feasibility of moving away from the dependence on fossil fuels to cleaner energy

Africa	Asia & Oceania	Europe	Central America	South & North America
Ethiopia Djibouti Kenya Malawi Morocco Rwanda Saudi Arabia South Africa Tanzania Tunisia Uganda Zambia	Australia China Indonesia Iran Japan Korea New Zealand Papua New Guinea Philippines Samoa Taiwan Vanuatu	Austria Croatia France Germany Hungary Iceland Italy Portugal Serbia Spain Switzerland Turkey	Costa Rica Dominica El Salvador Guatemala Honduras Martinique Mexico Nevis Nicaragua St Lucia St Vincent & Grenadines	Argentina Bolivia Chile Ecuador Peru Venezuela Canada (several provinces) USA (Alaska, California, Nevada)

Fig. 11.4 The pioneering Reykjanes Geothermal Power Station in Iceland started to operate in 2006, and since 2009 has the capacity to produce 100MWe[1] using reservoir steam at 290–320 °C. A planned expansion by 30MW is aiming to utilise the geothermal water, which is currently disposed of into the nearby sea

seems that most problems are manageable and can be overcome. For instance, the operation of geothermal power plants has the potential to negatively impact water quality and water consumption; however, this can be counteracted by using closed-loop systems, where the extracted hot water is re-injected into the reservoir after use. The practice of re-injection is also used to decrease carbon dioxide emissions to meet clean air standards as per requirement of the country where the geothermal energy production takes place and to reduce a potential impact on the overall state of the environment. Compared to other means of power generation, geothermal energy production provides an opportunity to

[1]Megawatts electric—electricity output capability.

harness a renewable resource without major environmental impacts if managed responsibly. Therefore, more incentives are needed to maximise the use of geothermal resources together with other ways to generate clean energy to replace fossil and nuclear fuels in the times ahead.

11.2.2 Geothermal Power—Research and Development

Over the last few decades many countries have contributed to the production of cleaner energy by using natural geothermal resources. To further advance the geothermal industry sector, numerous research projects are currently underway to explore the potential of geothermal reserves for the generation of clean[er] energy (Table 11.2). Research into new ground-breaking methods and the development of advanced technologies is ongoing; so is the development of improved drilling and wellbore materials for geothermal projects. Test productions are prepared as soon as funding is secured, and partnerships are formed to support the latest projects to generate geothermal energy with the aim to reduce the pressure on the natural environment caused by conventional methods of power generation.

In the year 2001 the Iceland Deep Drilling Project (IDDP) was launched to drill boreholes into supercritical zones to reach higher enthalpy fluids than those exploited for power production at the time. In 2017 the project was successfully completed after 176 days of drilling the IDDP-2 well to a depth of 4659 m with well bottom temperatures measured at 427 °C and fluid pressure of 340 bars (IDDP 2017). This was a significant milestone in the quest for renewable energy.

A review of current projects reveals substantial investments into new geothermal power plants and related infrastructure (Fig. 11.4), additional facilities and expansions as well as the inclusion of refurbishments of existing older geothermal power stations. Preliminary investigations are in progress worldwide, including feasibility studies and environmental impact assessments to establish the viability and sustainability of any future projects. However, when it comes to the exploitation of geothermal resources, the term 'sustainable' can have a double meaning; it may be a reference to the sustained availability of the natural resource, or it can refer to the environmental sustainability in relation to the exploitation technology of geothermal resources. To ensure the sustainability of geothermal resources to produce energy, it is essential to avoid excessive production rates which can lead to reservoir depletion. Taking into account the size of the geothermal field, the recharge rate, and the temperature at depth is vital for the long-term use of geothermal resources (Fridleifsson et al. 2008).

The large number of countries with a stake in geothermal energy development confirms the view held by scientists that the threat of climate change, exacerbated by the combustion

Fig. 11.5 Geothermal power plant in Terceira, one of the nine islands of the Azorean Archipelago

of fossil fuels, is starting to have an impact on energy production in the coming years. Although many projects are still at planning stage, it nevertheless indicates a growing interest into harnessing natural geothermal resources to replace fossil fuels. To illustrate the point, on the volcanic island of *Martinique* in the **Caribbean** currently 25% of electricity are generated from renewable sources like wind and sun or bagasse, a biofuel. At present, the installation of a geothermal power plant in *Petite Anse* is under investigation with the objective to raise the production of renewable energy on Martinique to 56% (RCI 2020).

On the other side of the Atlantic Ocean, the city of *Chaves* in the north of **Portugal** is in the process of creating a large urban geothermal heat network. In Chaves geothermal energy is already used for facilities including a thermal health complex, the municipal swimming pool and two hotels as well as supplying the city's retirement village. To replace other energy sources such as fossils fuels, thermal water normally used for health treatments and recreation is planned to be utilised in a pilot project, which aims to provide 25 public and private buildings with clean, renewable energy. The source for this energy is a large hydrothermal reservoir that discharges water at temperatures between 66° and 77 °C. The analysis of the project data indicates that this new geothermal development may lead to a cut in CO_2 (carbon dioxide) emissions from fossil fuels by up to 1300 tons per year (Caldas 2020). Geothermal energy is also increasingly developed on the volcanic Islands of the **Azores** with several smaller power plants in operation (Fig. 11.5). In 2017 the new 'Pico Alto' power plant in *Terceira* went online, with the objective to cover at least 10% of the island's electricity needs.

In **Chile** (South America), hydrothermal manifestations of volcanic origin occur in several regions with steam emitting from the ground, fumarolic activity and geysers spouting boiling water. For the purpose of generating geothermal energy, potential resources like the *Mariposa* geothermal system with reservoir temperatures of >250 °C have been located at depths between 700 and 1000 m with an inferred resource of 320 MWe (Hickson et al. 2011). The Mariposa Project is planned to be established in stages with the objective to contribute 100% renewable energy to the national electricity system (GRC 2019).

On the Asian continent in **Mongolia** a geophysical survey was conducted to investigate the possibility of constructing a geothermal power plant that combines the generation of heat and electricity. Funded by Switzerland, the National Renewable Energy Centre of Mongolia is exploring ways of developing geothermal energy in the *Hangai* Province to supply clean and cost effective energy from the geothermal reservoir near *Tsetserleg*, which is also the source of the volcanic hot springs in *Shivert* and *Tsenkher* (GEG 2018).

Compared to Mongolia, the country of **Japan** has abundant geothermal resources based on the vast number of active volcanoes, which could provide a renewable source for the generation of clean energy. Instead, high costs and strict regulations have not been favourable to the development of larger geothermal power plants in the past, although smaller plants contribute about 0.3% to Japan's total electricity generation. In the aftermath of the Fukushima nuclear disaster however, Japan's government has decided to speed up the development of geothermal energy and has rationalised the approval process for new projects. Paradoxically, Japan is one of the world's largest developers of geothermal power plants outside its own country (IRENA 2018). Current geothermal energy projects in Japan, either planned or already under construction, are located across the country with seven new power plants and 17 ongoing projects from past fiscal years (Think Geoenergy 2020a). On the other hand, concerns were raised previously about a potential reduction in the discharge capacity of natural hot spring water supplying Japan's large onsen tourism industry if geothermal resources are extracted near established spa towns. With over 27,000 hot springs providing the highly sought-after authentic hot mineral water for thousands of onsen throughout Japan, naturally onsen operators are not overly supportive of geothermal power plants that could potentially interfere with the quality and the availability of their most important resource—natural hot spring water.

In countries like **Australia** a trend towards using renewable resources has not been established either, with only minor progress made in the development of geothermal energy. Research has shown that the public perception was largely influenced by biased media reporting, which focused more on potential concerns than on the benefits of geothermal technology and therefore did not generate enough public and political support (Romanach et al. 2015). This is regrettable and disappointing, as Australia has the potential to provide a large percentage of its energy needs by using a combination of renewable natural resources (geothermal, solar, hydro, wind) in preference to conventional power plants that pollute the atmosphere by burning fossil fuels. Following the challenges of Australia's initial geothermal development, change was initiated with the formation of the new Australian Geothermal Association (AGA). Established in 2016, AGA's role is to 'promote and encourage the science, technology and development of geothermal energy in Australia' as a centre for information and communication about everything related to geothermal resources, technology and applications. On a positive note, a new geothermal power plant started operation in 2019 in the shire of Winton, Queensland (Think GeoEnergy 2020). While small-scale only (310 kW) compared to other countries' output measured in MW this is a first step in a promising direction with other similar projects reportedly at planning stage (Beardsmore et al. 2020).

11.2.3 Hot Spring Geosystems and Their Exploitation for Commercial Purposes

Geosystems, compared to ecosystems, are land formations and other natural physical features and phenomena either on the surface of the Earth or below, which combined make up the physical geography or physiography found on our planet. This includes systems made up of geothermal and hydrothermal resources, often closely connected to volcanic or magmatic systems.

Conventional Use of Geothermal Resources

Part of the geophysical heritage of natural high-temperature hot springs are by-products in various forms that are of commercial and industrial value such as ore mineral deposits, precious metals, or minerals. Sulphur in particular has long been used in the production of gunpowder, explosives and matches, as well as in manufacturing rubber (Sæmundsson and Gunnlaugsson 2002). As an extractable resource geothermal systems contribute towards various by-products, which represent a significant commercial asset, contingent on the volume of the deposits and how easy and economical these are to access.

In the cosmetics industry, natural hot springs are used as a resource for a range of products including geothermal mud packs, mineral bath salts or powders and mineral enriched lotions, soap and crèmes (Fig. 11.6a–c). Hot spring spas and resorts often develop and/or produce their own range of signature cosmetic and health products, e.g. Peninsula Hot Springs (Australia), the Polynesian Spa (New Zealand), or the Blue Lagoon (Iceland). In the field of complementary and alternative medicine (CAM), balneotherapy is an integrated part of the health and wellness sector and relies not only on thermal spring water, but also on related by-products like minerals and thermal mud (Erfurt-Cooper 2010), which are used to treat various medical conditions.

a

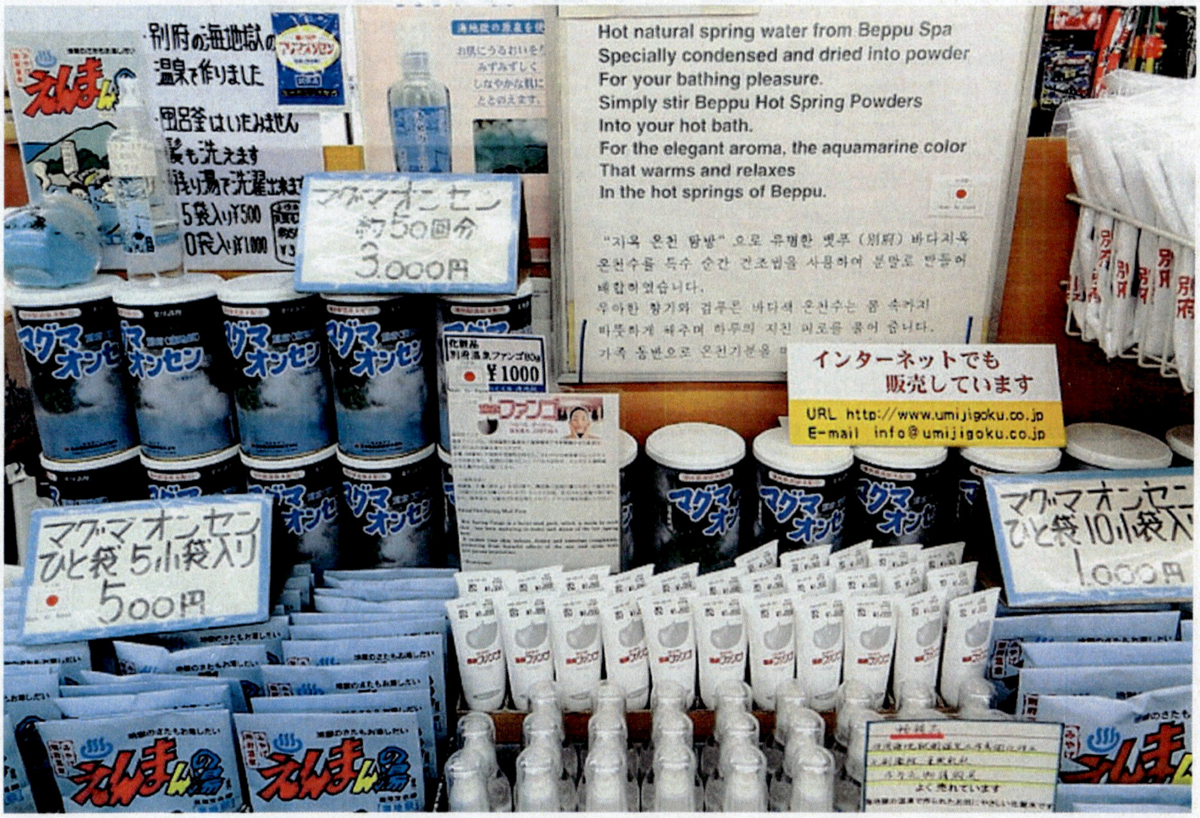

Fig. 11.6 **a** Beneficial minerals extracted from volcanic hot springs in Japan are used in a range of bath powders, soap, crèmes and lotions. **b** Cosmetic products including mud packs contain minerals from New Zealand's volcanic hot springs and are popular for natural skin care. **c** Bath additives in the form of mineral salts from the Dead Sea in Jordan

b c

Making use of natural resources, one of the most remarkable tourist attractions is geothermal cooking, a custom found on many volcanic island nations like the Azores, New Zealand, Iceland and Japan (Figs. 7.8 and 10.9). Heat from the Earth was also historically used by native American tribes at Yellowstone for the preparation of food. In fact, many other countries worldwide, including Kenya, the Philippines, Mexico and Indonesia, still use geothermal water and steam for cooking (Erfurt 2018). Restaurants with access to heat from volcanic activity have specialised in in geothermal cuisine or geyser cooking, which is exceedingly popular with visitors from non-volcanic countries and regions.

11.2.4 The Development of Innovative Technological Solutions to Extract Geothermal By-Products

Apart from the conventional use of hot springs for health and recreation and the generation of heat and electricity from geothermal fluids, novel technologies are aiming for more. The development of an interlinked process to co-produce energy and metals as a commercial by-product is presently investigated in more detail by researchers, who are studying 'combined heat, power and metal' extraction (CHPM), how to make it possible and how to optimise this process to meet future market demands. Initial research into such hybrid processes has shown that certain metals can be recovered through electrodeposition at temperatures up to 300 °C and under elevated pressure, whereby higher temperatures and pressure would lead to an increase in metal recovery. Other methods to recover metals from geothermal solutions investigated were gas-diffusion electro-precipitation as well as electro-crystallisation, the latter potentially allowing for nearly complete recovery of any metals of interest being present. While the implementation of combined heat, power and metal extraction currently is a technological challenge, the CHPM2030 research team is aiming for the scheme to be operational by 2030/2050 (CHPM2030 2019).

Of particular interest and receiving increased attention is the recovery of **lithium** in areas where high temperature fluids are leaching this metal from volcanic rocks like rhyolite (Parkway Minerals 2020). Projects aiming to extract lithium from geothermal brine are considered in several countries including the United States (New Mexico), Germany (Upper Rhine Area), the United Kingdom (Cornwall) and New Zealand. The growth in demand for lithium is partly due to the expected transition to electric vehicles on a global scale (Cariaga 2019; Cornish Lithium 2019).

But there are other geothermal by-products currently investigated for commercial use. In New Zealand, the construction of the first commercial-scale plant has started, where **silica** and other valuable minerals will be extracted from geothermal brine. In addition to the recovery of silica, studies are also conducted to further develop the technology to extract minerals such as lithium as well (Think Geoenergy 2020b). In Iceland, preparations including environmental assessments are underway to investigate the proposed **hydrogen** and **methane** production in conjunction with the *Hellisheiði* Power Station (HGPS) (ON 2019). These are relatively new geothermal by-products and differ from the conventional hydrothermal ore deposits, where metallic mineral masses have accumulated, formed by the precipitation of solids from hot mineral-laden fluids, and are able to be recovered without the need of processes including electrodeposition, gas-diffusion electro-precipitation or electro-crystallisation.

11.2.5 The Fragile Heritage of Hot Springs

As with all technologies that invade, and more often than not violate the natural environment, it is critical to apply best practice management to minimise any environmental and social impacts to avoid long term resource depletion and degradation of the groundwater quality. A definite disadvantage regarding the new technologies mentioned earlier is associated with the disposal of waste brine, which can be costly as well as complicated. Different stages of mineral extraction from geothermal fluids can lead to substantial quantities of high-concentration brine waste.

Coming back to the topic of hydraulic fracturing or fracking (see also Chap. 44), whether for the purpose of extracting geothermal heat or for the unconventional recovery of oil and gas, the principal process can put the surrounding environment under pressure. While the extraction of geothermal resources (hot water or steam) from subterranean reservoirs generally is thought of as not causing any chemical contamination of aquifers or artesian basins, concerns have been raised that the fracturing of rock using explosive charges is causing seismic tremors, which may lead to unpredictable responses of the subterranean hydrogeologic environment (Ndzelen and Reinhard 2016).

Far more concerning is the impact of the unconventional oil and gas (UOG) extraction, which is a controversial issue for several reasons with the main concern the potential for groundwater reservoir contamination. The use of large quantities of special water-based fracturing fluids, which are injected under high pressure into subterranean rock formations to open up additional cracks and fissures (Fig. 11.7), has the potential to pollute entire reservoirs. Besides water and sand, fracking fluids can contain a toxic cocktail of over 200 different chemicals (Chap. 4). Of these chemicals, at least a dozen substances were identified not only as hazardous and extremely toxic, but also as carcinogenic. Some reports are suggesting that groundwater reservoirs are already contaminated in several areas (DiGiulio and Jackson 2016). In addition to the potential for chemical contamination there is a risk of fugitive emissions of gases such as

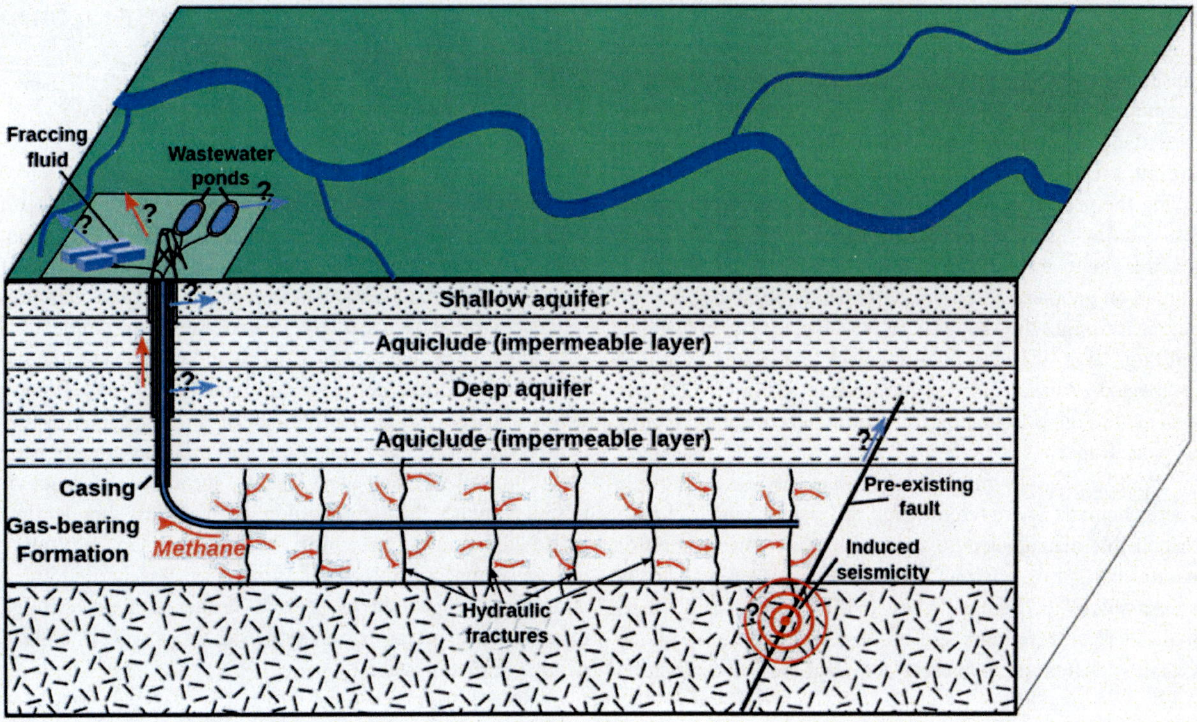

Fig. 11.7 Hydraulic Fracturing diagram. *Source* Norton (2013) (Public Domain)

methane, a powerful 'greenhouse gas', known to have a negative impact on the global climate.

To assess the potential quantity of escaped methane during coal seam gas (CSG) extraction, the former Australian federal Department of the Environment together with the Commonwealth Scientific and Industrial Research Organisation (CSIRO) measured methane emissions from 43 production wells in the states of Queensland and New South Wales (Chap. 4). The findings showed that only at three CSG wells no methane was detected, while at 40 CSG sites fugitive methane was measured. As this was only a pilot study, less than one percent of Australia's CSG wells were assessed for emissions, with further limitations due to the focus on well pads and nearby equipment only (CSIRO 2020).

Environmental Pressures

Hot spring areas and hydrothermal systems are indeed quite fragile and every form of pressure on these environments interferes with the natural balance needed to protect and conserve them for coming times. Once the damage is done and a hot spring or a hydrothermal system has been damaged, destroyed and/or contaminated, in most cases it cannot be reversed. Too many times the out-of-sight equals out-of-mind attitude towards our subterranean water sources has caused irreparable degradation, including the depletion of over-exploited aquifers.

External damage to hot springs is often caused through thoughtless actions by ignorant or inconsiderate tourists (Figs. 4.1 and 4.2), who are not aware of the importance to preserve natural resources from contamination and anthropogenic pollution. Other irresponsible activities are also carried out by tourism operators or tour companies, who exploit a certain area for their economic benefits without taking the necessary care to prevent degradation of a priceless natural resource. As discussed in Chap. 7, the trend towards large-scale waterparks is placing a considerable burden on geothermal resources, which are wasted to a large degree through the disposal of wastewater by constantly refilling large thermal pools (unless the water is either re-injected or used for other purposes, e.g. Figure 11.2), as well as through evaporation from thousands of square metres of waterscapes.

From a tourism management point of view, extreme visitor demand and overcrowding are issues, which require responsible management to stop any further deterioration of natural hot spring environments. Recommendations for the conservation of natural hot spring sites and their hydrothermal reservoirs are discussed in more detail in Chap. 4. After all, overused and degraded hydrothermal systems lead to decreased water quality as well as quantity, thereby causing the loss of economic and recreational value for the hot spring tourism industry, as well as affecting the sustainability of local ecosystems. To counteract the environmental challenges caused by human activity, the conservation of renewable natural resources and sustainable management are the only options to ensure that their geodiversity remains in balance.

11.2.6 Raising Awareness: Hot Springs and Environmental Concerns

The development of conservation strategies to protect the geodiversity of natural resources including hot springs is increasing on a global scale (Burek and Prosser 2008; Farsani et al. 2014; Henriques et al. 2011). To raise more awareness about conservation issues, the educational aspect can be an important factor for visitors of hot spring destinations, where the diversity of hydrothermal features and their geophysical heritage offers more than just sightseeing and photo opportunities for curious visitors. In many countries hydrothermal features such as hot springs, geysers and travertine terraces are already protected in national parks, geoparks and world heritage sites, where conservation is coordinated with promoting best-practice approaches and policies covering land use planning and sustainable resource management (Bentivenga and Geremia 2015; Erfurt 2012).

11.2.7 Sustainability vs Over-exploitation of Hot Springs

For conservation purposes it is essential to take a long-term view in the management of hot spring resources. This is not always the case, which has contributed to the deterioration of many hydrothermal features in the past. For conservation to be successful it is necessary to understand the value of a natural resource (e.g. hot springs) to appreciate their need to be preserved for coming generations (Burek and Prosser 2008; Prosser 2011). Unfortunately, the sustainability of natural hot springs as a renewable resource has already been compromised in many areas, mainly through human activities. To avoid the deterioration of countless hot springs worldwide it is our responsibility to ensure that any future use of natural hydrothermal systems is closely monitored and strictly regulated to prevent further damage through extreme commercial extraction processes as well as anthropogenic contamination.

Fig. 11.8 The valley of Geysers on the Kamchatka Peninsula (Russia) before many hydrothermal features were buried under a large mudslide in 2007. Since then many features have re-emerged. *Source* Nunn (2006) (Public Domain)

At the same time, natural events can also cause the deterioration or disappearance of hot springs; earthquakes for example can sometimes reduce or stop their flow altogether (Example 5.1). This applies especially to countries subject to tectonic instability with frequent volcanic eruptions and/or seismic tremors, which can trigger landslides or rockfalls that cover up existing springs. Sometimes the buried springs and other hydrothermal features re-emerge after a while, which has happened in the *Valley of Geysers* on the Kamchatka Peninsula in far-east Russia (Fig. 11.8), where in 2007 a large number of active hot springs and geysers disappeared under a mudslide (Erfurt 2012).

In some areas, it is necessary to drill wells into the ground and pump hot water to the surface to create a man-made hot spring that can be either commercially exploited or used like a natural hot spring (Erfurt 2012). This is usually monitored and regulated by local environmental and water legislation to avoid over-exploitation, especially during times of drought with little aquifer recharge.

11.3 Hot Spring Tourism—A Brief Summary

The growing interest in the geodiversity of unique landforms such as active hydrothermal areas has led tour operators to offer exclusive and/or adventure tours, which frequently include nearby volcanoes, to cater to groups who want to study these environments at close range. Countries rich in hot springs are taking advantage of these natural resources by exploiting them for destination development, predominantly for health and recreational tourism. Because of their popularity, natural hot spring areas are increasingly at risk from growing visitor pressure, which has a damaging effect on their fragile environment (Chap. 4; Erfurt 2012). Then again, compared to other practices disturbing and negatively impacting natural water resources including hydrothermal environments, hot spring tourism may only be a relatively minor contributor to any degradation of natural springs and their source reservoirs. A positive aspect of hot spring tourism, especially if integrated in the field of geotourism and ecotourism, is the increasing interest in learning about the geophysical background of hot springs and other related hydrothermal features (Cataldi et al. 1999; Erfurt 2012). This provides an opportunity for visitors to understand factors impacting on the environment and the need for conservation, as well as a chance to observe sustainable management approaches in natural hot spring areas.

11.3.1 Hot Springs as a Resource for Health, Wellness and Recreation

The use of hot springs for health and recreation is well established in the history of countries worldwide (see Chaps. 5, 6, 8). Natural springs were commonly considered as a health-giving divine contribution and a blessing for the people who were able to settle in their vicinity. Immortalised in legends and mythology is a thousands of years old practice of treating illness and injuries with hot mineral water. Passed on from one generation to the next, accounts of successful healing gradually developed into today's alternative medical field of balneotherapy or spa medicine, the natural approach to treating various forms of medical conditions.

One of the reasons for the continuing growth of hot spring tourism is the enduring trend that started in the early 1990s, which focuses on healthy living, improving fitness and maintaining overall wellbeing (Erfurt-Cooper 2009a, b; Erfurt 2012; Foster et al. 2011). Because natural hot springs have a long history of contributing a range of positive features and values (Fig. 11.9) for humankind worldwide, they are used and appreciated for reasons discussed in detail in previous chapters of this book.

Fig. 11.9 Hot springs contribute a wide range of natural features combined with cultural traditions, which are valued and appreciated, as well as commercially exploited by societies worldwide. *Source* Author

11.3.2 Health and Wellness—Research Update

Globally, the health, wellness and recreation industries continue to increase development investments in hot spring spas and resorts. New facilities are built, while others are modernised or extended to respond to the demand for better health and recreation, driven by the desire for improving and/or maintaining one's personal wellbeing (Fig. 11.10).

At the **Global Wellness Summit** in 2013, a 'Hydrothermal Forum' raised well-founded concerns about the lack of international standards and guidelines for the correct planning and architectural design of thermal bathing areas. In response, one year later the 'Global Wellness Institute's Hydrothermal Initiative' published the first edition of a 'Guide to Hydrothermal Spa and Wellness Development Standards' to provide a comprehensive overview of essential knowledge required for the design of thermal spa and wellness facilities (Cavanagh 2018). With regular updates every two years, by 2018 a third updated edition was published, which reflects the importance of global developments in the health and wellness industry with a focus on thermal treatment facilities, many of which are using natural hot spring water.

Continuous research and ongoing studies of the beneficial properties of hot mineral springs, together with a high success rate in treating a wide range of medical conditions and the apparent lack of side effects, strongly support the use of these 'medicinal waters' in holistic and alternative health therapies. Widely recognised as a natural source of health and wellbeing, hot mineral springs have maintained their reputation as a successful therapy also in the field of rehabilitation after accidents or injury trauma as well as post-surgery (Altman 2000; Bullard 2004; Cohen and Bodeker 2008; Erfurt 2012; Leavy and Bergel 2003).

Supported by medical associations based on research carried out under the supervision of qualified professional teams, the use of natural hot springs in spa medicine is integrated into the national health care system of many European countries. Especially in the field of rehabilitative care, balneotherapy administered at a suitable health clinic or medical spa is known for its success rates and therefore covered either completely or partially by statutory health insurance (Erfurt 2012; Chap. 8). For a hot spring treatment facility in Europe to be formally recognised and approved as a natural resource for health purposes, government standards as a rule require certain criteria to be covered. With variations in different countries, the major benchmarks generally include the type and amount of mineral and trace elements, the required minimum temperature to be classed as a warm or hot spring and the rate of discharge of hot mineral water from the source (Erfurt-Cooper 2009a, b; Erfurt 2012). To remain relevant and acknowledged as a

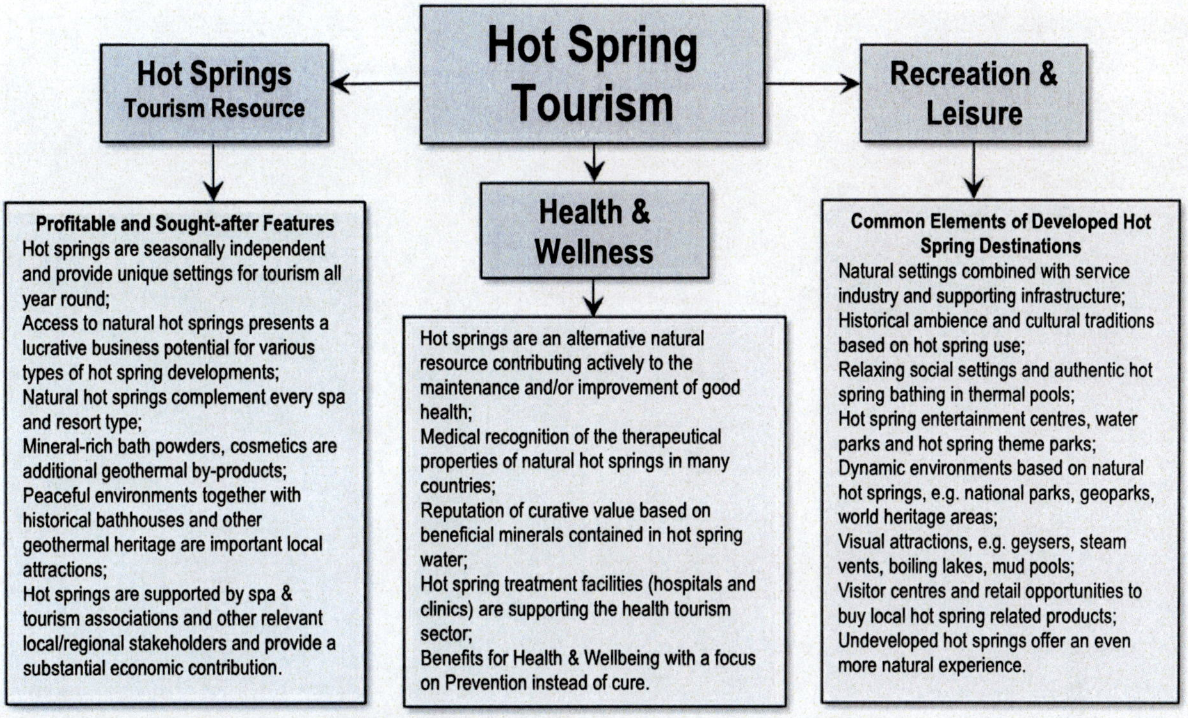

Fig. 11.10 The framework outlines the diversity of natural hot springs as a resource for health, wellness, recreation and leisure. *Source* Author

reliable method of treating health conditions without side effects, medical professionals, especially in Europe, Japan, China, New Zealand and Taiwan continue to research the connection between mineral-rich hot spring water and any associated health benefits. For further information, medical journals are reliable scientific sources that provide updated research findings of clinical trials, conducted under medical supervision, which evaluate the curative value of natural hot springs in the treatment of numerous medical conditions. An extensive review of the literature revealed the fact that, despite hot springs being a priceless resource for the tourism industry in general, there still is a scarcity of more recent research publications in the non-medical field of hot spring tourism. Occasionally, research articles about 'geomedicine' based on the use of hot springs for health and recreation are published in journals in the fields of Earth science or Environmental science. And yet, it seems the available research literature mainly concentrates on widely known hot spring spa towns, often applying a Eurocentric view, although with slowly increasing contributions from researchers based in countries in the Asia-Pacific part of the world.

11.3.3 Recreational Advantages of Hot Springs

From an economic point of view, the key factors in the development of hot spring tourist destinations include (a) the worldwide occurrence of hot springs, (b) the long-term sustainability as a renewable resource if responsibly managed, (c) the availability of a clean and unpolluted source of hot water, (d) the economic significance as a cost effective resource for domestic, commercial and industrial purposes and lastly, (e) a unique natural environment usually surrounding hot springs, another benefit for their recreational use. Many national parks and other protected sites include extensive areas with hydrothermal features, often linked to active volcanism nearby. Annual visitor numbers are large as people are drawn by the visual impact of geysers, boiling lakes and bubbling mud ponds, which play an important role in attracting tourists. Countries such Iceland, New Zealand and Japan are examples of a successful tourism industry based on an abundance of volcanic hot springs, used either for health and wellness or as unique recreational attractions. While people in Iceland enjoy their hot pools (Fig. 11.11),

Fig. 11.11 The Viking Springs pool in Leirubakki (southern Iceland) is built from volcanic rock, surrounded by a wall of lava blocks, fed by volcanic spring water and is close to the active volcano Hekla

the Japanese are devoted to their onsen (Fig. 11.12). New Zealand's hot spring spas, public baths and geothermal parks on the other hand are influenced by different cultures including Polynesian, Japanese and European (Fig. 11.13), all contributing to the diversity of visitor drawing power.

In many European countries the use of natural hot springs for health and recreation goes back several thousand years and hot spring tourism, despite being punctuated by centuries of decline, has repeatedly regained its momentum. By contrast, some countries' hot spring tourism is of a more contemporary nature. In **Iceland,** where geothermal activity is nearly everywhere under one's feet, it is safe to assume that most visitors will sooner or later participate in some form of hot spring tourism, if they visit the well-known *Blue Lagoon* the *Mývatn Nature Baths* (Fig. 11.14) or the recently established *GeoSea* in Húsavík (Appendix 11.1). The first two hot spring destinations are located in the middle of endless lava fields with geothermal power stations either as a backdrop (Svartsengi) or nearby (Bjarnarflag). Due to the country's geologic features, unique volcanic landforms as well as spouting geysers, hissing steam vents, boiling mud and even caves with hot springs (Fig. 11.15) can be explored within a reasonably short time; all surrounded by ever-present layers of ancient and not so ancient lava flows.

11.4 Hot Springs—Final Points of Reflection

Although cultural and social settings of hot spring destinations may vary from one country to another, they all share the common goal of seeking health benefits and recreation in a safe and peaceful environment. Combined with other outdoor activities, the pleasantly warm water of an undeveloped hot spring in the wilderness offers a relaxing soak

Fig. 11.12 This outdoor onsen (rotenburo) in Hokkaido (Japan) is designed with great care to integrate natural features and blend in with the environment

after a long hike and is an unforgettable experience (Figs. 2.3, 2.21, 4.8, 5.28, 9.14, 10.21). For geotourists, hot spring environments create an opportunity to learn more about their natural and cultural geoheritage and their vulnerable geosystems. Yet some of the most remarkable hot springs are not found on the Earth's surface, but are submerged on the bottom of lakes, rivers and oceans. As mentioned in Chap. 2, the geophysical diversity of hot springs also includes the multitude of submarine hydrothermal vents, known as black and white smokers (Figs. 2.19 and 11.16), which were first discovered in 1977 at the East Pacific Rise near the Galápagos Islands and are only accessible using a remote operated vehicle (ROV).

Geothermal water is used worldwide as a renewable resource to generate energy for commercial, agricultural and industrial processes, space heating, health and recreation as well as for drinking and cooking. From undeveloped to developed hot springs, they have a universal reputation for the curative properties of the mineral-rich waters. Hot springs and their surrounding natural environments are used in sustainable and sometimes less sustainable destination development and provide employment for local communities in many countries. The combination of health and recreational hot spring facilities with protected sites like national parks, geoparks or world heritage areas has been historically successful, and there is no reason why this would change in the future.

As a matter of concern however, the development and maintenance of large-scale waterparks should ultimately face stricter regulations; especially in countries, where clean water is becoming increasingly scarce and has turned into a valuable commodity. At this point in time the potential of a negative environmental impact of such facilities, while foreseeable, although not accurately predictable, requires more in-depth research to assess any effects that may have already manifested themselves and thereby contribute to the depletion of hot and cold water reservoirs (Chap. 7). Also, a

Fig. 11.13 Rock pools at the thermal complex in Hanmer Springs on New Zealand's South Island

Fig. 11.14 The lagoon of the Mývatn Nature Baths (36–40 °C) in the north-east of Iceland was created with excess water originating from the Námafjall high temperature field. The geothermal water is extracted for power generation at the nearby Bjarnarflag power plant

Fig. 11.15 Grjótagjá is a small cave with a hot spring near Lake Mývatn that was used for bathing in the past. During a period of volcanic activity between 1975 and 1984, the water temperature increased to the point where bathing in the cave's thermal pool became impossible

Fig. 11.16 Hydrothermal vents discharge particle-laden superheated water from the Earth's interior. On mixing with cold seawater, suspended minerals precipitate, causing 'smoke' columns to rise from chimney-like vents known as black or white smokers. This black smoker is located at a depth of 2,980 metres on the Mid-Atlantic Ridge. *Source* MARUM (2020)—Center for Marine Environmental Sciences, University of Bremen (CC-BY 4.0)

Fig. 11.17 This small cove is a natural thermal pool where hot spring water (61.8 °C) mixes with the cold seawater of the Atlantic Ocean. A popular, although somewhat adventurous hot spring experience at Ponta da Ferraria, São Miguel (Azores)

reminder that at many undeveloped hot springs (Fig. 11.17) as well as in active hydrothermal areas effective safety measures for visitors are either not always in place or simply inadequate; a situation that needs to be addressed and improvements made where necessary (Chap. 9).

One of the more challenging tasks while writing this book was to obtain reliable visitor numbers and accounts of the economic value of hot spring tourism to determine upcoming trends; but Chap. 9 presents some insightful data from representative destinations. Depending on their location, accessibility, and individual marketing strategies, hot spring areas are indisputably viable tourism destinations and continue to be developed in many countries around the world on account of their enduring success and in response to continued visitor demand (Erfurt 2012).

The rationale for this book was based in part on the intention to include topics that generate not just a bigger picture of the natural and cultural geoheritage of hot springs, but also focus on the connections with various industry sectors from health tourism and recreation to geothermal power generation and other commercial applications. Throughout the chapters the main objective was to present this information to an audience with an interest in natural hot springs and their geophysical origins, their cultural history, their reputation for curative properties and their contemporaneous use in many societies worldwide to the present day. Readers who have not visited a natural hot spring yet should add it to the list of things to do one day—there are many opportunities all over the world and there could be one not far away.

11.5 Appendix

Appendix 11.1 GeoSea in Húsavík, Iceland

A recent addition to Iceland's unique thermal bathing destinations is GeoSea, short for geothermal sea baths. The infinity pools at GeoSea in Húsavík are built on a cliff of ancient lava flows facing north, overlooking the Arctic Ocean. Magnificent sunsets and Northern Lights add to an unforgettable natural experience

References

Altman N (2000) Healing springs: the ultimate guide to taking the waters—from hidden springs to the world's greatest spas. Healing Arts Press, Rochester

Beardsmore G, Davidson C, Ricard L, Pujol M, Larking A, Bendall B (2020) Current directions for geothermal energy development in Australia. In: Proceedings world geothermal congress 2020, Reykjavík, Iceland, 26 April 2 May, 2020

Bentivenga M, Geremia F (2015) VII international symposium ProGEO on the conservation of the geological heritage 'geoheritage: protecting and sharing', Bari (Apulia, Italy), 24th to 28th September 2012. Geoheritage 7:1–3. https://doi.org/10.1007/s12371-015-0146-z

Bullard L (2004) Healing waters—Missouri's historical mineral springs and spas. University of Missouri Press, Columbia

Burek CV, Prosser CD (eds) (2008) The history of geoconservation. Geological Society, London, Special Publications 300, 1–5. https://doi.org/10.1144/SP300.1

Caldas D (2020) Maior rede de geotermia em Portugal Continental criada em Chaves. Diário Atual. Retrieved from https://tinyurl.com/tn2c9jg

Cariaga C (2019) Large Lithium resource reported at German geothermal site. Retrieved from https://tinyurl.com/vhbzt6u

Cataldi R, Hodgson SF, Lund JW (1999) Stories from a heated Earth—our geothermal heritage. Geothermal Resources Council, International Geothermal Association, Sacramento

Cavanagh C (ed) (2018) Guide to hydrothermal spa & wellness development standards, 3rd edn. Published by the Global Wellness Institute

CHPM2030 (2019) Combined heat, power and metal extraction (CHPM) impact factsheet. Retrieved from https://tinyurl.com/uhl6kqm

Cohen M, Bodeker G (2008) Understanding the global Spa industry: Spa management. Butterworth Heinemann, London

Cornish Lithium (2019) Cornish Lithium commences exploration in Gwennap. Retrieved from https://tinyurl.com/t3xpns2

CSIRO (2020) Hydraulic fracturing—What is unconventional gas? Retrieved from https://tinyurl.com/qprcp56

DiGuilio D, Jackson RB (2016) Impact to underground sources of drinking water and domestic wells from production well stimulation and completion practices in the pavillion. Environmental Science & Technology, Wyoming, Field. https://doi.org/10.1021/acs.est.5b04970

Erfurt P (2012) An assessment of the role of natural hot and mineral springs in health, wellness and recreational tourism. Unpublished doctoral dissertation, School of Business, James Cook University, Cairns, Queensland, Australia

Erfurt-Cooper P (2009a) Introduction: development of the health and wellness Spa industry. In: Erfurt-Cooper P, Cooper M (eds) Health and wellness tourism—Spas and hot springs. ChannelView, Bristol, pp 1–24

Erfurt-Cooper P (2009b) Geological background of natural hot and mineral springs. In: Erfurt-Cooper P, Cooper M (eds) Health and wellness tourism—Spas and hot springs. ChannelView, Bristol , pp 130–155

Erfurt-Cooper P (2010) The importance of natural geothermal resources in tourism. In: Proceedings world geothermal congress, Bali, Indonesia, 25–29 April 2010

Erfurt-Cooper P (2018) Active hydrothermal features as tourist attractions. In: Fearnley CJ, Bird DK, Haynes K, McGuire WJ, Jolly G (eds) Observing the volcano world: volcano crisis communication. Advances in volcanology series. Springer, Berlin, Heidelberg, pp 85–105. https://doi.org/10.1007/11157_2016_33

Farsani NT, Coelho COA, Costa CMM, Amrikazemi A (2014) Geo-knowledge management and geoconservation via geoparks and geotourism. Geoheritage 6:185–192. https://doi.org/10.1007/s12371-014-0099-7

Foster LT, Keller CP, McKee B, Ostry A (2011) British Columbia atlas of wellness, 2nd edn. Retrieved from https://dspace.library.uvic.ca/handle/1828/3838

Fridleifsson IB, Bertani R, Huenges E, Lund JW, Rybach L (2008) The possible role and contribution of geothermal energy to the mitigation of climate change. In: Hohmeyer O, Trittin T (eds) IPCC scoping meeting on renewable energy sources—proceedings. IPCC special report

GEG (Geothermal Energy and Geofluids) (2018) Development of a geoscientific framework for geothermal exploration and energy utilization in Mongolia. Retrieved from https://geg.ethz.ch/project-geothermal-mongolia/

GeoSea (2021) The Wonder of GeoSea. Retrieved from https://www.geosea.is

GNS Science (2020) Earth energy—posters & maps. With permission to use. Retrieved from https://tinyurl.com/ybnwd5jp

GRC (Global Resources Council (2019) Chile: first 50 MW at Mariposa geothermal power project expected in 2024. Retrieved from https://tinyurl.com/qo9w4hu

Henriques MH, Pena dos Reis R, Brilha J, Mota T (2011) Geoconservation as an emerging geoscience. Geoheritage 3:117–128. https://doi.org/10.1007/s12371-011-0039-8

Hickson C, Ferraris F, Rodriguez C, Sielfeld G, Henriquez R, Gislason T, Selters J, Benoit D, White P, Southon J, Yehia R (2011) The Mariposa geothermal system, Chile. Trans Geothermal Resourc Council 35:817–825

IDDP (2017) The drilling of the Iceland deep drilling project geothermal well at Reykjanes has been successfully completed. Retrieved from https://tinyurl.com/rx39c4n

IRENA (International Renewable Energy Agency) (2018) Unlocking geothermal potential in Japan through small-scale generation. Retrieved from https://tinyurl.com/sq3dgl3

Leavy HR, Bergel RR (2003) The Spa encyclopedia: a guide to treatments & their benefits for health & healing. Thomson Delmar Learning, New York

Lund JW, Boyd TL (2015) Direct utilization of geothermal energy 2015 worldwide review. In: Proceedings world geothermal congress 2015, Melbourne, Australia, 19–25 April 2015

MARUM (2020) Deep sea—hydrothermal vents. Research Faculty Universität Bremen. Retrieved from https://www.marum.de/en/Discover/Deep-Sea.html

Ndzelen R, Reinhard N (2016) Hydraulic fracturing and seismicity. Working paper. https://doi.org/10.13140/RG.2.1.3891.6082

Norton M (2013) Hydraulic fracture diagram. Public Domain. Retrieved from https://commons.wikimedia.org/wiki/File:HydroFrac2.svg

Nunn R (2006) Valley of the Geysers. Creative commons attribution-share alike 2.0 generic. Retrieved from https://tinyurl.com/qw8ngc9

ON (Orka náttúrunnar) (2019) Nýtt fyrirtæki í jarðhitagarðinn? (New company in the geothermal park?). Retrieved from https://tinyurl.com/vh74wqp

Parkway Minerals (2020) New Mexico Lithium Project (NMLP). Retrieved from https://tinyurl.com/tofzsyr

Prosser CD (2011) Principles and practice of geoconservation: lessons and case law arising from a legal challenge to site-based conservation on an eroding coast in eastern England, UK. Geoheritage 3:277–287. https://doi.org/10.1007/s12371-011-0042-0

Ragnarsson Á (2015) Geothermal development in Iceland 2010–2014. In: Proceedings world geothermal congress 2015, Melbourne, Australia, 19–25 April 2015

RCI (2020) Un projet de géothermie aux Anses-d'Arlet. Une concertation publique lancée. Retrieved from https://tinyurl.com/vvr8cwy

Romanach L, Carr-Cornish S, Muriuki G (2015) Societal acceptance of an emerging energy technology: how is geothermal energy portrayed in Australian media? Renew Sustain Energy Rev 42:1143–1150. https://doi.org/10.1016/j.rser.2014.10.088

Sæmundsson K, Gunnlaugsson E (2002) Icelandic rocks and minerals. Mal og mennig, Reykjavík

Think Geoenergy (2020a) 24 geothermal power projects chosen to receive funding by JOGMEC in Japan. Retrieved from https://tinyurl.com/tfqslz5

Think Geoenergy (2020b) Construction starts for geothermal silica extraction facility in New Zealand. Retrieved from https://tinyurl.com/tfxxc2k

Verkís (2020) Nesjavellir hot water main. Retrieved from https://tinyurl.com/y7m33rnb

Empty travertine terracettes before their controlled flooding with hot spring water. Pamukkale, Turkey

Geographical Index

P. Erfurt, *The Geoheritage of Hot Springs*, Geoheritage, Geoparks and Geotourism,
https://doi.org/10.1007/978-3-030-60463-9

Subject Index

P. Erfurt, *The Geoheritage of Hot Springs*, Geoheritage, Geoparks and Geotourism,
https://doi.org/10.1007/978-3-030-60463-9